FUNDAMENTALS OF ASTROMETRY

Astrometry encompasses all that is necessary to provide the positions and motions of celestial bodies. This includes observational techniques, instrumentation, processing and analysis of observational data, reference systems and frames, and the resulting astronomical phenomena. Astrometry is fundamental to all other fields of astronomy, from the pointing of telescopes, to navigation and guidance systems, to distance and motion determinations for astrophysics. In the past few decades, new observational techniques have facilitated improvements in accuracy by orders of magnitude. Starting from basic principles, this book provides the fundamentals for this new astrometry at milli- and microarcsecond accuracies. Topics include basics of general relativity; coordinate systems; vectors, tensors, quaternions, and observational uncertainties; determination and use of the celestial and terrestrial reference systems and frames; applications of new observational techniques; and present and future star catalogs and double-star astrometry. Examples of astronomical computations and a large glossary are also provided. This comprehensive text is an invaluable reference for graduate students and research astronomers.

JEAN KOVALEVSKY is an associate (retired) astronomer of the Centre d'Etudes et de Recherches Géodynamiques et Astronomiques (CERGA), at the Observatoire de la Côte d'Azur, France. He is President of the International Committee of Weights and Measures and is a member of the French Academy of Sciences.

P. KENNETH SEIDELMANN is a research professor in the Department of Astronomy, University of Virginia. Until 2000 he was Director of Astrometry at the US Naval Observatory, and a visiting professor at the University of Maryland.

FUNDAMENTALS OF ASTROMETRY

JEAN KOVALEVSKY
CERGA, Observatoire de la Côte d'Azur

P. KENNETH SEIDELMANN
Department of Astronomy, University of Virginia

CAMBRIDGE
UNIVERSITY PRESS

PUBLISHED BY THE PRESS SYNDICATE OF THE UNIVERSITY OF CAMBRIDGE
The Pitt Building, Trumpington Street, Cambridge, United Kingdom

CAMBRIDGE UNIVERSITY PRESS
The Edinburgh Building, Cambridge CB2 2RU, UK
40 West 20th Street, New York, NY 10011-4211, USA
477 Williamstown Road, Port Melbourne, VIC 3207, Australia
Ruiz de Alarcón 13, 28014 Madrid, Spain
Dock House, The Waterfront, Cape Town 8001, South Africa

http://www.cambridge.org

First published 2004

Printed in the United Kingdom at the University Press, Cambridge

Typeface Times 11/14 pt.   *System* LATEX 2_ε [TB]

A catalog record for this book is available from the British Library

Library of Congress Cataloging in Publication data

Kovalevsky, Jean.
Fundamentals of astrometry / Jean Kovalevsky and P. Kenneth Seidelmann.
    p.   cm.
Includes bibliographical references and index.
ISBN 0 521 64216 7 (hardback)
1. Astrometry.   I. Seidelmann, P. Kenneth.   II. Title.
QB807.K682 2003
522–dc21   2003053227

ISBN 0 521 64216 7 hardback

In memory of the late Jeannine Kovalevsky and to Bobbie Seidelmann in appreciation for their continuing support, encouragement, and love during the preparation of this book by their respective husbands

Contents

Preface

Astrometry is undergoing fundamental changes. The celestial reference frame based on bright optical stars is being replaced by the extragalactic reference frame observed at radio wavelengths by Very Long Baseline Interferometry (VLBI); Hipparcos has proven the capabilities of astrometry from space; photographic plates are being replaced by charge coupled devices (CCDs) as the detectors of choice; optical interferometers are replacing transit circles and astrolabes; accuracies of tenths of arcseconds are being replaced by milli- and microarcseconds; the dynamical reference frame is being replaced by a kinematic reference frame; Global Positioning System observations are changing the determinations of Earth Orientation Parameters; and the theory of general relativity is required as the basis of astrometry.

This book is designed to provide the fundamentals for the new astrometry at the milli- and microarcsecond accuracies. The intent is to start from basic principles, without approximations, and develop the fundamentals of astrometry at microarcsecond accuracies. It is based on the general theory of relativity and the concepts introduced by the International Astronomical Union (IAU) in the past decade. The book provides the definitions and applications of the International Celestial Reference System (ICRS) to astronomy and astrometry. It is also designed to provide the philosophy and concepts of astrometry for the future, the principles behind the algorithms, the reasons for astrometry and its relationships with astronomy, geophysics, planetary sciences, astrophysics, cosmology, and celestial mechanics. It defines and describes the reference systems, reference frames, their relationships and realizations. Definitions, astronomical values, and examples are given in Appendixes.

Other volumes provide the methods for almanacs (*Explanatory Supplement to the Astronomical Almanac*), and observational techniques in detail (Kovalevsky, *Modern Astrometry*). New developments in astronomical data and star catalogs, algorithms and software, latest values of constants, ephemerides, and Solar System bodies can be found in specialized documents like the *IERS Conventions*, astronomical data banks and SOFA software. It is indeed expected that the latest information, data, and software will be available by the world wide web and so some web sites are given in the book as sources of the latest data.

It is our pleasure to acknowledge and thank Dr Thomas Corbin, who reviewed all the chapters of the book and provided suggestions concerning content and language. We also thank Dr Serguei Kopeikin who has reviewed and given very useful suggestions in the parts concerning general relativity. We have also profited from discussions with Dr Nicole Capitaine, Dr Dennis McCarthy, and Dr Patrick Wallace.

Abbreviations

AC	*Astrographic Catalogue*
ACRS	Astrographic Catalogue Reference Stars
ACT	*Astrographic Catalog Tycho*
ADC	Astronomical Data Center at Goddard Space Flight Center
AGK1	*Katalog der Astronomischen Gesellschaft*
AGK2	*Zweiter Katalog der Astronomischen Gesellschaft*
AGK2A	Astronomischen Gesellschaft Katalog Reference stars
AGK3	*Dritter Astronomischen Gesellschaft Katalog*
AGK3-R	Astronomischen Gesellschaft Katalog Reference stars
AGU	American Geophysical Union
BCRS	Barycentric Celestial Reference System
BGI	Bureau Gravimétrique International (at Toulouse)
BIH	Bureau International de l'Heure
BIPM	Bureau International des Poids et Mesures
BdL	Bureau des Longitudes (now IMCCE)
BPT	binary pulsar time scale
CCD	charge coupled device
CDS	Centre de Données stellaires (in Strasbourg)
CEO	Celestial Ephemeris Origin
CEP	Celestial Ephemeris Pole
CERGA	Centre d'Etudes et de Recherches Géodynamiques et Astronomiques (in OCA)
CHARA	Center for High Angular Resolution Astronomy
CIP	Celestial Intermediate Pole
CIRF	Celestial Intermediate (or True) Reference Frame
CNES	Centre National d'Etudes Spatiales
COAST	Cambridge Optical Aperture Synthesis Telescope

CPC	*Cape Photographic Catalogue*
CTE	charge transfer efficiency
CTI	charge transfer inefficiency
CTRF	Conventional Terrestrial Reference Frame
CTRS	Conventional Terrestrial Reference System
DGFI	Deutsches Geodatisches Forschungsinstitut
DIVA	Double Interferometer for Visual Astrometry
DMA	Defense Mapping Agency (now NIMA)
DUT1	UT1-UTC
EOP	Earth Orientation Parameters
ERS	European Remote Sensing Satellite
ESA	European Space Agency
ET	Ephemeris Time
FAME	Full-sky Astrometric Mapping Explorer
FCN	Free Core Nutation
FGS	Fine Guidance Sensors
FK3	*Fundamental Katalog 3*
FK4	*Fundamental Catalogue 4*
FK5	*Fundamental Catalogue 5*
GAIA	Global Astrometric Instrument for Astrophysics
GC	*General Catalogue*
GCRF	Geocentric Celestial Reference Frame
GCRS	Geocentric Celestial Reference System
GEM	Goddard Earth Model
GI2T	Grand interféromètre à deux télescopes
GMST	Greenwich Mean Sidereal Time
GPS	Global Positioning System
GRGS	Groupe de Recherches de Géodésie Spatiale
GRIM	GRGS – Institute of Munich (Earth gravity model)
GRS	Geocentric Reference System
GRS	Geodetic Reference System
GSC	*Guide Star Catalog*
GSD	Greenwich Sidereal Date
GST	Greenwich Sidereal Time
GTRF	Geocentric True Reference Frame
HCRS	Hipparcos Catalogue Reference System
HST	Hubble Space Telescope
IAG	International Association for Geodesy
IAU	International Astronomical Union
ICRF	International Celestial Reference Frame

ICRS	International Celestial Reference System
IERS	International Earth Rotation Service
IGS	International GPS Service
IHRF	International Hipparcos Reference Frame
IOTA	Infrared-optical Telescope Array
IRS	*International Reference Star catalog*
ISI	Infrared Spatial Interferometer
ITRF	International Terrestrial Reference Frame
ITRS	International Terrestrial Reference System
IUGG	International Union of Geodesy and Geophysics
JD	Julian Date
JDN	Julian Day Number
JGM	Joint Gravity Model
K	kelvin (degree)
KI	Keck Interferometer
LBT	Large Binocular Telescope
LHA	local hour angle
LOD	length of Day
MACHOS	Massive Compact Halo Objects
mas	milliarcsecond
MJD	Modified Julian Day
nas	nanoarcsecond
NASA	National Aeronautical and Space Agency
NPMP	Northern Proper Motion Program
NPOI	Navy Protoype Optical Interferometer
OCA	Observatoire de la Côte d'Azur
pdf	probability density function
PGC	*Preliminary General Catalogue*
POSS	Palomar Optical Sky Survey
p.p.b.	parts per billion
PPM	*Positions and Proper Motions catalog*
psf	point spread function
PT	pulsar time scale
PTI	Palomar Testbed Interferometer
RGC	reference great circle
r.m.s.	root mean square
SAO	Smithsonian Astrophysical Observatory
SDSS	Sloan Digital Sky Survey
SI	Système International (d'Unités)
SIMBAD	main data base of the CDS

SPMP	Southern Proper Motion Program
STScI	Space Telescope Science Institute
SUSI	Sydney University Stellar Interferometer
TAI	International Atomic Time
TCB	Barycentric Coordinate Time
TCG	Geocentric Coordinate Time
TDB	Barycentric Dynamical Time
TDI	time delay integration
TEO	Terrestrial Ephemeris Origin
TT	Terrestrial Time
UCAC	*USNO CCD Astrographic Catalog*
USNO	US Naval Observatory
UT	Universal Time
UTC	Coordinated Universal Time
UT1	Univeral Time 1
VLA	Very Large Array
VLBA	Very Large Baseline Array
VLBI	Very Long Baseline Interferometry
VLTI	The Very Large Telescope Interferometer
WGD	World Geodetic Datum
WGD2000	World Geodetic Datum 2000
WGS	World Geodetic System
2MASS	2 Micron Astronomical Sky Survey
μas	microarcsecond

1

Introduction

Astrometry is positional astronomy. It encompasses all that is necessary to provide the positions and motions of celestial bodies. This includes observational techniques, instrumentation, processing and analysis of observational data, positions and motions of the bodies, reference frames, and the resulting astronomical phenomena.

The practical side of astrometry is complemented by a number of theoretical aspects, which relate the observations to laws of physics and to the distribution of matter, or celestial bodies, in space. Among the most important are celestial mechanics, optics, theory of time and space references (particularly with regards to general relativity), astrophysics, and statistical inference theory. These scientific domains all contribute to the reduction procedures which transform the observed raw data acquired by the instruments into quantities that are useable for the physical interpretation of the observed phenomena. The goals of this book are to present the theoretical bases of astrometry and the main features of the reduction procedures, as well as to give examples of their application.

Astrometry is fundamental to, and the basis for, all other fields of astronomy. At minimal accuracy levels the pointing of telescopes depends on astrometry. The cycle of days, the calendar, religious cycles and holidays are based on astrometry. Navigation and guidance systems are based on astrometry, previously for nautical purposes and now primarily for space navigation.

Astronomy and astrophysics are also strongly dependent on astrometry. The dynamics in the Solar System and in other gravitationally linked celestial bodies, stellar motions as obtained from the determination of proper motions, masses of double and multiple star systems, distances of stars in the vicinity of the Sun, apparent dimensions and shapes of planets, stars and other celestial objects, are all derived from astrometric observations.

Improvements in accuracies of observational techniques have been accompanied by increases in knowledge and improvements in theories. This process should continue in the future.

A very touchy but essential feature of astrometry is the expression of the uncertainties of the observations. It is customary, and actually very important, to distinguish precision, which describes the capability of an instrument to reproduce the observations, and accuracy, which represents how much these results deviate from what are thought to be the true values (which are by nature not attainable). The precision is improved by increasing the power, the resolution, or the sensitivity of the instrument. The accuracy is improved by measuring, or modeling, all effects that possibly can bias the results. This is done during the reduction process, even though it may mean additional measurements (temperature or humidity of air, optical distortions of the instrument, etc.). A given instrument's observations of stars may have a precision of 20 milliarcseconds (mas), but when compared with a catalog from many instruments, one may infer that its accuracy is 50 mas (assuming, however, some evaluation of the accuracy of the reference catalog itself). Evaluating the accuracy of observations is one of the most difficult and fundamental tasks of astrometry.

1.1 Classical astrometry

Astrometry began with the first naked-eye observations of the sky. These observations led to the discovery of *wanderers* (planets) among the stars, knowledge of the cycles of days and years, predictions of eclipses, the concept of almanacs and calendars, and the catalogs of observations. The naked eye limited both the faintness of objects and the accuracies of the positions observed. The observations of phenomena, such as eclipses and occultations, along with the location on Earth of the observer are the most accurate observations from ancient times.

From the beginning, astrometric observations were the basis for theories and discoveries. In 150 BC, Hipparchus discovered precession, the long period (26 000 year) motion of the Earth's pole with respect to the fixed stars, from observations with a precision of 1200″ (Dick, 1997). This was possible due to observations of Spica from 160 years earlier. Hipparchus may have prepared the first star catalog, but certainly Ptolemy's (AD 150) star catalog, as in his *Almagest*, is the first passed down from ancient times. Ptolemy developed theoretical explanations of the motions of the wanderers around the Earth. In 1496 Copernicus concluded that the wanderers moved around the Sun, not the Earth.

The Islamic culture produced the star catalogs of Al-Sufi (960), using armillary spheres, and Ulugh Beg (1430), who used a huge sextant, of precision of the order of 5–10′. Tycho Brahe (1546–1601) constructed improved instruments using sights and scales on quadrant- and sextant-type instruments. His observations, at a precision of 15–35″, had a wide impact on astronomy. However, even these improved observations were unable to measure stellar parallax, the angle through which a star seems to be displaced due to the orbital motion of the Earth, and this raised questions about the Copernican theory, since the parallactic displacement due to the motion of the Earth around the Sun was not detected. On the other side, the observations of the planets, in particular Mars, which was the only planet whose observations were of the necessary precision, led Kepler to develop his laws of motion, which in turn led to Newton's Universal Law of Gravity. The invention of the telescope permitted Galileo to discover the satellites of Jupiter and make astrometric observations of those objects, the first astrometric observations made with a telescope.

The application of astrometry for navigation and time-keeping led to the establishment of national observatories in Paris (1667), Greenwich (1675), Berlin (1701), and St Petersburg (1725). Telescopes were mounted on instruments similar to those used by Tycho Brahe. At the Royal Observatory at Greenwich, John Flamsteed produced the *Historiae Coelestis* (1725), the first great star catalog based on telescopic observations. In 1718 Edmond Halley, who would succeed Flamsteed as Astronomer Royal, showed that the bright stars, Aldebaran, Sirius, and Arcturus, were displaced by many minutes of arc from their positions in antiquity. Thus, the star positions are not fixed, but they move, perpendicular to the line of sight, by stellar proper motions. In 1728 James Bradley, Halley's successor, discovered a stellar aberration of 30″ due to the Earth's orbital motion. Aberration is the displacement of the angular position of an object due to the finite speed of light, in combination with the motion of the observer and the observed object. In 1748 Bradley detected the periodic motion of the Earth's celestial pole with respect to the stars due to nutation, an effect which can amount to 18″. William Herschel, who discovered Uranus in 1781, detected the motion of the Sun toward the constellation Hercules from an analysis of stellar proper motions.

Improvements in the precision of astrometric observations were sufficient that in 1838–40 Fredrich Wilhelm Bessel, Wilhelm Struve, and Thomas Henderson independently announced observations of a few tenths of an arcsecond precision showing that star positions shifted as the observer changed position in the Earth's orbit. Thus, the effect of parallax was observed, the truth of the Copernican theory solidified, and a method of directly measuring accurate stellar distances was established. In 1844 at Pulkovo, Bessel, from positional observations of 0″.7 and proper motions of 0″.5/year precision, announced unseen stellar companions, which were producing variations in proper motions of Procyon and Sirius. Direct telescope observations confirmed the companions later in the century. In 1885 Chandler detected, from stellar measurements, the motion of the Earth's axis of rotation with respect to the Earth's crust. This is called polar motion and amounts to 0″.5/year.

For most of the twentieth century, the primary instrument for large-angle astrometry was the transit circle. The transit instrument, which could observe only right ascensions, was first used by Ole Roemer in Copenhagen in 1689. The transit circle, which has a circle to measure declination, was succesfully introduced by Troughton in 1806 for Groombridge. By the mid nineteenth century most observatories had transit circles and for 150 years they were the prime instruments of large-angle astrometry. The observations from transit circles were the bases for the positions and proper motions of the fundamental catalogs through the FK5, which provided the astronomical reference frames.

The transit circle is a specialized telescope that can be moved only along the meridian, a North–South arc passing through the observer's zenith. This restriction of motion and the rigid mounting provides the stability necessary for precise observations. Observations are made in the East–West coordinate (right ascension) by timing when the star crosses, or transits, the meridian due to the Earth's rotation, while the North–South coordinate (declination) is measured on a very finely divided circle.

While the transit circle was good for accurate measurements of individual stars on the hour circle of the instrument all over the sky, the photographic plate could observe many stars in a small field of view. David Gill began a photographic survey of the southern sky from Cape Observatory at the end of the nineteenth century. The plates were measured and the positions analyzed by J. C. Kapteyn. The 1887 International Astrophotographic Congress in Paris coordinated the first international photographic survey of the sky, the *Carte du Ciel*, and the resulting *Astrographic Catalogue* (AC). This sky survey reached about 12th magnitude, while subsequent photographic surveys using Schmidt telescopes have gone as deep as 21st magnitude. Schmidt telescopes achieve wide fields of view by a correcting glass at the center of curvature and a concave focal plane, which the plate must be bent to form.

Long-focus refractors were being used at the end of the nineteenth century in small fields for observations of double stars, asteroids and satellites. The advantage of long-focus astrometry is that it provides a large scale on the plate, proportional to the focal length. One gets better precision in positions of star images, namely a few hundreds of arcseconds. A large number of stars are thought to be in multiple star systems. In some cases the components can be detected individually; others can be detected photometrically by changes in magnitude, some by astometry due to their nonlinear proper motions, some by periodicities in their radial velocities and others by multiple spectral characteristics (Chapter 12). At the beginning of the twentieth century Frank Schlesinger at Yerkes, Allegheny and Yale Observatories began the application of photography on long-focus refractors for the determination of parallaxes (Section 11.2).

1.2 New astrometry

The last decades of the twentieth century have experienced a revolution in astrometry. This was because of the charge coupled device (CCD) as a detector replacing the photographic plate and measuring machine, optical interferometers able to observe at milliarcsecond levels, space astrometry pioneered by the Hipparcos project, and radio astrometry by Very Long Baseline Interferometers (VLBI, Section 2.4.2) producing a celestial reference frame of extragalactic sources at milliarcsecond accuracy levels.

The CCD is a silicon chip that converts photons into electrons. The CCD directly provides a digital output, eliminating the need for the measuring machine. It is a small-size detector compared with the photographic plate, but it is extremely efficient, more than 50%, dependent on the filter used, compared with 1% for photographic plates (see Sections 2.1 and 14.3). Also, the data can be processed to eliminate backgrounds, subtract bright stars, etc. The small pixel sizes and the point spread function across many pixels provides the means of measuring the centroids of star positions very accurately. As a result, parallaxes of faint stars with respect to the mean parallax of background stars (relative parallaxes) can be determined with accuracies of better than 0.5 milliarcseconds (mas); and relative positions with 20 mas accuracies are possible from ground-based observations.

Albert A. Michelson, in the late nineteenth century, developed interferometry, which uses the interference of light waves for its measurement. The first stellar interferometer was not built until 1920 and only in the 1970s was it used for astrometry, other than double-star observations. Optical interferometers (Section 2.3) are currently being developed for astrometric positional measurements at the mas level, double-star measurements, and imaging of individual stars.

The Hubble Space Telescope (Section 2.5.2) is capable of astrometric observations with the Wide Field Planetary Camera, and the Fine Guidance Sensors. Double stars, parallaxes, proper motions in clusters, planetary satellite positions, and planetary nebulae motions have been measured (Section 2.4). The Hipparcos satellite (Section 2.5.1) was developed as an astrometric satellite to determine parallaxes. It has determined the positions and proper motions of 120 000 stars to accuracies of 1 mas and 1.1 mas/year for stars of 9th magnitude and brighter and parallaxes for 60 000 stars to 1 mas accuracies.

There are now plans for interferometer and astrometric satellite missions that could reach microarcsecond accuracies for stars as faint as 20th magnitude. The Space Interferometer Mission (SIM) and the Global Astrometric Instrument for Astrophysics (GAIA) are two such missions under study (Section 2.6).

In parallel with these developments in optical astrometry, the field of radio interferometric techniques (Section 2.4.2) applied to astrometry was developed primarily for geodesy and Earth orientation purposes, but the accuracies and the wealth of data have had a significant impact on astrometry via the recently adopted International Celestial Reference System (ICRS).

Observations at a few microarcsecond precision imply that, in parallel, one must model the light path between the source and the observer with a similar accuracy (Chapter 6). This means, in particular, that one uses general relativity to describe the space-time (Chapter 5). Another consequence is that sophisticated mathematical methods must be used for the treatment of astrometric data (Chapter 4).

1.3 Time

From early times people kept track of time based on the apparent diurnal motion of the Sun. This is called apparent solar time and undergoes seasonal variations, because of the obliquity of the ecliptic and the eccentricity of the Earth's orbit. This is the time from sundials. Clocks and chronometers are not subject to the variations of the Earth's orbit, so mean solar time was introduced based on a fictitious mean Sun with a constant rate of motion and Earth rotation. It was known that the diurnal rotation with respect to the stars was different from that with respect to the Sun. Sidereal time is defined by the apparent diurnal motion of the catalog equinox, and, hence, is a measure of the rotation of the Earth with respect to the stars, rather than the Sun. The rotation of the Earth with respect to the Sun is called Universal Time (UT), and is related to sidereal time (now, stellar angle), by means of a numerical formula (see Section 10.5).

For ephemerides of Solar System bodies' motions, the tabular times are the values of a uniform measure of time. Discrepancies between observed and computed positions were most evident for the Moon, because its motion is both complicated and rapid. Adams (1853) showed that the observed secular acceleration of the mean motion of the Moon could not be produced by gravitational perturbations. Ferrel (1864) and Delaunay (1865) showed that the tides exert a retarding action on the rotation of the Earth, accompanied by a variation of the orbital velocity of the Moon in accordance with the conservation of momentum. Newcomb (1878) considered the possibility of irregular variations of the Earth's rotation as an explanation of lunar residuals, but he could not find collaboration from planetary data (Newcomb, 1912). De Sitter (1927) and Spencer Jones (1939) correlated irregularities of the motions of the inner planets with those of the Moon, thus proving the irregularity of the rotation of the Earth.

Danjon (1929) recognized that mean solar time did not satisfy the need for a uniform time scale, and suggested a time scale based on the Newtonian laws of planetary motion. Unaware of Danjon's paper, Clemence (1948) proposed a uniform fundamental standard of time which was adopted by the International Astronomical Union (IAU) in 1952 and designated as *Ephemeris Time* (ET). The tropical year of 1900.0, the period of one complete revolution of the longitude of the Sun with respect to the dynamical equinox, was the basis for the definition of Ephemeris Time.

Until 1960, the unit of time, the second, was defined as a specific fraction (1/86 400) of the mean solar day. With the adoption of ephemeris time, the second was defined in the Système International (SI) of units as a specific fraction of the tropical year. This definition was more accurate, but its determination was less precise, because the motions of the Sun and Moon being slow, one could not transform the observations of their positions, made to a few tenths of an arcsecond, into time measurements of better than a few hundredths of a second. So, in 1967, a new definition of the second was adopted, following a proposal by Markowitz *et al.* (1958), in terms of a cesium beam frequency standard. Since then, it has been the fundamental unit of time. Whenever another unit is used, it has to be specified (e.g. ET second, etc.).

The second is defined as the duration of 9 192 631 770 periods of radiation corresponding to the transition between two hyperfine levels of the ground state of the cesium-133 atom. International Atomic Time (TAI) is a practical time scale determined by the Bureau International des Poids et Mesures from time and frequency standards worldwide and conforms as closely as possible to the definition of the second (Section 5.5.1). The second is the unit that is the most accurately and precisely realized at present, and thus, with the conventional value of the speed of light, is the basis for the definition of the meter. Coordinated Universal Time (UTC) is based on TAI and kept within 0.9 s of Universal Time (UT1) by the introduction of leap seconds as needed. UTC is the standard time scale defined for the zero meridian and used with appropriate time differences for time zones all over the Earth (Section 10.5.3).

Ephemeris Time was difficult to determine and its definition was not consistent with the theory of relativity. In 1976 and 1991 the IAU adopted resolutions clarifying the relationships between space-time coordinates. Terrestrial Time (TT), previously named Terrestrial Dynamical Time (TDT), is the time reference for apparent geocentric ephemerides, has a unit of measurement so that it agrees with the SI second on the geoid, and is offset from TAI by +32.184 s. Geocentric Coordinate Time (TCG) is the coordinate time appropriate for a coordinate system with its spatial origin at the center of mass of the Earth (Section 5.5.1). TCG differs from TT by a scaling factor. Barycentric Coordinate Time (TCB, section 5.5.2) is the coordinate time appropriate for a coordinate system with its spatial origin at the barycenter of the Solar System. The relationship between TCB and TCG involves a full four-dimensional transformation involving the position and velocity of the center of mass of the Earth, a scaling factor, and periodic terms. Thus, the resulting space-time coordinates are consistent with the theory of relativity.

The time scale Barycentric Dynamical Time (TDB) was introduced in 1976 to be a time scale for the barycenter of the Solar System such that it differed from TDT only by periodic terms. This results in a definitional ambiguity dependent upon the length of time being considered. Also, the definition results in different values of some constants, such as the unit of length, to be used with the two time scales. These time scales can be precisely determined from atomic clocks and the theory of relativity. They do not consider the variability of the rotation of the Earth as needed for actual observations.

1.4 Earth orientation and reference frames

During the first three quarters of the twentieth century, the rotation of the Earth and polar motion were determined by observations with visual or photographic zenith tubes, transit circles, and astrolabes. Then the techniques of lunar and satellite laser ranging (Section 2.4.3), radio interferometry (Section 2.4.2), and the Global Positioning System (Section 2.4.1) became available and much better accuracy could be achieved. The International Earth Rotation Service (IERS) was founded to coordinate these activities and to provide international values for polar motion and other Earth rotation parameters.

In determining Earth orientation parameters by means of Very Long Baseline Interferometry (VLBI) many observations of radio sources have been made. The accuracies of the positions are about a milliarcsecond and the sources are very distant, so they have no appreciable proper motions. Thus, it became apparent that an accurate, fixed, stellar reference frame could be developed, improving the preceding FK5 reference frame.

A list of stable, point-like radio sources was needed. The ties between these generally optically faint sources and optical and dynamical reference frames were necessary. Such a list of sources was prepared and investigated for structural motions so that an International Celestial Reference Frame (ICRF) could be designated (Section 7.2). Optically this reference frame is represented by the *Hipparcos Catalogue* (Section 11.6.1). The reference frame is independent of the dynamical reference frame, defined by the equator and the ecliptic planes which are dependent on the Solar System bodies. The origin of the ICRF has been selected to be as close as possible to the FK5 origin and the dynamical equinox of J2000.0. However, in principle, the ICRF is based on observations from the Earth and the kinematics of the Earth's equator. A new Celestial Intermediate Pole (CIP), and Celestial Ephemeris Origin (CEO) were adopted in 2000 by the IAU. Solar System observations can be made on this reference frame and the equinox and ecliptic determined on that system.

The Hipparcos satellite has provided milliarcsecond accuracies for optical sources. VLBI is determining radio sources at sub-milliarcsecond levels and detecting motions of 25 microarcseconds per year. Optical interferometers in space are planned for microarcsecond accuracies. Thus, astrometry must be developed fundamentally on an extragalactic reference frame to microarcsecond accuracies.

2

New observational techniques

What characterizes the new observational techniques is, of course, not only that they appeared recently – say in the past couple of decades – but also that they strive to reach one or two orders of magnitude better accuracy than the classical techniques. The unit for describing the new astrometric capabilities is no longer one tenth of a second of arc, but one thousandth of a second of arc, a subunit that we shall designate throughout this book by the abbreviation mas (milliarcsecond). Plans are underway to reach microarcsecond (μas) accuracies. Such a gain in precision impacts directly on all aspects of astronomy and, particularly, on the reduction procedures. One may divide these new techniques into three major groups: interferometry, time-measuring techniques, and space astrometry.

2.1 New detectors

Many astrometric techniques take advantage of the development of much more sensitive new detectors. Among them, one must mention the following.

- *The CCD detector.* The CCD (for charge coupled device) has become a major tool in astronomy since the 1970s (Monet, 1988). It consists of a semiconductor device, where a photoelectric effect takes place when light reaches it, producing an electronic image. This image is preserved by arrays of small positive electrodes, which attract photoelectrons and keep them in a similar array of potential wells, providing the possibility of long exposure times by adding photoelectrons in the same pack (Kovalevsky, 2002). Once the exposure is over, one shifts, by a periodic change of the potential of electrodes, the electronic image along the array. It is then recovered pack by pack at the edge of the array, digitized, and registered in a computer. There, software can analyse it and produce images, determine relative positions or magnitudes of the objects, or deduce any other information that is needed. It is also possible to synchronize the transfer speed with the image shift due to the rotation of the Earth. This is the drift–scan mode, which allows registration of a long strip of the sky with a fixed telescope. The CCD technique has completely superseded astronomical photography. For more information, see Martinez and Klotz (1999).

- *Infrared detectors.* Infrared astronomy and astrometry are now gaining importance. Photon detectors have the property to absorb photons, which cause electron transitions and alter the electrical properties of the substrate. The photons may change the electrical conductivity (photoconductive detectors) or generate electron pairs, which are separated and produce a detectable potential difference (photovoltaic detectors). Now, they are often associated with a CCD, which ensures the readout. For instance, the NICMOS receiver (Near Infrared Camera for Multi Object Spectrograph), has an array dimension of 1024×1024 pixels. These, and a few other infrared detectors, are presented by Léna (1996).

2.2 Basics of interferometry

The principle of interferometry is based upon the propagation properties of electromagnetic waves, whatever their domain (optical, radio, microwaves, etc.). The propagation of an electro-magnetic wave along a certain direction Ox is represented by its instantaneous amplitude U at time t:

$$U = a \cos \omega (t - t_0), \tag{2.1}$$

where a is the amplitude of the wave, t_0 is a time at which U is maximum, and ω is the angular frequency. It is related to the frequency by

$$f = \omega/2\pi,$$

to the period of the wave

$$T = 1/f = 2\pi/\omega,$$

and to the wavelength λ' by

$$\lambda' = \frac{v}{f}, \tag{2.2}$$

where v is the speed of light in the medium. In vacuum, $v = c = 299\,729\,458$ m/s exactly. This actually defines the meter as the length of the path traveled by light in vacuum during a time interval of $1/299\,729\,458$ of a second. The quantity c/v depends upon properties of the medium. It is the index of refraction of the medium

$$n = \frac{c}{v}. \tag{2.3}$$

It is always larger than one. For vacuum, $n = 1$ and is, therefore, often neglected in the equations. Generally, one considers only the wavelengths λ in a vacuum, so that

$$\lambda = \frac{c}{f} = \frac{2\pi c}{\omega},$$

and

$$\lambda' = \frac{c}{f} = \frac{2\pi nc}{\omega}. \tag{2.4}$$

Let us now assume that, at time t_0, the light has already propagated along the Ox axis by a distance x. Then $t_0 = x/v$ and using the previous equations, (2.1) transforms into

$$U = a \cos\left(\omega t - \frac{2\pi nx}{\lambda}\right), \tag{2.5}$$

where the quantity nx is called the optical path. For simplicity, one sometimes introduces a phase ϕ

$$\phi = \frac{2\pi nx}{\lambda}, \tag{2.6}$$

and U becomes

$$U = a \cos(\omega t - \phi).$$

These two results are fundamental for interferometry.

2.2.1 The superposition principle

This principle states that when two (or several) waves with the same wavelength are superposed and propagate along the same direction, they can be considered as being a single wave whose instantaneous amplitude is the geometric sum of the individual instantaneous amplitudes of the separate waves.

As an example, let us consider two monochromatic waves with the same amplitude, but different phases, ϕ and ϕ',

$$U = a \cos(\omega t - \phi); \quad U' = a \cos(\omega t - \phi').$$

The combined instantaneous amplitude is

$$V = U + U' = a[\cos(\omega t - \phi) + \cos(\omega t - \phi')]$$

$$= 2a \cos\frac{\phi - \phi'}{2} \cos\left(\omega t - \frac{\phi + \phi'}{2}\right). \tag{2.7}$$

The new amplitude,

$$A = 4a^2 \cos^2 \frac{\pi}{\lambda}(nxnx')$$

$$= 2a^2 \left(1 + \cos \frac{2\pi n(x - x')}{\lambda}\right), \tag{2.8}$$

depends upon the difference of optical paths of each wave as expressed in units of wavelengths. Thus, there can be both constructive and destructive interference between the two waves, and the resulting amplitude can be anything between 0 and $2a$.

2.2.2 Illumination

The intensity of light (or the illumination) is proportional to the square of the instantaneous amplitude of the wave

$$I \cong \frac{1}{T} \int_0^T U^2 \mathrm{d}t = \frac{a^2}{2}.$$

Applying this to the amplitude (2.7), one gets

$$I \cong 2a^2 \cos \frac{2\pi}{\lambda}(nx - n'x). \tag{2.9}$$

If we assume that the resulting wave is observed in such a way that the difference of optical paths varies with time, the illumination will change with time. If different paths correspond to different points on a screen, one will see periodic patterns of illuminated and dark regions; these are the interference fringes. The most classical example is Young's experiment in which two small holes in a diaphragm allow only a plane wavefront (hence with the same phase) to cross. By diffraction, each of them illuminates a large surface of the screen (Fig. 2.1). In some places, the difference of optical paths is exactly an integer number of wavelengths, and the illumination is maximum. In other places, when this difference is equal to an integer number of wavelengths plus half a wavelength, then $I = 0$. The pattern of illumination is a succession of dark and light regions called interference fringes.

2.2.3 Coherence length and bandwidth

Natural light is never strictly monochromatic. Even with the most narrow filters, a light beam contains rays of different wavelengths, or frequencies,

$$\lambda_0 - \Delta\lambda \le \lambda \le \lambda_0 + \Delta\lambda$$
$$f_0 - \Delta f \le f \le f_0 + \Delta f.$$

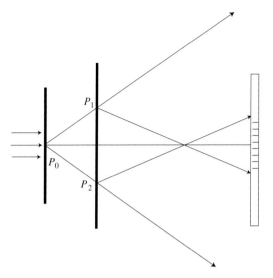

Fig. 2.1. Scheme of Young's experiment and the structure of fringes.

If such a beam is allowed to interfere, for instance in Young's experiment, each ray system with a given wavelength, λ, will produce a fringe pattern of its own and the ensemble will mix. In points for which $x = x'$, for all λ, I is maximum. It is called white, or *central fringe*. But the next ones will be slightly displaced, resulting in a blurring which increases when $x' - x$ increases. It increases more quickly for large $\Delta\lambda$ than for small $\Delta\lambda$. At a certain limit, fringes become indistinguishable. There is a mathematical definition of this limit. It is called the *coherence length*, whose value is

$$\Delta x = \lambda^2/\Delta\lambda. \tag{2.10}$$

Transforming this into frequencies one gets the *bandwidth*

$$\Delta f = c\Delta\lambda/\lambda^2. \tag{2.11}$$

The time taken by the light to travel Δx is called the *coherence time*

$$\Delta t = 1/\Delta f. \tag{2.12}$$

Examples. For $\lambda = 0.5\,\mu$m and a filter (bandwidth) of 1 nm, the coherence length is $\Delta x = 0.25$ nm.

For $f = 10$ GHz ($\lambda_0 = 3$ m) and a bandwidth of 2 MHz, the coherence time is $0.5\,\mu$s.

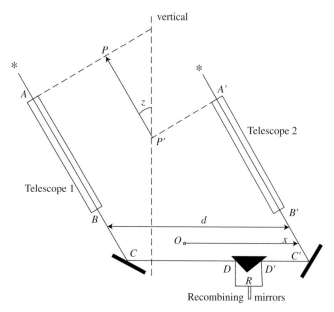

Fig. 2.2. Principle of a Michelson North–South interferometer. In this figure, it is assumed, for simplicity, that the star is in the meridian plane containing the two telescopes.

2.3 Stellar interferometry

2.3.1 Principle of the Michelson interferometer

Let two identical telescopes point at some given star. Let us first assume that it is a unique point-like source and all rays in a plane perpendicular to the direction of the star have the same phase ϕ. Let us place some mirrors in such a way that the two beams are recombined into a single one. Figure 2.2 presents a schematic view of such an interferometer at a moment when the star and the two axes of the telescopes are in the same vertical (meridian) plane. If R is the receiver, the optical path from the wavefront P, for telescope 1, is

$$x_1 = AB + BC + CD + DR.$$

For telescope 2, it is

$$x_2 = PP' + A'B' + B'C' + C'D' + D'R.$$

We assume that the segments directly attached to the interferometer are well known or can be calibrated.

If the mirror DRD' is mobile along CC', one can put it in such a way that $x_1 = x_2$, so that R receives the central fringe of the interferometric pattern built up by the two beams.

If now DD' is servo controlled by R in such a way that R always receives the central fringe, the measurement of $CD - C'D'$ equals PP' and its variations with time, while the direction of the star changes because of the rotation of the Earth. The quantity PP' is the optical path difference of the rays at the entrance of each telescope or, at the arrival at the baseline $CC' = d$ of the interferometer. In the case described by Fig. 2.2, this path difference $\Delta x = d \sin z$ where z, is the angle of the zenith distance of the star.

This property can be extended to any configuration of the interferometer, and if, in some local system of coordinates, the unit vector of the direction of the star is $\mathbf{S}$ and $\mathbf{B}$ is the base vector of the interferometer, one has

$$\Delta x = \mathbf{S} \cdot \mathbf{B}. \tag{2.13}$$

In other terms, measuring Δx gives a relation between the apparent direction of the star and the base vector attached to the Earth.

The main limitations in Michelson astrometric interferometry are the stability of the baseline throughout the night, the accuracy of the laser interferometers that monitor the positions of vital parts of the optical design, and the evaluation of the refraction in a consistent manner throughout the night. Another difficulty comes from the instability of the images due to atmospheric turbulence. More critical is the fact that this turbulence deforms the wavefront differently in each telescope, so that a stochastic additional error in the optical path difference is introduced. The region of the sky within which the refraction effect is sufficiently uniform to permit interferometry is called the isoplanatic patch. Its importance varies depending on the seeing during the observation. In any case, it represents a natural limitation to the precision of the instrument, generally a few milliarcseconds. To get rid of it, one should use active, or adaptive, optics, which are now commonly in use in large telescopes, but are very expensive and would represent an additional complication to the instrument.

2.3.2 *Other uses of optical interferometry*

Since A. Labeyrie (1975) obtained the first fringes on a two-telescope interferometer, many other optical interferometers have been built and become operational. But they have had an objective that is less difficult to reach than determination of star positions over wide angles. The objective was to determine the apparent diameters of stars.

Each point of a star's surface emits independently of the others, so that it produces an independent interference pattern. The superposition principle does not apply, and the illumination produced is the sum of the individual illuminations by elementary fringes.

If the baselines are small, the differences of optical paths between various points on the star are small, and the interference patterns are very similar, resulting in a blurring, which is measured by the fringe visibility

$$V = \frac{I_{\max} - I_{\min}}{I_{\max} + I_{\min}}.$$

One can show that, in the case of a North–South horizontal interferometer observing in the meridian plane,

$$V = \frac{J_1(2\pi DR \cos\phi \cos\delta)}{\pi DR \cos\phi \cos\delta}, \tag{2.14}$$

where R is the radius of the star expressed in radians, δ is its declination, and ϕ is the latitude of the observatory. J_1 is the first-order Bessel function. For different directions, the formula is somewhat more complicated, but the principle remains. Interferometers designed to determine R can observe with different baselines, D, so that by fitting the observed visibilities to formula (2.14), one determines R. Various existing interferometers have baselines ranging from 12 to 60 m or more, and the precision with which radii are determined is of the order of a few milliarcseconds. The result, however, may be biased, if the star has a noticeable limb darkening which, in this case, has to be modeled.

The same instruments are also able to determine the separation, the position angle, and the magnitude difference between the components of a double star. The observations must be performed in different directions of the sky using the Earth's rotation to get a variety of viewing orientations. The modeling of the illumination is that of two separate point sources. In this case, the precision of a few milliarcseconds is also attained. Of course, an astrometric optical interferometer can also be used to measure visibilities, or interferometric responses, of a double star in what is called the imaging mode.

2.3.3 The main interferometers

The technology has advanced to the state that optical interferometers are now built in many places. Specifically, computer-controlled delay lines, laser metrology systems, and detectors are now possible at the accuracies and speeds necessary for optical interferometric astrometry measurements. Thus, a number of optical interferometers have been designed and are in stages between construction and routine operation. Space-based optical interferometers are under development. Short descriptions and references to these interferometers follow. All interferometers feature one or more actively controlled optical elements to compensate for atmospheric turbulence and image motion. Several have vacuum feed systems to avoid the effects

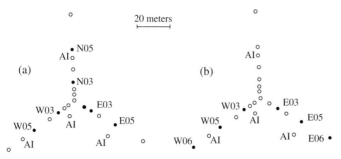

Fig. 2.3. General scheme of the main elements of the NPOI interferometer. Black dots indicate two possible observing configurations; (a) and (b) give different coverages of imaging planes.

of ground turbulence. All the smaller-aperture interferometers are Michelson interferometers, where the combination of the afocal beams takes place at one or more beam splitters. All interferometers employ optical delay lines (with the exception of ISI, which uses electronic delays) to remove the path difference in order to combine the beams of light within their coherence length. Note the large step between the classic interferometers and the new giant large-aperture interferometers. The latter require adaptive optics to co-phase the apertures.

(i) *Cambridge Optical Aperture Synthesis Telescope (COAST)*
COAST incorporates five 40 cm telescopes in a Y configuration with a maximum separation of 48 m. Up to four telescopes can be used simultaneously. It is designed to have telescopes up to 100 m apart. This is primarily an imaging system and they plan to image the surfaces of giant stars. More information is at http://www.mrao.cam.ac.uk/telescopes/coast/status.html.

(ii) *Navy Prototype Optical Interferometer (NPOI)*
NPOI is a large optical interferometer for both astrometry and imaging. There are three 250 m legs in a Y configuration with four astrometric huts within the inner 38 m baselines and stations for imaging siderostats providing baselines between 2 m and 440 m as shown in Fig. 2.3. Six fast vacuum delay lines for astrometry can be augmented by long delay lines for imaging. The siderostats are 50 cm feeding image compressors and vacuum light pipes to the delay lines and detectors. The beam combiners will accept up to six beams. This is a joint project of the US Naval Observatory, Naval Research Laboratory, and Lowell Observatory, being built on Anderson Mesa in Arizona. More information is available at http://usno.navy.mil/projects/npoi.

(iii) *Sydney University Stellar Interferometer (SUSI)*
SUSI has been designed to operate with North–South baselines from 5 m to 640 m, with intermediate baselines forming a geometric progression increasing in steps of $\sim 40\%$. This gives resolutions from 20 mas to 70 μas at wavelengths of 450 nm. There is one East station for a potential East–West arm of the interferometer.

This interferometer has the highest resolving power of all on the longest baseline, making it suitable for the determination of the diameters of early type dwarfs. More information is available at

http://www.physics.usyd.edu.au/astron/susi/susi_baselines.html.

(iv) *Infrared-optical Telescope Array (IOTA)*

IOTA is a Michelson stellar interferometer located on Mt Hopkins, Arizona, with three 45 cm collectors which can be located on different stations on each arm of an L-shaped array of 15 m by 35 m, with a maximum baseline of 38 m. At visible and near-infrared (near-IR) wavelengths, combination occurs at a beam splitter. Also at near-IR wavelengths, the FLUOR experiment combination occurs in single-mode fibers. Detection is done in the near-IR with an HgCdTe NICMOS3 array and in the visible with a CCD. Organizations operating this array include the Smithsonian Astrophysical Observatory, Harvard University, and the University of Massachusetts. More information is at

http://cfa-www.harvard.edu/cfa/oir/IOTA/Description.html.

(v) *Infrared Spatial Interferometer (ISI)*

ISI is a two-telescope interferometer operated by the University of California at Berkeley on Mt Wilson Observatory. The telescopes are 65 inch fixed parabolic mirrors and 80 inch flat mirrors, operating as siderostats. Observations are made at mid-infrared wavelengths of 11 μm using the heterodyne technique with a CO laser as a local oscillator. The telescopes, electronics, and computer control systems are contained in trailors, which can be moved to reconfigure the baselines. A third telescope is being added to the interferometer and three baselines will be operated simultaneously. More information is at http://isi.ssl.berkeley.edu/system.html.

(vi) *Palomar Testbed Interferometer (PTI)*

The PTI is a K-band 110 m maximum baseline stellar interferometer located at Palomar Observatory. Two different baselines are available. This interferometer was conceived, designed, and constructed primarily as an engineering facility to demonstrate ground-based differential astrometry for the detection of faint companions and planets. More information is available at

http://huey.jpl.nasa.gov/palomar/.

(vii) *Center for High Angular Resolution Astronomy (CHARA) Array*

The CHARA Array has six 1 m telescopes dispersed in a two-dimensional layout with a maximum baseline of 350 m on Mt Wilson in California. The light from the six telescopes is conveyed through vacuum tubes to a central Beam Synthesis Facility. More information about CHARA can be found at

http://www.chara.gsu.edu/CHARA/array.html.

(viii) *The Grand interferomètre à deux télescopes (GI2T)*

This optical and infrared interferometer in France allows siderostats to be placed along a North–South baseline with separations between 12 m and 65 m. The telescopes have 1.5 m diameter primaries and feature a unique alt-azimuth mount based on a concrete sphere. More information can be found at

http://www.obs-nice.fr/fresnel/gi2t/en/.

(ix) *Keck Interferometer (KI)*

The Keck Interferometer will combine the two 10 m Keck telescopes with four proposed 1.8 m outrigger telescopes as an interferometric array. A Michelson beam combination will be used for the 85 m baseline of the Kecks and the 25–140 m baselines of the outriggers. The Kecks have adaptive optics and the outriggers will have fast tip–tilt correction. The co-phasing system uses active delay lines in the beam combining room and dual-star modules at each telescope. Back-end beam combiners will be two-way beam combiners at 1.5–3.4 μm for fringe tracking, astrometry, and imaging, a multi-way combiner at 1.5–5 μm for imaging, and a nulling combiner for high dynamic range observations at 10 μm. The first interferometric fringes were observed in March 2001. More information is available at http://huey.jpl.nasa.gov/keck/publicWWW/techdesc/index.html.

(x) *Large Binocular Telescope (LBT)*

The LBT is two 8.4 m $F/1.142$ primary mirrors on a common mount on Mt Graham in Arizona. This gives the light-gathering power of a single 11.8 m telescope and the resolving power of a 22.8 m telescope. The focal stations will be infrared, dual $F/15$, Gregorian, and phased combined, reimaged $F/15$, center. The instruments will be interferometric imaging at 0.4–400 μm, infrared imaging/photometry at 2.0–30 μm, wide-field multi-object spectroscopy at 0.3–1.6 μm, faint-object/long-slit spectroscopy at 0.3–30 μm, and high-resolution spectroscopy at 0.3–30 μm. More information is at

http://medusa.as.arizona.edu/ibtwww/tech/lbtbook.html.

(xi) *The Very Large Telescope Interferometer (VLTI)*

This array will ultimately combine all four 8 m telescopes of the European Southern Observatory's VLT on Cerro Paranal in Chile. Five auxiliary telescopes with 1.8 m primaries are planned to provide more baselines with a maximum length of about 200 m. Interference fringes were observed between two of the 8 m telescopes in October of 2001. The instruments include near- and far-infrared beam combiners, as well as a camera for phase-referenced imaging and microarcsecond astrometry. More information is available at http://www.eso.org/projects/vlti/.

2.3.4 Fizeau interferometer

The major drawback of Michelson interferometry is that the useful field of view is very small and does not exceed a few seconds of arc. There are essentially two reasons for this. One is that the collimated light has to travel for at least one half of the baseline B. A small displacement in the field of view by an angle α displaces the ray by $B\alpha$, assuming no magnification. In practice, it is much larger unless the intermediary optics are very large. The second reason is that there is no geometric relation between the images produced by the two telescopes, so that the superposition of rays disappears very quickly from the recombination center and no fringes are built.

The solution to suppress these problems is that a single telescope is used to produce the two interfering rays. These are separated by occulting the whole mirror with the exception of two entry pupils. This is the principle of the Fizeau interferometer: both entrance pupils produce the same image, and therefore the rays interfere uniformly all over the field of view, limited only by the aberrations of the optics in places where images are distorted differently. Therefore, the useful field can be as large as 1°, as large as the field of view of the full telescope.

A Fizeau interferometer is much more difficult to build and the baselines are necessarily small, of the order of the diameter of a telescope mirror. Actually, it is not necessary to have the full telescope built, but the two areas must be shaped and solidly placed, one with respect to the other, as if they are parts of a unique, large mirror. Such systems have not yet been built, but they have been proposed to be flown in several future astrometric space missions. The advantage would have been that interferences would be produced for all stars present in the field of view, allowing simultaneous determinations of their relative positions. But actually, the Fizeau interferometer design has not been retained.

2.3.5 Speckle interferometry

This new observational method was introduced by A. Labeyrie (1970). The turbulent atmosphere near the ground consists of cells 5–20 cm across, depending on the wind and the thermal response of the ground. The light coming from a point-like source and the crossing cells interfere and produce irregular patterns at the focus of the telescope called speckles, each corresponding in size to the resolving power of the instrument. Hence, the speckles contain all the angular information that the instrument might have provided were the atmosphere ideally calm. The speckles move and produce fuzzy and spread out images, whose dimensions depend on the *seeing*. But if the exposure time is as short as 1/100th of a second, they do not have the time to move, or deform significantly, and each is a noisy image of the star's Airy diffraction disk. The theory of the speckle interferometry was made first by Korff (1973) and improved by Roddier and Roddier (1975). In practice, it consists of combining the information of these speckles, recorded on a fast CCD (see Section 2.1), and transferred to a computer which computes an autocorrelation function of a large number of images taken during the observation. Finally, one has to smooth out the autocorrelations and subtract the background. One way is to compute the Fourier transform of the autocorrelation function and interpret the information contained in the Fourier space.

This technique recovers much of the resolution power of telescopes. For instance, a 4 m telescope with the best seeing gives images larger than $0\overset{''}{.}5$, while the resolving power of the speckle interferometry is $0\overset{''}{.}03$. This is illustrated in Section 12.3.2 as applied to observations of double stars.

2.4 Applications of time measurements

Several new astrometric techniques are not based on direct measurements of angles, but on an indirect method. The observation consists either of a measure of time intervals, or involves synchronization of very stable clocks.

Time or frequencies are, at present, the physical quantities that are measured with the greatest accuracy. The time scale universally adopted is the international atomic time (TAI: Temps Atomique International), which is based on cesium frequency standards operated by some primary metrological laboratories. At present, the most accurate is the atomic fountain based upon cold – and consequently with very small Doppler broadening – cesium atoms producing the spectral line selected to define the second, and set by definition as having a frequency $f_0 = 9\,192\,631\,770$ Hz. The first ever built operates regularly at the Paris Observatory with an uncertainty of 10^{-15}. It is now approaching 5×10^{-16}. Several other such *cesium fountain clocks* have been built using the same technology, and have similar accuracies.

2.4.1 The Global Positioning System and TAI

In principle, the Global Positioning System (GPS) is a ranging system that allows the determination of the position of a point on the Earth from measured distances and Doppler shifts to several satellites, whose orbits are very well known with respect to a terrestrial reference frame (see Chapter 9). It is essentially a navigation and geodetic tool, but it is used as an off-line tool in many astrometric applications, so that it is natural to present it here. A detailed description of the system and its uses can be found in Hofmann-Wallenhoff *et al.* (1992).

All satellites have on-board clocks that are synchronized to a unique time scale, called the GPS time scale, which is, in turn, well defined with respect to the TAI. They emit signals at some specified instant, t_s, of the on-board clock, and the ground stations record the arrival time, t_g, in its own scale, which has to be corrected to fit the GPS time scale. The range is derived from $R = c(t_g - t_s)$ after corrections for ionospheric and atmospheric refraction and relativistic effects.

The GPS satellites transmit the orbit ephemerides, but more precise orbital elements are available after a few days from the International GPS Service (IGS), which determines them anew from observations made at about 40 sites around the world. In the satellite transmission, a formula to transform the on-board time into the GPS time is given. The signal is emitted at two frequencies, which allow determination of the first-order propagation refraction correction.

By using the improved IGS orbit, and accumulating data over several days, it is possible to locate the receiving station with an uncertainty of 1 cm in the dual-frequency mode.

GPS also plays a major role in establishing TAI. If a satellite is observed simultaneously by two stations with known positions on the Earth, one gets the difference between the times provided by the two ground clocks.

The TAI unit interval is set to agree with the observations of atomic standards, but the stability is ensured by the comparison of more than 200 atomic clocks dispersed around the world and intercompared using the GPS-based technique. The clocks on board the satellites are intermediary between the users and master clocks in the US Naval Observatory, well connected to TAI. The relative accuracy of TAI is of the order of one in 10^{14}, and its stability is evaluated to be of a few units in 10^{15}. Many clocks are built industrially and have an accuracy and stability better than one in 10^{13} for periods ranging from one hour to one year. A much better short-range stability is achieved by hydrogen masers (down to one part in 10^{15} during a few hours, 10^{14} for up to one month), but the link to the actual definition of the second is not yet as good. The frequency standards are very useful in some astronomical applications, such as Very Long Baseline Radio Interferometry, in which a very good, short-term stability of frequencies is needed.

2.4.2 VLBI (Very Long Baseline Interferometry)

The resolving power of an interferometer is of the order of λ/D. So, for $D = 20$ m at $\lambda = 0.5$ μm, it is about 10 mas, a result that can be somewhat improved by adding up many observations. Unless D becomes very large, the milliarcsecond seems to be a limiting boundary. But for observations in the radio domain, with wavelengths of a few centimeters, to reach analogous precision necessitates a very large D, namely a few thousand kilometers. While there exist connected radio interferometers, which are based on the same principles as optical, as for instance the Very Large Array (VLA; Thomson *et al.*, 1980), their dimensions hardly reach 230 kilometers (MERLIN in England; Thomasson, 1986). For better accuracy, a completely different approach is taken. This is the case of VLBI. A basic book on VLBI is by Felli and Spencer (1989). Its applications to astrometry are discussed in Walter and Sovers (2000).

The basic idea is to record signals that are received in each radio telescope, and then correlate them in a computer. An antenna transforms the electro-magnetic field of a radio wave into an electric potential, v, proportional to the instantaneous amplitude of the wavefront

$$v = b\cos(\omega t - \phi),$$

where ϕ is the phase of the wave to which a constant, instrumental dephasing is added. Remembering that 3 cm wavelength corresponds to a 10 GHz frequency, such a frequency cannot be continuously analyzed. A local oscillator, generally a hydrogen maser, produces a stable reference frequency,

$$v' = b' \cos \omega_0 t,$$

where ω_0 is close to ω.

These frequencies are mixed in a special heterodyne circuit. After filtering out high-frequency terms, there remains a periodic electrical intensity proportional to b and b', an angular frequency $\omega - \omega_0$, and the same phase ϕ

$$I \cong bb' \cos ((\omega - \omega_0)t - \phi). \tag{2.15}$$

Choosing ω_0 in such a way that the frequency of $\omega - \omega_0$ is of the order of 1–2 MHz, it is now possible to sample this frequency and represent numerically on a tape the time variations of (2.15). The timing is accurately marked on the same tape.

At a second radio telescope the same procedure is followed on the same radio source. Then both recordings are brought together and compared. The better the local clocks are synchronized, the better one can recognize that the two records correspond to the same wavefront. An approximate reckoning of the path difference produces a first approximation to the time delay, Δt_0, between the two wavefronts. Let Δt be the correction to t_0 that gives the exact time delay. Then the correlation between the two signals is a maximum. So, the special computer, called a correlator, computes the value of Δt such that the normalized correlation function

$$R(\Delta t) = \frac{\int_{t_1}^{t_2} I_1(t)I_2(t + t_0 + \Delta t)dt}{\int_{t_1}^{t_2} I_1^2 dt \int_{t_1}^{t_2} I_2^2 dt} \tag{2.16}$$

is a maximum.

The actual determination of the time delay at a reference instant is much more complex. One has first to correct for the unknown ionospheric and tropospheric refractions. This is done by observing simultaneously at two frequencies, the S band (2300 MHz) and the X band (8400 MHz). One has also to take into account the continuous variation of the received frequency due to the Doppler shift produced by the rotation of the Earth.

In order to increase the precision of the result, two techniques are simultaneously used. First one increases the time interval $t_1 - t_2$ present in formula (2.16), but this implies a very strict identification of the time markers in both tapes. Independent of the stability provided by hydrogen masers, the station clocks must indicate the same time scale. This is done through frequent synchronization using the GPS system. The other technique is to separate the observed band into a number of adjacent narrow bands and make a simultaneous reduction on a parallel computer. The Mark III receiver and correlator work simultaneously on the S and X bands, subdivided

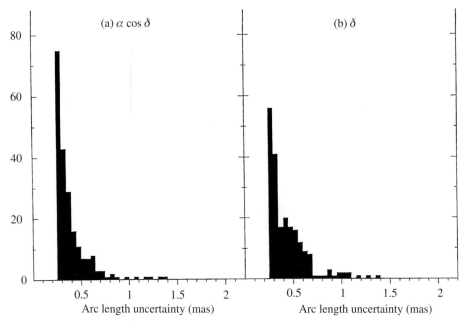

Fig. 2.4. Histograms of uncertainties of the coordinates of the 212 defining radio sources of the ICRF (Ma *et al.*, 1998).

into 28 channels of 2 MHz bandwidth. Even with a two-point sampling per period, more than 100 million bits have to be processed for each second of observation.

The fundamental equation applied to interpret the data is formula (2.13). So VLBI can be used to determine the baseline, assuming the celestial coordinates of the sources, usually quasars, are known. This gives information both on the geodetic position of the antennae and the parameters of the rotation of the Earth. Conversely, knowing the position of antennae and the elements of the Earth's rotation, one can determine the directions, **S**, of sources.

Actually, there are too many parameters to be determined from a single baseline. So, networks of several radio telescopes are operated for any observing program. The uncertainty of a single observation of a source is about 5 mas, but synthetic treatment of many baselines gives the parameters of the rotation of the Earth with uncertainties of about 0.2 mas. A combined solution for source positions has been iterated by IERS (International Earth Rotation Service). It uses results from the several existing VLBI networks. A total of 606 sources constitutes the final solution, the International Celestial Reference Frame (ICRF). But most of them have changing structures, and must be continuously monitored. Only 212 have very good observing records, and are considered as primary reference sources (see Fig. 2.4, and Section 7.2).

Furthermore, the positions of about 20 radio telescopes contributing to the VLBI networks are determined with uncertainties of the order of 0.5 mm. They are the primary net of positions upon which the International Terrestrial Reference Frame (ITRF) is based (see Chapter 9).

A difficulty with VLBI – in addition to its very complex and costly operations – is that the great majority of observable radio sources are optically faint extragalactic objects. This is optimum for defining a non-rotating reference system. But such a reference frame is necessary also for stars, and the problem is that there are almost no stars emitting radio waves with a sufficient intensity to be observable by VLBI. The link between radio and optical reference coordinates relies at present on only 11 radio stars observed with VLBI and a few dozens of other stars which have been observed with connected interferometry (VLA, MERLIN), but, in those cases, with only a few milliarcsecond uncertainty.

2.4.3 *Laser ranging*

Another use of timing is measuring distances. Since the meter is defined as a function of the velocity of light in vacuum, measuring the time during which a light pulse travels from a terrestrial source to and back from a target is actually a measure of twice the distance, provided the necessary corrections for velocity through air, relative motions, and the instrument delays are applied. In addition to artificial satellites, widely used as targets for geodetic purposes (see Chapter 9), the astronomical application is laser ranging to the Moon. Let us describe first the lunar laser ranging, which is the most difficult to realize and operate.

The main part of laser ranging equipment is a Q-switched laser which provides very short light pulses. The most common type is a crystal containing neodymium (YAG) which emits in near infrared (1.064 μm). The light beam then crosses a frequency-doubling system, so that the actual emission includes both 1.064 and 0.532 μm. The width of a pulse can now be as short as 0.1 ns with a repetition rate of 10 Hz or more. Note that the length of a 0.1 ns pulse is only 3 cm, so that the uncertainty of the distance due to the scatter of photons around the center of the pulse is only 0.75 cm.

The light pulse is collimated towards the Moon by a large telescope (1.50 m aperture in the case of the CERGA lunar laser, which has the best performance) guided by an ephemeris of one of the retroreflectors deposited on the Moon by three US Apollo and one Soviet Lunokhod missions (Kovalevsky, 2002). These return the received light beam in the direction of its source. On the ground, the same telescope directs the return photons to a photomultiplier. The return time is determined by an event timer that receives signals from the outcoming laser pulse and the receiver.

The main difficulty is to recognize the return signal from noise generated by Moon light. This is done using a very narrow band filter in front of the photomultiplier, which is opened during only 50 ns around the theoretical return time as computed from the ephemeris. Because of the tremendous loss of photons due to atmospheric turbulence, the diffraction in the telescope and the retroreflector, and the transmission factors of the optics, the atmosphere, and the receiver, a real return is only occasionally detected. Generally an observing session lasts 10–15 min, representing something like 6000 pulses, out of which about 10–40 are recorded in good nights. They are recognized from the background by computing the residuals from comparing with the ephemerides; they all have the same residual whereas noise photons have a random distribution in time.

At present, after various instrumental and calibration corrections, the uncertainty achieved is of the order of 5 mm and the accuracy is probably about 1 cm. This is a major tool for the study of the Earth–Moon dynamics, including the rotations of both bodies. It also provides information on general relativity parameters and dynamical reference frames.

Satellite laser ranging systems are built under the same scheme. However, the satellites being much closer to the Earth, the efficiency of the illumination is much larger than in the case of the Moon. This allows the recognition of the returns by the fact that two or more photoelectrons are received within the pulse time. This is achieved in spite of the fact that artificial satellite ephemerides are not as accurate. It is necessary to broaden the outcoming beam to several tens of arcseconds. Many of the satellite laser ranging systems in existence have accuracies of the order of 2 cm.

2.4.4 Planetary radars

Although optical ranging is limited to the neighborhood of the Earth and where retroreflectors are located, it is possible to use large radio telescopes to send pulses to planets, minor planets, or satellites, and then to receive return signals, so that, with the same principles, the distance between a point on the Earth and the closest portion of the surface of the planet may be obtained. The terrestrial planets, Jupiter satellites, and many asteroids have been ranged by this technique. Difficulties arise from the fact that the points on the surface are located at different distances from the center of mass of the planet, so that the main source of error is not the timing – which is made to a microsecond precision – but the uncertainty of the shape of the satellite or the planet, which can be of the order of a few kilometers. Nevertheless, ranging provides the most accurate measurements of Solar System positions for the construction of planetary ephemerides.

In the case of minor planets, the radar technique has been used to determine the shapes of some of these bodies and their rotation periods. Additional information is provided, in this case, by the broadening of the Doppler width of the signal. Radar observations of asteroids are also essential to the preparation of space missions to these bodies and to the control of the operations of the space probe around the encounter.

2.4.5 Pulsar timing

This is a very important development of timing techniques applied to astronomy. It consists of observing the time variations of the emissions of a pulsar, and studying their periods and the evolution of the period with time. These observations are useful in discussing the reference frames and are used as a test of Solar System ephemerides accuracies. Their contribution as a time source has been suggested, but is still uncertain.

The various aspects of the interpretation of these observations are discussed in Section 14.7.

2.5 Launched space astrometry satellites

Among the most limiting factors in ground-based astrometric observations are the atmospheric refraction and turbulence, the deformations of the mechanical parts of an instrument, and the fact that from no single point on the Earth can one observe the whole sky. These drawbacks are suppressed for an astrometric instrument in space. However, other difficulties then arise: the cost of space instrumentation, launch, and operations, and the instability of the satellite which is subject to a number of disturbing forces, while the Earth is a very stable platform whose motions are very well monitored and modeled.

Space astrometry was born in 1989 with the launch of Hipparcos and, in 1990, of the Hubble Space Telescope. While the first satellite was essentially an astrometric satellite – although it also obtained important results in photometry – the second is a multi-purpose space telescope which has, among others, astrometric capabilities. We will briefly describe them, referring the reader to more complete descriptions of the instruments and their reduction procedures, which are much too complex to be summarized in a few pages.

2.5.1 Hipparcos

The main principle of Hipparcos is to scan the sky along great circles in order to determine, on a reference great circle (RGC), the abscissae of stars situated in a small ring around the RGC.

Then, by varying the position of the RGCs following a certain scanning law, which allows all the parts of the sky to be successively observed, one gets, for each

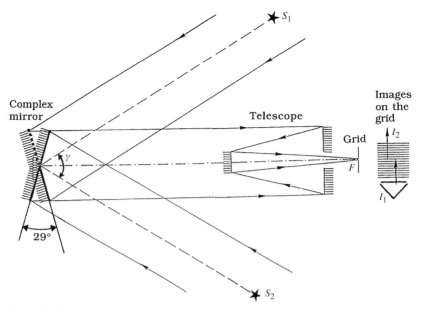

Fig. 2.5. Schematic view of the Hipparcos instrument. The images I_1 and I_2 of stars S_1 and S_2 are formed on the same grid which modulates the illumination while the satellite rotates around an axis perpendicular to the plane of the figure.

star, a certain number of values of abscissae in different directions at different times. From this one can determine the five astrometric parameters of the stars, namely two positional coordinates, two components of the proper motion, and the parallax. Additional information on double and multiple stars is also obtainable from the data.

The instrument essentially consists of a Schmidt telescope receiving light from two regions of the sky separated by a *basic angle* of 58°. This is achieved by two glued mirrors at an angle of 29°, as shown in Fig. 2.5. As the satellite rotates around an axis parallel to the intersection of the mirrors, the stellar images transit through a double grid formed of a star-mapper and a main grid. The star-mapper consists of four chevron slits and four slits perpendicular to the motion of the images. Its primary task is to recognize the stars whose positions have been determined before launch from ground-based astrometry. They are contained in the *Input Catalogue* of more than 118 000 stars of the observing program of the satellite (Turon *et al.*, 1992a). This recognition provides an equation to the continuous determination of the orientation of the satellite as a function of time. It is indeed of the utmost importance to the project that, at each instant, the orientation is known. It is actually represented either by spline functions or by trigonometric functions associated with linear functions of time defined between successive gas jet thrusts used to maintain the scanning law. The on-board determined orientation is used to compute at every instant the position of a star image on the main grid.

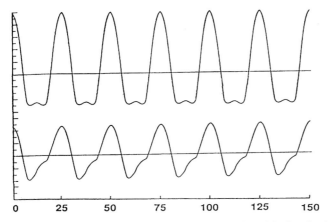

Fig. 2.6. Example of modulations by the main grid of a single star (upper curve) or a double star (lower curve). The vertical scale is the intensity in an arbitrary scale. The abscissae represent time.

The main grid is composed of 2688 parallel slits with a separation of $1''.208$. An image dissector tube collects the light in a small field of $30''$ around the assumed position of the image, and registers the photon counts while the star light is modulated by the grid. Several stars can be observed successively during an observing frame of about 2.13 s. The modulation is represented by a constant term and the first two harmonics, the phase of which represents the fraction of a grid step where the star was located at a given reference time t_i within each observing frame i. This gives the grid coordinate at time t_i provided that the positions given in the *Input Catalogue* and the orientation are precise enough to recognize on what grid step the star was located. The uncertainty for each grid coordinate is of the order of 15 mas. In the case of double stars, the modulations of the two components are combined and the phases of the harmonics are not equal (see Fig. 2.6). It is still possible, by comparing the modulation curves obtained at different RGCs, to reconstruct the relative positions and magnitudes of the components and correct the phases to represent that of the primary, or of the photocenter. From grid coordinates, it is possible to determine, for most of the 1500 stars observed during about 8 h, their abscissae on a fixed RGC and, in addition, to improve the orientation along the scan. This involves modeling of the instrument and the along-scan orientation, writing equations of condition for each star present in each observing frame, and solving about 40–50 thousand equations with about 3000 unknowns for abscissae, orientation spline functions, and calibration parameters. Each star is observed between 15 and 40 times, so that the average uncertainty of abscissae falls to about 3 mas.

During the mission, observations were made on about 2400 RGCs. A synthesis of all the results on the RGCs is made, each star having between 20 and 50 abscissae on independent RGCs. The principle of this synthesis is sketched in Fig. 2.7; the

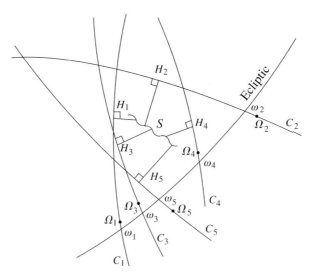

Fig. 2.7. Apparent motion of a star, S, and its projections $H_1, \ldots, H_5$ on reference great circles $C_1, \ldots, C_5$.

star moves along a curve representing the combination of its proper motion and its parallax, and the abscissa points are projections at different times of a point on the curve on the RGC. Descriptions of the reduction procedures of Hipparcos data are sketched in Chapter 8 of Kovalevsky (2002), or in van Leeuwen (1997). Full details appear in Volume 3 of the *Hipparcos Catalogue* (ESA, 1997). As a final result of the reductions, uncertainties in positions, yearly proper motions and parallaxes range between 0.5 and 1.5 mas for stars of magnitude brighter than 10 depending on their positions on the sky and the efficiency for these positions of the adopted scanning law (see also Section 11.6.1).

2.5.2 Hubble Space Telescope

The Hubble Space Telescope (HST) is a 2.4 m aperture Ritchey–Chrétien telescope with a field of view of $14'$ radius. There are six instruments that share the focal surface and two of the initial instruments can be used for astrometry: the Fine Guidance Sensors and the Wide Field/Planetary Camera.

The Fine Guidance Sensors (FGS) occupy three $90°$ segments of an annulus of inner and outer radii $11'$ and $14'$. They are essentially meant to be the guidance system of the telescope, but two are sufficient, so that the third one is available for astrometry (Fig. 2.8). A general description of the design, mode of operation and calibration were presented by Duncombe *et al.* (1991). The position of a point on the FGS is defined by two star selectors, which are two beam deflectors of fixed lengths, a and b, whose rotation angle is measured, and that bring the light

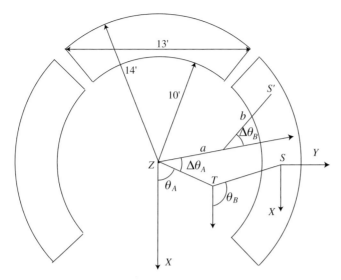

Fig. 2.8. Fine Guidance Sensors and the positioning of star selectors.

from the star image to a detector assembly. The detector consists of two Koester interferometric prisms which produce different intensities, if the selectors are not exactly centered on the star image. The difference controls the positioning of the star selectors until the pointing is correct. The angles, θ_A and θ_B, together with the nominal angular projections of a and b on the sky, suffice to define the position of the star in some local system of coordinates. Pointing successively to several objects, while the telescope remains fixed through the other two FGSs, gives the relative positions of the observed objects in the sky with uncertainties of the order of 3–4 mas, essentially limited by the general jitter of the telescope.

The Wide Field/Planetary Camera is a combination of two cameras each based on four 800×800 CCD detectors. The planetary camera has a field of view of $68''$ while for the wide field camera it is $160''$. The reduction of data for these cameras is based on the same principle as ground-based photography or CCD camera scanning (see Chapter 14). Let us note that while the Hubble repair mission did not improve the FGS, the Wide Field/Planetary Camera was replaced with the Wide Field Planetary Camera 2, with four CCDs and correcting optics in the re-imaging system, so the expected point-spread functions were achieved. More details are given in Seidelmann (1991).

2.6 Proposed space missions

The successes of Hipparcos and the astrometric use of the Hubble Space Telescope have led astrometrists to propose more sophisticated future missions that would

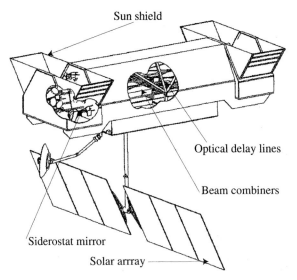

Fig. 2.9. Schematic view of the SIM configuration.

greatly improve the results already obtained. Three projects are Hipparcos-like scanning missions allowing complete sky surveys: DIVA, FAME and GAIA. Another type of mission, SIM, consists of observations of individually selected stars as with the HST. All of them have been thoroughly studied and designed, but none of them has yet reached the stage of building. The first two have even been, possibly temporarily, descoped by the respective space agencies. Anyhow, we describe them all, because their principles will be the bases of future space astrometry.

2.6.1 Space interferometry

The US Space Interferometry Mission (SIM) is an orbiting stellar interferometer consisting of two siderostats with equal apertures of 30 cm collecting the light with a distance of 10 m (Fig. 2.9). In addition, each pod has two steerable mirrors which are the collectors of two guide interferometers directed towards bright stars and are used to determine the orientation in space of the baseline during the observation period. The collected light is directed towards delay lines and a beam combiner. An accurate positioning of the collecting mirrors with respect to the beam combiner will be performed by infrared stabilised interferometers with an absolute accuracy of 1 μm/m. It is expected that SIM will be launched on an Earth-trailing solar orbit about 2010. During its five years of operation, about 10 000 objects as faint as magnitude 20 should be observed, with an uncertainty of 4 μas for wide-angle astrometric observations (3 μas for parallaxes). This will need an integration time of 1 s for stars of magnitude smaller than 10, 200 s at magnitude 15, and more than 4 h for the faintest objects. For narrow-angle ($<1°$) astrometry, one expects to reduce uncertainties to 0.5 μas. More information is at http://sim.jpl.nasa.gov.

2.6.2 DIVA

DIVA was a proposed German project (Bastian *et al.*, 1996; Röser, 1999) which has unfortunately been delayed. The scanning law and the rotation period were very similar to those of Hipparcos. A solid-angle beam combiner assembles two fields of view separated by 100°. The apertures are two rectangles of 22.5×11 cm. The equivalent focal length is 11.2 m. The detection is done by four identical CCD mosaics, each consisting of ten 1024×2048 pixel CCDs.

A star image first crosses two star-mapper mosaics used for the recognition of the position of the incoming star and collecting data for astrometry and attitude determination. Then, the two science mosaics receive light by a diffraction grid. The scan direction is perpendicular to the fringes. Because the fringe separation is a function of the wavelength, the star-mapper is used to evaluate the color of the star, while the dispersed fringes provide spectrophotometric information.

The DIVA project was expected to produce a star map of 40 million stars to magnitude 15, a spectroscopic survey of 12 million stars, and a UV survey of 30 million stars. More information is at www.ari.uni-heidelberg.de/diva/diva.html.

2.6.3 FAME

FAME (Full-sky Astrometric Mapping Explorer) is another project that was expected to be launched on a geosynchronous orbit with a total expected lifetime of five years (Urban *et al.*, 2000) but, like DIVA, it has been stopped. Based on Hipparcos principles with two fields of view separated by 84°3, it scans the sky with a 35° separation of the spin axis from the Sun, but with faster rotation and scanning periods. The aperture sizes are 40×9 cm and the effective focal length is 10.5 m.

The receiving system consists of thirteen 2048×4096 pixel CCD arrays, two of which are covered with filters, providing data in three photometric SLOAN Digital Sky Survey bands (Section 11.4.3). Eleven CCDs are used for astrometry, with the brighter stars, of 4–8 magnitudes, being observed through neutral density filters on three CCDs, by means of a start–stop technique, or by centroiding on extended point-spread functions of saturated images. The read-out is in drift scanning mode, the drift speed being determined from the on-board attitude determination.

It was designed in such a way that 40 million stars would have been observed with, for a five-year mission, accuracies in astrometric parameters of 0.050 mas,

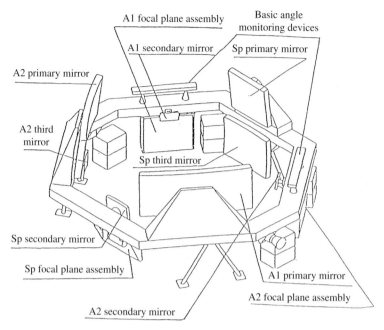

A1 focal plane assembly

Basic angle monitoring devices

A1 secondary mirror

Sp primary mirror

A2 primary mirror

A2 third mirror

Sp third mirror

Sp secondary mirror

Sp focal plane assembly

A1 primary mirror

A2 focal plane assembly

A2 secondary mirror

Fig. 2.10. Tentative GAIA payload configuration: A1 and A2 are the astrometric mirrors and Sp is the photometric collector.

and 0.5 mas, respectively for stars of magnitude 9 and 15. More information is at www.usno.navy.mil/fame/.

2.6.4 GAIA

GAIA (Global Astrometry Instrument for Astrophysics) is being studied by the ESA and is approved for a launch in 2011–2012. As with the previous missions, it is based upon the Hipparcos main principles with two fields of view separated by 106° and a rotation duration of 6 h (Kovalevsky *et al.*, 2000; ESA, 2000). It is intended to be placed on an orbit around the Lagrange point, L_2, of the Earth–Sun system, designed in such a way that it avoids eclipses of the Sun by the Earth.

After some reduction in size, the astrometric instrumentation now consists of two telescopes A1 and A2 with a rectangular entrance pupil whose dimensions are 140×50 cm with a 46.7 m equivalent focal length (Fig. 2.10). The common focal plane covers a field of view of $0°66 \times 0°66$ filled with some 180 CCDs operating in drift scanning mode. The first two columns of the mosaic are star-mappers which detect images above some given threshold and determine their positions and speeds in the focal plane. Then the image reaches the astrometric CCDs and the five broad-band photometers, but only small pixel windows around the predicted path are read and transmitted to the ground after some on-board treatment.

The expected accuracies of positions and parallaxes for a five-year mission range from $4\,\mu$as for stars up to magnitude 12, degrading to $11\,\mu$as at magnitude 15, $27\,\mu$as at magnitude 17, and $160\,\mu$as for magnitude 20. The accuracy of yearly proper motions should be about three quarters of these numbers.

A third telescope, placed midway between the two astrometric instruments, has a square 50 cm entrance pupil with a 2.1 m equivalent focal length. The field of view is split into a central part devoted to the radial velocity measurements, and two outer regions, for medium-band photometry recording in 10 different spectral bands. More information is at http://astro.estec.esa.nl/SA-general/Projects/GAIA/gaia.html.

3

Basic principles and coordinate systems

Astronomy and astrometry began with observations of directions, or spherical angles, without knowledge of distances. So the concept of a celestial sphere and the use of spherical trigonometry have long traditions. With the increased knowledge and improved accuracies of distances, or parallaxes, and radial velocities, the use of space coordinates and vectors and matrices have become appropriate. In this chapter, the development of vectors can be used as the basis for the equations of spherical trigonometry and matrix algebra. In addition, tensors and quaternions are now used in the Theory of Relativity and artificial satellite applications. Then, differential coordinates, coordinate frames and the various coordinate systems used in astronomy will be discussed. In the process the changes in concepts being introduced with the new International Celestial Reference System (ICRS) will be presented.

3.1 Vectors

While a scalar has magnitude and a sign, a vector has a length and a direction. A scalar k multipling a vector $\mathbf{AB}$ changes its length.

$$\mathbf{AB} + \mathbf{BA} = 0$$
$$k\mathbf{AB} = \mathbf{AB}k.$$

Vectors can be added (see Fig. 3.1) and the commutative and associative laws apply.

$$\mathbf{AB} + \mathbf{BC} = \mathbf{AC}$$
$$\mathbf{a} + \mathbf{b} = \mathbf{b} + \mathbf{a} \quad \text{(commutative law)}$$
$$(\mathbf{a} + \mathbf{b}) + \mathbf{c} = \mathbf{a} + (\mathbf{b} + \mathbf{c}) \quad \text{(associative law)}.$$

A vector can be specified in components. If $\mathbf{i}$ and $\mathbf{j}$ are not parallel unit vectors and $\mathbf{i}, \mathbf{j}$, and $\mathbf{r}$ are coplanar, then unique scalars x, y exist so that

$$\mathbf{r} = x\mathbf{i} + y\mathbf{j}.$$

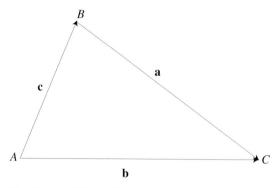

Fig. 3.1. Addition of vectors.

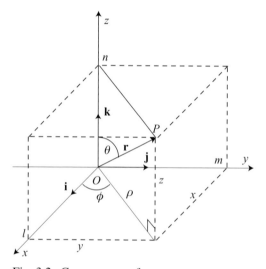

Fig. 3.2. Components of a vector.

If we add **k**, not in the plane, then **r** can be three dimensional and specified by:

$$\mathbf{r} = x\mathbf{i} + y\mathbf{j} + z\mathbf{k}. \tag{3.1}$$

r is then resolved along directions **i**, **j**, **k** by its components x, y, z.

In general, **i**, **j**, **k** are mutually perpendicular unit vectors in a right-hand triad with positive directions as shown in Fig. 3.2. We shall generally use such a triad for reference in space. The components are written without ambiguity as $\mathbf{r} = (x, y, z)$. Let r be the length of **r** and, if α is the angle between **OP** and **i**, then

$$\cos \alpha = x/r = l.$$

With m and n similarly defined with respect to the j and k axes, l, m, n are the direction cosines of **r** with respect to the triad **i**, **j**, **k**, and

$$l^2 + m^2 + n^2 = 1.$$

If we denote $\mathbf{u}$ as a unit vector along the direction of the vector $\mathbf{r}$, then $\mathbf{r}$ is written as

$$\mathbf{OP} = \mathbf{r} = x\mathbf{i} + y\mathbf{j} + z\mathbf{k} = (x, y, z) = (rl, rm, rn) = r(l, m, n) = r\mathbf{u}.$$

If $\mathbf{r}_1 = (x_1, y_1, z_1)$ and $\mathbf{r}_2 = (x_2, y_2, z_2)$, then

$$\mathbf{r}_1 + \mathbf{r}_2 = (x_1 + x_2, \ y_1 + y_2, \ z_1 + z_2). \tag{3.2}$$

In addition to the above representation of a vector in Cartesian coordinates, two other types of coordinate system are in use (see Fig. 3.2).

(i) *Cylindrical coordinates* (ρ, ϕ, z), where ρ and ϕ are the polar coordinates of the projection of the vector $\mathbf{r}$ on the x–y plane.
(ii) *Spherical coordinates* (r, θ, ϕ) where $r = |\mathbf{r}|$, ϕ has the same definition as in cylindrical coordinates, and θ is the angle between Oz and $\mathbf{r}$, with its origin along Oz.

The components in these three coordinate systems are related by the following equations.

$$
\begin{aligned}
x &= r \sin \theta \cos \phi & &= \rho \cos \phi \\
y &= r \sin \theta \sin \phi & &= \rho \sin \phi \\
z &= r \cos \theta & &= z \\
r &= \sqrt{(x^2 + y^2 + z^2)} & &= \sqrt{(\rho^2 + z^2)} \\
\theta &= \arccos(z/\sqrt{(x^2 + y^2 + z^2)}) & &= \arcsin(\rho/z) \\
\phi &= \arctan(y/x) & &= \phi \quad \text{if } x > 0 \\
\phi &= \arctan(y/x) + 180^\circ & &= \phi \quad \text{if } x < 0 \\
\rho &= \sqrt{(x^2 + y^2)} & &= r \sin \theta
\end{aligned}
$$

3.1.1 Dot product

Taking the vectors $\mathbf{a}$ and $\mathbf{b}$ and the angle θ between them, the scalar, or dot, product is

$$\mathbf{a} \cdot \mathbf{b} = ab \cos \theta. \tag{3.3}$$

By the commutative law, one has

$$\mathbf{a} \cdot \mathbf{b} = ab \cos \theta = ba \cos \theta = \mathbf{b} \cdot \mathbf{a}.$$

This can be interpreted as a times the projected length of $\mathbf{b}$ on $\mathbf{a}$ ($b \cos \theta$).

Since $\mathbf{a} \cdot (\mathbf{b} + \mathbf{c})$ is the projection of the sum of separate projections on $\mathbf{a}$, it follows that

$$\mathbf{a} \cdot (\mathbf{b} + \mathbf{c}) = \mathbf{a} \cdot \mathbf{b} + \mathbf{a} \cdot \mathbf{c}.$$

Also

$$m(\mathbf{a} \cdot \mathbf{b}) = (m\mathbf{a}) \cdot \mathbf{b} = \mathbf{a} \cdot (m\mathbf{b}).$$

If $\mathbf{i}, \mathbf{j}, \mathbf{k}$ are a unit triad, the components of $\mathbf{r}$ on the orthogonal triad are $\mathbf{r} \cdot \mathbf{i}, \mathbf{r} \cdot \mathbf{j}, \mathbf{r} \cdot \mathbf{k}$, and

$$\mathbf{i} \cdot \mathbf{j} = \mathbf{i} \cdot \mathbf{k} = \mathbf{j} \cdot \mathbf{k} = 0.$$

The square of a vector $\mathbf{r}$ is

$$\mathbf{r} \cdot \mathbf{r} = rr = r^2.$$

If $\mathbf{r} = \mathbf{x} \cdot \mathbf{i} + \mathbf{y} \cdot \mathbf{j} + \mathbf{z} \cdot \mathbf{k}$, then

$$r^2 = x^2 + y^2 + z^2.$$

3.1.2 Vector product

Remark. We shall now need to orient the three-dimensional space and its triad. We shall assume the right hand convention. It is defined in such a way that screws advance in the direction $\mathbf{k}$ when rotated from $\mathbf{i}$ toward $\mathbf{j}$ (see Fig. 3.2).

Let us now define the vector, or cross, product

$$\mathbf{a} \times \mathbf{b} = (ab \sin \theta)\mathbf{n}, \tag{3.4}$$

where $\mathbf{n}$ is the unit vector perpendicular to the $\mathbf{a}$–$\mathbf{b}$ plane, and such that the triad $(\mathbf{a}, \mathbf{b}, \mathbf{n})$ is right handed.

Note that $\times$ is the accepted symbol for the cross product and care must be taken not to confuse this with a variable x. The vector product is distributive, but it is not commutative.

$$\mathbf{a} \times (\mathbf{b} + \mathbf{c}) = \mathbf{a} \times \mathbf{b} + \mathbf{a} \times \mathbf{c}.$$
$$\mathbf{a} \times \mathbf{b} = -\mathbf{b} \times \mathbf{a}.$$

In a right-hand triad $\mathbf{i}, \mathbf{j}, \mathbf{k}$, one has

$$\mathbf{i} \times \mathbf{j} = \mathbf{k}, \quad \mathbf{j} \times \mathbf{i} = -\mathbf{k},$$
$$\mathbf{j} \times \mathbf{k} = \mathbf{i}, \quad \mathbf{k} \times \mathbf{j} = -\mathbf{i},$$
$$\mathbf{k} \times \mathbf{i} = \mathbf{j}, \quad \mathbf{i} \times \mathbf{k} = -\mathbf{j}.$$

Note that $\mathbf{r} \times \mathbf{r} = 0$.

If the components of a and b are respectively (a_x, a_y, a_z) and (b_x, b_y, b_z), then

$$\mathbf{a} \times \mathbf{b} = (a_y b_z - a_z b_y, \; a_z b_x - a_x b_z, \; a_x b_y - a_y b_x),$$

or in determinant form:

$$\mathbf{a} \times \mathbf{b} = \begin{vmatrix} \mathbf{i} & \mathbf{j} & \mathbf{k} \\ a_x & a_y & a_z \\ b_x & b_y & b_z \end{vmatrix}.$$

3.1.3 Triple scalar product

The triple scalar product is defined by $\mathbf{a} \cdot (\mathbf{b} \times \mathbf{c})$. It can be noted $[\mathbf{a}, \mathbf{b}, \mathbf{c}]$.

If $\mathbf{a}, \mathbf{b}, \mathbf{c}$ are considered in their components, then

$$\mathbf{a} \cdot (\mathbf{b} \times \mathbf{c}) = \begin{vmatrix} a_x & a_y & a_z \\ b_x & b_y & b_z \\ c_x & c_y & c_z \end{vmatrix}. \tag{3.5}$$

Note that

$$[\mathbf{a}, \mathbf{b}, \mathbf{c}] = [\mathbf{b}, \mathbf{c}, \mathbf{a}] = [\mathbf{c}, \mathbf{a}, \mathbf{b}] = -[\mathbf{a}, \mathbf{c}, \mathbf{b}].$$

One may also define the triple vector product as

$$\mathbf{a} \times (\mathbf{b} \times \mathbf{c}) = (\mathbf{a} \cdot \mathbf{c})\mathbf{b} - (\mathbf{a} \cdot \mathbf{b})\mathbf{c}. \tag{3.6}$$

3.2 Vector derivatives

A vector can vary as a function of a scalar, such as time, or as a function of another vector such as position. We assume the variation is continuous, so it can be differentiated.

If a particle moves along the path AB, P is a position at time t, and P' is a position at time $t + \delta t$. If δt is a small time interval, then PP' is δr (Fig. 3.3). Letting δt tend to zero, one can define the derivative of a vector, $\mathbf{V}(t)$ $(x(t),\ y(t),\ z(t))$,

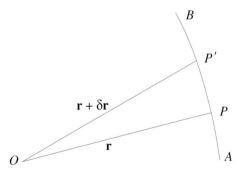

Fig. 3.3. Derivative of a vector.

as:

$$\frac{dV}{dt} = \left(\frac{dx}{dt}, \frac{dy}{dt}, \frac{dz}{dt} \right).$$

If $\mathbf{u}$ is a unit vector along PP' and s is the distance along the curve,

$$\delta \mathbf{r} = \mathbf{u} \delta s,$$

so that,

$$\lim_{\delta \to 0} \frac{PP'}{\delta t} = \lim_{\delta \to 0} \frac{\delta \mathbf{r}}{\delta t} = \frac{d\mathbf{r}}{dt} = \mathbf{u} \frac{ds}{dt}.$$

This is the velocity of a particle at P, commonly written as $\dot{\mathbf{r}}$ or $\mathbf{v}$. It has magnitude and direction. The components of $\dot{\mathbf{r}}$ are $(\dot{x}, \dot{y}, \dot{z})$ and those of $\mathbf{u}$ are $(dx/ds, dy/ds, dz/ds)$.

In modern conventions, velocity is a vector $\mathbf{v}$; speed, v, is a scalar value of velocity, ds/dt.

3.2.1 Rotating axes

Rotating axes is a common phenomenon that must be commonly dealt with. Let $\mathbf{p}$ be the unit vector along $\mathbf{r}$. Writing $\mathbf{r} = r\mathbf{p}$ and differentiating

$$\dot{\mathbf{r}} = \dot{r}\mathbf{p} + r\frac{d\mathbf{p}}{dt}, \tag{3.7}$$

where the first term is the change along $\mathbf{r}$, or the change in length, and the second term is the change perpendicular to $\mathbf{r}$, or a rotation. Suppose a vector $\mathbf{r}$ has a rate of change of $d\mathbf{r}/dt$ with respect to a fixed frame F_1, and a change $\partial\mathbf{r}/\partial t$ with respect to a frame F_2, which is rotating with respect to F_1 with an angular velocity $\vec{\omega}$ (which is not necessarily constant). In the case where F_2 is rotating with $\mathbf{r}$, then $\partial\mathbf{r}/\partial t$ can be written instead of $\dot{r}\mathbf{r}$. Then (3.7) can be written in the form

$$\frac{d\mathbf{r}}{dt} = \frac{\partial \mathbf{r}}{\partial t} + \vec{\omega} \times \mathbf{r}. \tag{3.8}$$

This is a general result, which will be illustrated. Let $\mathbf{i}, \mathbf{j}, \mathbf{k}$ be an orthogonal triad, rigidly attached to F_2. Then,

$$\mathbf{r} = (\mathbf{r} \cdot \mathbf{i})\mathbf{i} + (\mathbf{r} \cdot \mathbf{j})\mathbf{j} + (\mathbf{r} \cdot \mathbf{k})\mathbf{k} = \sum_{\mathbf{i}, \mathbf{j}, \mathbf{k}} (\mathbf{r} \cdot \mathbf{i})\mathbf{i}, \tag{3.9}$$

with a summation over $\mathbf{i}, \mathbf{j}, \mathbf{k}$. Then (3.7) can be written as

$$\frac{d\mathbf{r}}{dt} = \sum_{\mathbf{i}, \mathbf{j}, \mathbf{k}} \left(\frac{d(\mathbf{r} \cdot \mathbf{i})}{dt} \mathbf{i} \right) + \sum_{\mathbf{i}, \mathbf{j}, \mathbf{k}} \left((\mathbf{r} \cdot \mathbf{i}) \frac{d\mathbf{i}}{dt} \right). \tag{3.10}$$

Since $d\mathbf{i}/dt = \vec{\omega} \times \mathbf{i}$, then

$$\sum_{\mathbf{i,j,k}} \left((\mathbf{r} \cdot \mathbf{i}) \frac{d\mathbf{i}}{dt} \right) = \vec{\omega} \times \sum_{\mathbf{i,j,k}} (\mathbf{r} \cdot \mathbf{i})\mathbf{i} = \vec{\omega} \times \mathbf{r}. \tag{3.11}$$

If we treat $\mathbf{i}, \mathbf{j}, \mathbf{k}$ as constant non-rotating vectors, then

$$\frac{d\mathbf{r}}{dt} = \sum_{\mathbf{i,j,k}} \left(\frac{d(\mathbf{r} \cdot \mathbf{i})}{dt}\mathbf{i} \right). \tag{3.12}$$

Hence, the derivative of the vector $\mathbf{r}$, with a rotation can be given in the general form

$$\frac{d\mathbf{r}}{dt} = \frac{\partial \mathbf{r}}{\partial t} + \vec{\omega} \times \mathbf{r}. \tag{3.13}$$

3.3 Gradient

Consider, defined and continuous in space, a scalar function of a position which is written $f(x, y, z)$ or $f(\mathbf{r})$. The move from $f(x, y, z)$ to $f(x + \delta x, y, z)$ produces a change in f which can be written $(\partial f/\partial x)\delta x$, where $(\partial f/\partial x)$ is the conventional partial differentiation with respect to x, while y and z are assumed constant. This could be written

$$\frac{\partial f}{\partial x}\delta x = \left(\frac{\partial f}{\partial x}\mathbf{i} + \frac{\partial f}{\partial y}\mathbf{j} + \frac{\partial f}{\partial z}\mathbf{k} \right) \cdot \mathbf{i}\delta x$$

$$\frac{\partial f}{\partial x}\delta x = \nabla f \cdot \mathbf{i}\delta x. \tag{3.14}$$

The differential operator $\nabla = \frac{\partial}{\partial x}\mathbf{i} + \frac{\partial}{\partial y}\mathbf{j} + \frac{\partial}{\partial z}\mathbf{k}$ is called del or nabla.

Likewise the changes from y to $y + \delta y$ and z to $z + \delta z$ take the same form. In general, the change from r to $r + \delta r$ is $\nabla f \cdot \mathbf{r}$. The vector with the elements $(\partial f/\partial x, \partial f/\partial y, \partial f/\partial z)$ is called the "gradient" of f, and is represented by grad f or ∇f.

Consider a curve C and the values that f assumes along the curve. The rate of change of f along C with respect to the arc length is

$$\frac{\partial f}{\partial s} = \frac{\partial f}{\partial x}\frac{\partial x}{\partial s} + \frac{\partial f}{\partial y}\frac{\partial y}{\partial s} + \frac{\partial f}{\partial z}\frac{\partial z}{\partial s} = \mathbf{u}\nabla f.$$

Consider a surface with $f(x, y, z) = $ constant. Along a line on the surface one has $df/ds = 0$, so $\mathbf{u} \cdot \nabla f = 0$.

Thus, ∇f is perpendicular to the surface. In general, ∇f is a field vector that is a function of position.

3.4 Matrices

A matrix is an array of $m \times n$ numbers, a_q^p, set out in m rows and n columns. The superscript p indicates the row and the subscript q indicates the column. It is a matrix of order $m \times n$. In this book, we shall essentially use 3×3 matrices, exceptionally 2×2. A shortened notation of an $m \times n$ matrix is

$$A\left(a_q^p\right) = \left\| a_q^p \right\|.$$

Matrices are mathematical objects that obey a certain number of transformation and combination laws. For instance, a linear function of matrices of the same structure is a matrix whose components are the same combination of the components of the given matrices; for instance:

$$A\mathcal{M}\left(a_q^p\right) + B\mathcal{M}'\left(b_q^p\right) = \mathcal{S}\left(Aa_q^p + Bb_q^p\right).$$

An important transformation is the transposition, which inverts the roles of rows and columns and transforms an $m \times n$ matrix into an $n \times m$ matrix:

$$\mathcal{M}^{\mathsf{T}}\left(a_q^p\right) = \mathcal{M}\left(a_p^q\right). \tag{3.15}$$

The product of two matrices can be defined only if the number n of columns of the first matrix, A, is equal to the number m' of rows of the second matrix B ($n = m' = N$). The result is a matrix, C, with m rows and n' columns, whose general element of row i and column j is obtained by multiplying the rows of the first matrix into the columns of the second matrix, element by element, and summing the products algebraically

$$c_j^i = \sum_{k=1,\,N} a_k^i b_j^k.$$

As an example, the product of two 3×3 matrices

$$A\left(a_q^p\right) = \begin{pmatrix} a_1^1 & a_2^1 & a_3^1 \\ a_1^2 & a_2^2 & a_3^2 \\ a_1^3 & a_2^3 & a_3^3 \end{pmatrix}$$

and

$$B\left(b_q^p\right) = \begin{pmatrix} b_1^1 & b_2^1 & b_3^1 \\ b_1^2 & b_2^2 & b_3^2 \\ b_1^3 & b_2^3 & b_3^3 \end{pmatrix}$$

is

$$C(c_j^i) = \begin{pmatrix} a_1^1 b_1^1 + a_2^1 b_1^2 + a_3^1 b_1^3 & a_1^1 b_2^1 + a_2^1 b_2^2 + a_3^1 b_2^3 & a_1^1 b_3^1 + a_2^1 b_3^2 + a_3^1 b_3^3 \\ a_1^2 b_1^1 + a_2^2 b_1^2 + a_3^2 b_1^3 & a_1^2 b_2^1 + a_2^2 b_2^2 + a_3^2 b_2^3 & a_1^2 b_3^1 + a_2^2 b_3^2 + a_3^2 b_3^3 \\ a_1^3 b_1^1 + a_2^3 b_1^2 + a_3^3 b_1^3 & a_1^3 b_2^1 + a_2^3 b_2^2 + a_3^3 b_2^3 & a_1^3 b_3^1 + a_2^3 b_3^2 + a_3^3 b_3^3 \end{pmatrix}.$$

(3.16)

A vector in space may be considered as a 1×3 matrix:

$$\mathbf{V} = (x \quad y \quad z).$$

The product of a vector, $\mathbf{V}$, represented as a row matrix, $\mathcal{V}$, by a square 3×3 matrix, $\mathcal{A}$, is the vector, $\mathbf{V}'$, represented by $\mathcal{V}'$:

$$\mathcal{V}' = (x \quad y \quad z) \times \begin{pmatrix} a_1^1 & a_2^1 & a_3^1 \\ a_1^2 & a_2^2 & a_3^2 \\ a_1^3 & a_2^3 & a_3^3 \end{pmatrix} = (x' \quad y' \quad z'), \quad (3.17)$$

with

$$x' = a_1^1 x + a_1^2 y + a_1^3 z,$$
$$y' = a_2^1 x + a_2^2 y + a_2^3 z,$$
$$z' = a_3^1 x + a_3^2 y + a_3^3 z.$$

One may also represent a vector, $\mathbf{V}$, by a column matrix, the transpose of the row matrix. Conversely, the row matrix is the transpose of the column matrix:

$$\mathcal{V} = (x \quad y \quad z) = \begin{pmatrix} x \\ y \\ z \end{pmatrix}^{\mathsf{T}}.$$

The equivalent of Equation (3.17) with the column matrix is

$$\mathcal{V}' = \mathcal{A}^{\mathsf{T}} \times \mathcal{V}^{\mathsf{T}} = \begin{pmatrix} a_1^1 & a_1^2 & a_1^3 \\ a_2^1 & a_2^2 & a_2^3 \\ a_3^1 & a_3^2 & a_3^3 \end{pmatrix} \times \begin{pmatrix} x \\ y \\ z \end{pmatrix} = \begin{pmatrix} x' \\ y' \\ z' \end{pmatrix}.$$

Note that the multiplication rule imposes that in the horizontal representation of a vector, the latter must be placed before the matrix. In the case of vertical representation, it has to be placed after the matrix.

An important case of a 3×3 matrix is the unit matrix, $\mathcal{I}$,

$$\mathcal{I} = \begin{pmatrix} 1 & 0 & 0 \\ 0 & 1 & 0 \\ 0 & 0 & 1 \end{pmatrix}.$$

It is such that

$$\mathcal{M} \times \mathcal{I} = \mathcal{I} \times \mathcal{M} = \mathcal{M}.$$

3.4.1 Rotation matrices

Let us now describe a special class of 3×3 matrices, the rotation matrices. A first property of a square matrix is that one can associate a numerical value to a matrix, the value of the determinant having the same elements:

$$D = \det(\mathcal{A}) = \begin{vmatrix} a_1^1 & a_2^1 & a_3^1 \\ a_1^2 & a_2^2 & a_3^2 \\ a_1^3 & a_2^3 & a_3^3 \end{vmatrix}.$$

Rotation matrices are characterized by $D = 1$. They are called *orthogonal*. The product of such a matrix by a vector transforms the vector to another one having the same length. This transformation is therefore a rotation. If the matrix is the unit matrix, the transformation is an identity. We shall generally use, in this book, column representations of vectors.

There are three fundamental rotation matrices in the $Oxyz$ orthogonal right-handed triad:

(i) rotation of a vector **V** about the Ox axis positively through θ, operated by the matrix $\mathcal{R}_1$

$$(x' \quad y' \quad z') = (x \quad y \quad z) \times \begin{pmatrix} 1 & 0 & 0 \\ 0 & \cos\theta & -\sin\theta \\ 0 & \sin\theta & \cos\theta \end{pmatrix}$$

or

$$\begin{pmatrix} x' \\ y' \\ z' \end{pmatrix} = \begin{pmatrix} 1 & 0 & 0 \\ 0 & \cos\theta & \sin\theta \\ 0 & -\sin\theta & \cos\theta \end{pmatrix} \times \begin{pmatrix} x \\ y \\ z \end{pmatrix} = \mathcal{R}_1 \times \begin{pmatrix} x \\ y \\ z \end{pmatrix};$$

(ii) rotation of a vector **V** about the Oy axis positively through ω, operated by the matrix $\mathcal{R}_2$

$$(x' \quad y' \quad z') = (x \quad y \quad z) \times \begin{pmatrix} \cos\omega & 0 & \sin\omega \\ 0 & 1 & 0 \\ -\sin\omega & 0 & \cos\omega \end{pmatrix}$$

or

$$\begin{pmatrix} x' \\ y' \\ z' \end{pmatrix} = \begin{pmatrix} \cos\omega & 0 & -\sin\omega \\ 0 & 1 & 0 \\ \sin\omega & 0 & \cos\omega \end{pmatrix} \times \begin{pmatrix} x \\ y \\ z \end{pmatrix} = \mathcal{R}_2 \times \begin{pmatrix} x \\ y \\ z \end{pmatrix};$$

(iii) rotation of a vector **V** about the Oz axis positively through ϕ, operated by the matrix $\mathcal{R}_3$

$$(x' \quad y' \quad z') = (x \quad y \quad z) \times \begin{pmatrix} \cos\phi & -\sin\phi & 0 \\ \sin\phi & \cos\phi & 0 \\ 0 & 0 & 1 \end{pmatrix}$$

or

$$\begin{pmatrix} x' \\ y' \\ z' \end{pmatrix} = \begin{pmatrix} \cos\phi & \sin\phi & 0 \\ -\sin\phi & \cos\phi & 0 \\ 0 & 0 & 1 \end{pmatrix} \times \begin{pmatrix} x \\ y \\ z \end{pmatrix} = \mathcal{R}_3 \times \begin{pmatrix} x \\ y \\ z \end{pmatrix}.$$

An example is the Eulerian coordinate transformation defined in Section 3.8.

3.4.2 *Cartesian coordinate transformation*

If **r** and **r**′ are the Cartesian coordinates of a vector in two different coordinate systems, they can be related by

$$\mathbf{r}' = \mathcal{A}\mathbf{r} + \mathbf{a}, \tag{3.18}$$

where **a** is the translation of the origin of the unprimed system in the primed system, and matrix $\mathcal{A}$ represents a rotation. The matrix $\mathcal{A}$, or direction cosine matrix, can be obtained from the product of matrices of successive rotations about the three coordinate axes. The elements of $\mathcal{A}$ are direction cosines of the primed axes in the unprimed axes, and satisfy the orthogonality condition. Since $\mathcal{A}$ is an orthogonal matrix, the inverse transformation is the transposed matrix. Thus,

$$\mathcal{A}^{-1} = \mathcal{A}^{\mathsf{T}}.$$

The direction cosine matrix, in terms of the orthogonal unit vectors (**i**, **j**, **k**) and (**i**′, **j**′, **k**′) in the two coordinate systems, can be given by

$$\mathcal{A} \equiv \begin{pmatrix} \mathbf{i}'\cdot\mathbf{i} & \mathbf{i}'\cdot\mathbf{j} & \mathbf{i}'\cdot\mathbf{k} \\ \mathbf{j}'\cdot\mathbf{i} & \mathbf{j}'\cdot\mathbf{j} & \mathbf{j}'\cdot\mathbf{k} \\ \mathbf{k}'\cdot\mathbf{i} & \mathbf{k}'\cdot\mathbf{j} & \mathbf{k}'\cdot\mathbf{k} \end{pmatrix}. \tag{3.19}$$

Euler's theorem states that any finite rotation of a rigid body can be expressed as a rotation through some angle about some fixed axis. Therefore, the most general transformation matrix is a rotation by some angle, Φ, about some fixed axis, **e**. This axis is unaffected by the rotation and must have the same components in both primed and unprimed systems. Denoting the components by e_1, e_2 and e_3, the matrix

$\mathcal{A}$ is

$$
\mathcal{A} = \begin{pmatrix} a_1^1 & a_1^2 & a_1^3 \\ a_2^1 & a_2^2 & a_2^3 \\ a_3^1 & a_3^2 & a_3^3 \end{pmatrix},
$$

(3.20)

with

$$
a_1^1 = \cos\Phi + e_1^2(1 - \cos\Phi)
$$
$$
a_1^2 = e_1 e_2(1 - \cos\Phi) + e_3 \sin\Phi
$$
$$
a_1^3 = e_1 e_3(1 - \cos\Phi) - e_2 \sin\Phi
$$
$$
a_2^1 = e_1 e_2(1 - \cos\Phi) - e_3 \sin\Phi
$$
$$
a_2^2 = \cos\Phi + e_2^2(1 - \cos\Phi)
$$
$$
a_2^3 = e_2 e_3(1 - \cos\Phi) + e_1 \sin\Phi
$$
$$
a_3^1 = e_1 e_3(1 - \cos\Phi) + e_2 \sin\Phi
$$
$$
a_3^2 = e_2 e_3(1 - \cos\Phi) - e_1 \sin\Phi
$$
$$
a_3^3 = \cos\Phi + e_3^2(1 - \cos\Phi).
$$

The inverse transformation matrix may be obtained by Equation (3.15) or replacing Φ by $-\Phi$ in (3.20), that is a rotation of the same amount in the opposite direction.

3.5 Tensors

Tensors, like matrices, are tables of numbers or functions. They are mathematical tools that are particularly linked with changes in coordinate systems. Their existence is independent of coordinate systems, although their expression depends on them.

In a n-dimensional space, a tensor has n^q components, where q is the order, or rank, of the tensor. So, in classical three-dimensional space:

- a rank-zero tensor is a scalar,
- a rank-one tensor is a vector,
- a rank-two tensor is a matrix,
- a rank-three tensor has 27 components.

Although tensor analysis is a generalization of vector analysis, one cannot, in general, visualize tensors. In physics, they are important insofar as they are used to express physical laws that remain valid after coordinate system transformations. The rules on tensor calculus depend on rank and dimensions of space. We shall consider only those cases that are useful in the present book.

3.5.1 Rank-two tensors in three-dimensional space

An example in mechanics is the moment of inertia tensor. It is an order-two tensor in three-dimensional space (x, y, z). The 3×3 table expressing the tensor is

$$[I] = \begin{bmatrix} I_1^1 & I_2^1 & I_3^1 \\ I_1^2 & I_2^2 & I_3^2 \\ I_1^3 & I_2^3 & I_3^3 \end{bmatrix},$$

where the diagonal terms I_i^i are the inertia moments,

$$I_1^1 = I_{xx} = \sum_{i=1}^{3} m_i \left(y_i^2 + z_i^2 \right),$$

and the other terms are the products of inertia.

$$I_1^2 = I_2^1 = I_{xy} = \sum_{i=1}^{3} m_i x_i y_i,$$

so that the expression of the moment of inertia tensor is

$$[I] = \begin{bmatrix} I_{xx} & I_{xy} & I_{xz} \\ I_{xy} & I_{yy} & I_{yz} \\ I_{xz} & I_{yz} & I_{zz} \end{bmatrix}. \tag{3.21}$$

It is a symmetrical tensor. Let us now apply the following most general transformation of the system of coordinates,

$$\begin{aligned} x' &= a_1^1 x + a_1^2 y + a_1^3 z, \\ y' &= a_2^1 x + a_2^2 y + a_2^3 z, \\ z' &= a_3^1 x + a_3^2 y + a_3^3 z. \end{aligned} \tag{3.22}$$

Let it satisfy the condition that it is a transformation which, when followed by its inverse, is unity. In this particular case, it is a rotation,

$$\sum_i a_i^j a_i^k = 0 \quad \text{for} \quad j \neq k, \quad \text{and} \quad \sum_i a_{ij}^2 = 1.$$

One then finds that the current element of the transformed tensor $[I']$ is

$$I'_{ij} = \sum_{l,m} a_{il} a_{jm} I_m^l. \tag{3.23}$$

Let us now consider a linear transformation, whose coefficients are the components of the tensor I. A vector, $\mathbf{u}$, of components, (x, y, z), is transformed into a vector,

$\mathbf{u}'$, of components, (x', y', z'), by

$$x' = I_1^1 x + I_1^2 y + I_1^3 z,$$
$$y' = I_2^1 x + I_2^2 y + I_2^3 z,$$
$$z' = I_3^1 x + I_3^2 y + I_3^3 z. \tag{3.24}$$

These equations are formally identical to Equations (3.17). One can write these equations symbolically as

$$\mathbf{u}' = [I] \times \mathbf{u}.$$

This can be shown as follows. The components of $\mathbf{u}$ are transformed by Equations (3.22), and the components of the tensor by Equations (3.23). Computing the transform of $\mathbf{u}$, using in (3.22), as coefficients, the components of $[I]$, one reproduces exactly Equations (3.22). Actually, it is a consequence of the condition imposed on the components that the transformation, followed by its inverse, is an identity.

An application of this is the determination of the principal axes of inertia. One defines a proper vector of a tensor, any vector, $\mathbf{u}$, that is colinear with the product $[I] \times \mathbf{u}$. It is therefore given by the equations

$$x' = I_1^1 x + I_1^2 y + I_1^3 z = \lambda x,$$
$$y' = I_2^1 x + I_2^2 y + I_2^3 z = \lambda y,$$
$$z' = I_3^1 x + I_3^2 y + I_3^3 z = \lambda z. \tag{3.25}$$

To get the solution, one must determine the values of λ, for which the characteristic equation of the tensor is equal to zero,

$$\begin{vmatrix} I_{xx} - \lambda & I_{xy} & I_{xz} \\ I_{xy} & I_{yy} - \lambda & I_{yz} \\ I_{xz} & I_{yz} & I_{zz} - \lambda \end{vmatrix} = 0. \tag{3.26}$$

This equation has three solutions λ_i. The moment of inertia with respect to a principal axis i is

$$I_{\lambda_i} = \mathbf{u} \times [I] \times \mathbf{u} = \mathbf{u} \times \lambda_i \mathbf{u} = \lambda_i.$$

3.5.2 Tensors in a four-dimensional space

In the framework of the Theory of Relativity (Chapter 5), the reference is a four-dimensional space-time, and tensors of rank two that are introduced are square symmetric tables of 4×4 components. The most general form is:

$$\begin{bmatrix} g_{00} & g_{01} & g_{02} & g_{03} \\ g_{10} & g_{11} & g_{12} & g_{13} \\ g_{20} & g_{21} & g_{22} & g_{23} \\ g_{30} & g_{31} & g_{32} & g_{33} \end{bmatrix}. \tag{3.27}$$

It represents the relationship between the square of the distance of two generalized events separated by $ds = (dx_0, dx_1, dx_2, dx_3)$ (Section 5.3.3),

$$ds^2 = \sum_{i=0}^{3} \sum_{j=0}^{3} g_{ij} dx_i dx_j, \qquad (3.28)$$

with $g_{ij} = g_{ji}$.

Another application of four-dimensional rank-two tensors is the energy–momentum tensor. Let us consider a collection of non-interacting particles, n per volume unit, each having an energy m. The energy density is, evidently,

$$T_{00} = nm.$$

But this is not all the energy of the system. One must consider also the energy flux, which has three components. For instance, the x component is the amount of energy transported through a unit yz area, if the velocity is v_x. So, the energy flux density has three components,

$$T_{0i} = nmv_i \qquad \text{with } i = 1 \text{ to } 3.$$

It can also be regarded as the density of momentum.

Finally, one must also consider the momentum flux. The x–y momentum flux is the amount of x momentum that flows in the y direction. It is expressed by $mnv_x v_y$. Again, there are three components,

$$T_{ji} = nmv_i v_j = T_{ij} \qquad \text{with } i, j = 1 \text{ to } 3, \quad i \neq j.$$

These three types of energy can be expressed in a single entity, the energy–momentum tensor,

$$\begin{pmatrix} T_{00} & T_{01} & T_{02} & T_{03} \\ T_{10} & 0 & T_{12} & T_{13} \\ T_{20} & T_{21} & 0 & T_{23} \\ T_{30} & T_{31} & T_{32} & 0 \end{pmatrix}. \qquad (3.29)$$

The application to electro-magnetism is more complex (see, for instance, Mould, 1994). We shall only give the result. Let $\mathbf{E}(E_x, E_y, E_z)$ and $\mathbf{B}(B_x, B_y, B_z)$ be, respectively, the electrical and the magnetic fields and c is the speed of light, the field tensor is, in complex notation:

$$\begin{pmatrix} 0 & B_z & -B_y & -iE_x/c \\ -B_z & 0 & B_x & -iE_y/c \\ B_y & -B_x & 0 & -iE_z/c \\ iE_x/c & iE_y/c & iE_z/c & 0 \end{pmatrix}. \qquad (3.30)$$

3.6 Quaternions

Quaternions were invented by Hamilton (1866) in the mid nineteenth century. They can be considered to be four-vectors that can be used in place of a normal 3×3 matrix, so they have some computational efficiency advantage and are commonly used in space attitude and video game applications. They are based on Euler's theorem that any arbitrary rotation can be described as a rotation about a single axis, which is the eigenvector of the transformation, and has the same components in the rotated and unrotated systems.

3.6.1 Properties of quaternions

A quaternion can be thought of as the combination of a scalar, s, and a three-vector, $\mathbf{v}$, and can be written $(s, \mathbf{v})$. The vector represents the axis direction multiplied by the sine of half of the rotation angle, and the scalar is the cosine of half of the rotation angle (Section 3.6.2).

Quaternions are also an extension of complex numbers into the three-dimensional realm. In this case, the scalar is the real quantity, and the vector components are three imaginary quantities. This permits the definition of operations like addition, subtraction, multiplication, inverse, normalization, etc., as straightforward extensions of the same operations on complex numbers. The quaternions can be expressed in ordinary vector notation. So there is an entire mathematical formalism for quaternions, and for rotational dynamics cast in quaternions. Two successive rotations are represented by the multiplication of two quaternions. The advantage of using quaternions over rotation matrices is that only four quantities are required, not nine. Quaternion representation of rigid-body rotations leads to kinematical expressions involving the Euler symmetric parameters.

The most important properties of quaternions are summarized here. Let the four parameters (q_1, q_2, q_3, q_4) form the components of the quaternion $\bar{q}$ as follows:

$$\bar{q} \equiv q_4 + iq_1 + jq_2 + kq_3, \tag{3.31}$$

where i, j, k are the hyperimaginary numbers satisfying the conditions

$$i^2 = j^2 = k^2 = -1,$$
$$ij = -ji = k,$$
$$jk = -kj = i,$$
$$ki = -ik = j. \tag{3.32}$$

The conjugate, or inverse, of $\bar{q}$ is defined as

$$\bar{q}^{\star} \equiv q_4 - iq_1 - jq_2 - kq_3.$$

The quantity q_4 is the real, or scalar, part of the quaternion, and $iq_1 + jq_2 + kq_3$ is the imaginary, or vector, part.

A vector in three-dimensional space, **U**, having the components (U_1, U_2, U_3) is expressed in quaternion notation, with a scalar part equal to zero, as

$$\mathbf{U} = iU_1 + jU_2 + kU_3. \tag{3.33}$$

If the vector **q** corresponds to the vector part of $\bar{q}$ $(\mathbf{q} = iq_1 + jq_2 + kq_3)$, then an alternate representation of $\bar{q}$ is

$$\bar{q} = (q_4, \mathbf{q}). \tag{3.34}$$

Quaternion multiplication is performed in the same manner as the multiplication of complex numbers, or algebraic polynomials, except that the order of the operations must be taken into account, because they are not commutative. For example, the product of two quaternions is

$$\bar{q}'' = \bar{q}\bar{q}' = (q_4 + iq_1 + jq_2 + kq_3)(q_4' + iq_1' + jq_2' + kq_3'). \tag{3.35}$$

Using Equation (3.31), this reduces to

$$\begin{aligned}
\bar{q}'' = \bar{q}\bar{q}' &= (-q_1q_1' - q_2q_2' - q_3q_3' + q_4q_4') \\
&= +i(+q_1q_4' + q_2q_3' - q_3q_2' + q_4q_1') \\
&= +j(-q_1q_3' + q_2q_4' + q_3q_1' + q_4q_2') \\
&= +k(q_1q_2' - q_2q_1' + q_3q_4' + q_4q_3').
\end{aligned} \tag{3.36}$$

If $\bar{q} = (q_4, \mathbf{q})$, then (3.36) can alternatively be expressed as

$$\bar{q}'' = \bar{q}\bar{q}' = (q_4q_4' - \mathbf{q} \cdot \mathbf{q}' + q_4\mathbf{q}' + q_4'\mathbf{q} + \mathbf{q} \times \mathbf{q}'). \tag{3.37}$$

The length, or norm, of $\bar{q}$ is defined as

$$|\bar{q}| \equiv \sqrt{\bar{q}^*\bar{q}} = \sqrt{\bar{q}\bar{q}^*} = \sqrt{q_1^2 + q_2^2 + q_3^2 + q_4^2}. \tag{3.38}$$

3.6.2 Euler symmetric parameters

The Euler symmetric parameters (q_1, q_2, q_3, q_4), used to represent finite rotations, are defined by the following equations

$$\begin{aligned}
q_1 &\equiv e_1 \sin \Phi/2, \\
q_2 &\equiv e_2 \sin \Phi/2, \\
q_3 &\equiv e_3 \sin \Phi/2, \\
q_4 &\equiv \cos \Phi/2.
\end{aligned} \tag{3.39}$$

Then,

$$q_1^2 + q_2^2 + q_3^2 + q_4^2 = 1. \tag{3.40}$$

The transformation matrix, $\mathcal{A}$, in terms of the Euler symmetric parameters, is

$$\mathcal{A} = \begin{pmatrix} q_1^2 - q_2^2 - q_3^2 + q_4^2 & 2(q_1q_2 + q_3q_4) & 2(q_1q_3 - q_2q_4) \\ 2(q_1q_2 - q_3q_4) & -q_1^2 + q_2^2 - q_3^2 + q_4^2 & 2(q_2q_3 + q_1q_4) \\ 2(q_1q_3 + q_2q_4) & 2(q_2q_3 - q_1q_4) & -q_1^2 - q_2^2 + q_3^2 + q_4^2 \end{pmatrix}. \tag{3.41}$$

The inverse transformation matrix may be obtained by replacing q_1, q_2 and q_3 by $-q_1$, $-q_2$ and $-q_3$, respectively, in (3.41) and leaving q_4 unaltered.

The Euler symmetric parameters may be regarded as components of a quaternion $\bar{q}$ defined by

$$\bar{q} \equiv q_4 + iq_1 + jq_2 + kq_3$$

where i, j, k are hyperimaginary numbers as specified in (3.32).

3.6.3 *Quaternion rotation*

If a set of four Euler symmetric parameters, corresponding to the rigid-body rotation defined by the transformation matrix, $\mathcal{A}$, are the components of the quaternion $\bar{q}$, then $\bar{q}$ is the representation of the rigid-body rotation. If $\bar{q}'$ corresponds to the rotation matrix, $\mathcal{A}'$, then the rotation described by the product $\mathcal{A}' \times \mathcal{A}$ is equivalent to the rotation described by $\bar{q}\bar{q}'$. Note the inverse order of quaternion multiplication as compared with matrix multiplication.

The transformation of a vector, **U**, corresponding to multiplication by matrix $\mathcal{A}$,

$$\mathbf{U}' = \mathcal{A} \times \mathbf{U}$$

is effected in quaternion algebra by the operation

$$\mathbf{U}' = \bar{q}^\star \mathbf{U} \bar{q}. \tag{3.42}$$

For computation purposes, it is convenient to express quaternion multiplication in matrix form. Let the components of $\bar{q}$ form a four-vector as follows,

$$\bar{q} = \begin{pmatrix} q_1 \\ q_2 \\ q_3 \\ q_4 \end{pmatrix}. \tag{3.43}$$

This procedure is analogous to expressing the complex number $c = a + ib$ in the form of the two-vector

$$\mathbf{c} = \begin{pmatrix} a \\ b \end{pmatrix}.$$

In matrix form, (3.36) then becomes

$$\begin{pmatrix} q_1'' \\ q_2'' \\ q_3'' \\ q_4'' \end{pmatrix} = \begin{pmatrix} q_4' & q_3' & -q_2' & q_1' \\ -q_3' & q_4' & q_1' & q_2' \\ q_2' & -q_1' & q_4' & q_3' \\ -q_1' & -q_2' & -q_3' & q_4' \end{pmatrix} \times \begin{pmatrix} q_1 \\ q_2 \\ q_3 \\ q_4 \end{pmatrix}. \tag{3.44}$$

Given the quaternion components corresponding to two successive rotations, (3.44) gives the quaternion components corresponding to the total rotation.

In the same manner as elementary rotation matrices, there are elementary quaternions. The quaternion for rotation about the x-axis through an angle θ is

$$q_x = \cos\theta/2 + \mathbf{i}\ \sin\theta/2,$$

and similarly about the y-axis and z-axis through angles ω and ϕ respectively, the quaternions are

$$q_y = \cos\omega/2 + \mathbf{j}\ \sin\omega/2,$$

and

$$q_z = \cos\phi/2 + \mathbf{k}\ \sin\phi/2.$$

3.7 Spherical trigonometry

Take the surface of a sphere with unit radius. All lines of interest are parts of great circles. There is only one great circle through two points, that are not on opposite ends of a diameter. A spherical triangle ABC is defined by three points on the surface of the sphere, the sides being parts of great circles. O is the center of the sphere, $(OA, OC) = b$, $(OB, OC) = a$, and $(OA, OB) = c$ (see Fig. 3.4). The angles between the planes are $A = (OAC, OAB)$, $B = (OBA, OBC)$, $C = (OCB, OCA)$. Let $\mathbf{i}, \mathbf{j}$ and $\mathbf{k}$ be non-mutually perpendicular unit vectors along OA, OB and OC. Then $\mathbf{i} \times \mathbf{j}$ is a vector with magnitude $\sin c$ perpendicular to AOB and $\mathbf{i} \times \mathbf{k}$ is a vector with magnitude $\sin b$ perpendicular to AOC. Then

$$(\mathbf{i} \times \mathbf{j}) \cdot (\mathbf{i} \times \mathbf{k}) = \sin b \sin c \cos A.$$

Using the triple scalar product (3.5) and then the triple vector product (3.6),

$$(\mathbf{i} \times \mathbf{j}) \cdot (\mathbf{i} \times \mathbf{k}) = \mathbf{i} \cdot [\mathbf{j} \times (\mathbf{i} \times \mathbf{k})]$$
$$= \mathbf{i} \cdot [\mathbf{i} \cdot (\mathbf{j} \cdot \mathbf{k}) - \mathbf{k} \cdot (\mathbf{i} \cdot \mathbf{j})] = (\mathbf{j} \cdot \mathbf{k}) - (\mathbf{i} \cdot \mathbf{k}) \cdot (\mathbf{i} \cdot \mathbf{j}),$$

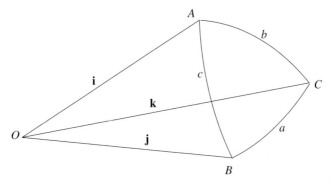

Fig. 3.4. Angles in a spherical triangle.

and, replacing the unit vector products by their values as functions of angles,

$$\sin b \sin c \cos A = \cos a - \cos b \cos c.$$

Hence,

$$\cos a = \cos b \cos c + \sin b \sin c \cos A, \tag{3.45}$$

and two other relations obtained by permutation of the letters that are known as the cosine law of spherical trigonometry.

By the definition of the vector product, the quantities in parentheses give vectors perpendicular to the planes and their cross product is the sine of the angle A. The denominator eliminates the magnitude from the numerator. Thus,

$$\sin A = |(\mathbf{i} \times \mathbf{j}) \times (\mathbf{i} \times \mathbf{k})|/|\mathbf{i} \times \mathbf{j}| \, |\mathbf{i} \times \mathbf{k}|.$$

Applying the formula for the triple vector product on $\mathbf{i} \times \mathbf{j}$, $\mathbf{i}$ and $\mathbf{k}$, one gets

$$\sin A = |(\mathbf{i} \times \mathbf{j}) \cdot \mathbf{k} - (\mathbf{i} \times \mathbf{j}) \cdot \mathbf{j}|/\sin b \sin c.$$

The second product is null, because $\mathbf{j}$ is perpendicular to $\mathbf{i} \times \mathbf{j}$, so that

$$\sin A = [\mathbf{i}, \mathbf{j}, \mathbf{k}]/\sin b \sin c,$$

and, finally, dividing both sides by $\sin a$,

$$\frac{\sin A}{\sin a} = [\mathbf{i}, \mathbf{j}, \mathbf{k}]/\sin a \sin b \sin c.$$

The same approach can be used for $\sin B$ and $\sin C$, so the result is the sine law of spherical trigonometry:

$$\frac{\sin A}{\sin a} = \frac{\sin B}{\sin b} = \frac{\sin C}{\sin c}. \tag{3.46}$$

Two other important relations are obtained as follows. Eliminate $\cos b$ in (3.45) using the similar expression of $\cos b$, and one gets

$$\cos a = \cos^2 c \cos a + \cos c \sin c \sin a \cos B + \sin b \sin c \cos A,$$

and dividing all by $\sin b$, one gets

$$\sin b \cos A = \cos a \sin c - \cos c \sin a \cos B, \tag{3.47}$$

and, by permutation, five other equalities.

Now, let us divide (3.47) by either $\sin b \sin A$ or $\sin a \sin B$, which are equal following (3.46) and, rearranging it, one finally gets

$$\cos c \cos B = \sin c \cot a - \sin B \cot A, \tag{3.48}$$

and, by permutation, five other relations.

3.8 Coordinate systems

Astronomy has used and continues to use a number of different coordinate systems, reference frames, origins and transformations depending on the observations, purposes and accuracies required. The dynamical reference system at an epoch, based on the moving equinox and mean equator for a fixed time, has been the primary reference frame of the past. A new, fixed fundamental reference frame based on very distant radio sources is being introduced with sub-milliarcsecond accuracy. This change is from a dynamical system, based on the orbital motions of the planets, to a kinematic system, based on the rotations and motions of the Earth. While this reference frame is fixed at an epoch, the motions of the Earth, a common observing platform, will still be involved. Thus, the transformations between coordinate systems must be considered as necessary. The relationships or transformations may be given in spherical trigonometry expressions, or matrices and vectors. Also, there are necessary transformations for the appropriate relativistic time scales for the different reference frames (see Section 5.5).

The standard coordinate frame is defined as a right-handed system usually denoted in terms of x, y, z. Any positions can be designated in terms of this coordinate system. The origin and two of the axes are arbitrary. Thus, a plane, defined by two axes, and an origin, a fiducial direction defining one axis in the plane, are sufficient to define a coordinate system.

Combinations of rotations can be accomplished by matrix multiplications. Care must be exercised concerning the signs, or directions, of the rotations. Let us take, as an example, the transformation using Eulerian angles (see Fig. 3.5).

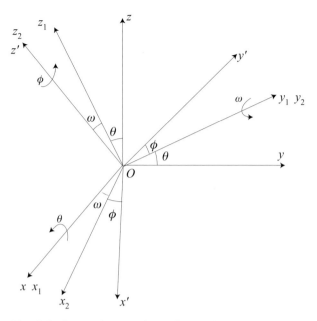

Fig. 3.5. Successive rotations of axes about Ox, Oy_1 and Oz_2.

We start from a coordinate system (x, y, z) and transform it by three successive rotations:

    (i) a rotation about the Ox axis through an angle θ transforms the coordinates x, y, z of a point P into $x_1 = x$, y_1, z_1;

    (ii) a rotation about the new Oy_1 axis through an angle ω transforms the coordinates x_1, y_1, z_1 of the point P into x_2, $y_2 = y_1$, z_2;

    (iii) a rotation about the new Oz_2 axis through an angle ϕ transforms the coordinates x_2, y_2, z_2 of the point P into x', y', $z' = z_2$.

Following Section 3.4.1, the result of these three rotations can be obtained by applying successively three matrix multiplications. If the horizontal vector representation is chosen then,

$$(x' \quad y' \quad z') = (x \quad y \quad z) \times \begin{pmatrix} 1 & 0 & 0 \\ 0 & \cos\theta & -\sin\theta \\ 0 & \sin\theta & \cos\theta \end{pmatrix}$$

$$\times \begin{pmatrix} \cos\omega & 0 & \sin\omega \\ 0 & 1 & 0 \\ -\sin\omega & 0 & \cos\omega \end{pmatrix} \times \begin{pmatrix} \cos\phi & -\sin\phi & 0 \\ \sin\phi & \cos\phi & 0 \\ 0 & 0 & 1 \end{pmatrix}.$$

The multiplications must be performed from left to right.

If the vertical vector representation is chosen, then, using the notations defined in Section 3.4.1,

$$
\begin{pmatrix} x' \\ y' \\ z' \end{pmatrix} = \mathcal{R}_3(\phi) \times \mathcal{R}_2(\omega) \times \mathcal{R}_1(\theta) \times \begin{pmatrix} x \\ y \\ z \end{pmatrix},
$$

or

$$
\begin{pmatrix} x' \\ y' \\ z' \end{pmatrix} = \begin{pmatrix} \cos\phi & \sin\phi & 0 \\ -\sin\phi & \cos\phi & 0 \\ 0 & 0 & 1 \end{pmatrix} \times \begin{pmatrix} \cos\omega & 0 & -\sin\omega \\ 0 & 1 & 0 \\ \sin\omega & 0 & \cos\omega \end{pmatrix}
$$

$$
\times \begin{pmatrix} 1 & 0 & 0 \\ 0 & \cos\theta & \sin\theta \\ 0 & -\sin\theta & \cos\theta \end{pmatrix} \times \begin{pmatrix} x \\ y \\ z \end{pmatrix} = \mathcal{D} \times \begin{pmatrix} x \\ y \\ z \end{pmatrix}.
$$

In this case, the multiplications must be performed from right to left.

The direction cosine elements of the rotation matrx $\mathcal{D}$ would then be as follows.

$$
\begin{aligned}
a_1^1 &= \cos\phi\cos\theta - \sin\phi\sin\theta\cos\omega \\
a_2^1 &= -\cos\phi\sin\theta - \sin\phi\cos\theta\cos\omega \\
a_3^1 &= \sin\phi\sin\omega \\
a_1^2 &= \sin\phi\cos\theta + \cos\phi\sin\theta\cos\omega \\
a_2^2 &= -\sin\phi\sin\theta + \cos\phi\cos\theta\cos\omega \\
a_3^2 &= -\cos\phi\sin\omega \\
a_1^3 &= \sin\theta\sin\omega \\
a_2^3 &= \cos\theta\sin\omega \\
a_3^3 &= \cos\omega.
\end{aligned} \tag{3.49}
$$

3.9 Differential coordinates

In many applications, one is confronted with the problem of determining the coordinates of an object B with respect to another object A whose position in a global coordinate system is known ($\alpha = \alpha_0, \delta_0$). Hence the coordinates of B are $\alpha_0 + \Delta\alpha$ and $\delta = \delta_0 + \Delta\delta$. Actually, in many applications, in particular in double-star astrometry, it is customary to use spherical polar coordinates ρ and θ as shown in Fig. 3.6. Here, $\rho = \text{Arc}AB$ and θ is the position angle reckoned eastward from the direction of the North pole P.

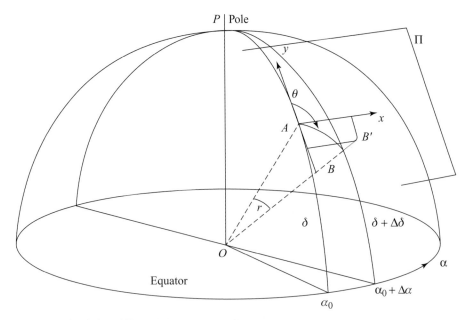

Fig. 3.6. Differential and standard coordinates.

3.9.1 Standard coordinates

In most of the cases, the observation is performed at the focal plane of a telescope, which is a geometric transform of the projection of the field onto the plane tangent to the celestial sphere at A. This transformation is called gnomonic transformation. It transforms the differential coordinates r, θ into rectangular coordinates, also called "*standard coordinates*" Ax, Ay in the tangent plane Π. Ay is directed towards the North pole, tangent to the local celestial meridian, and Ax is perpendicular and directed eastward, parallel to the celestial equator. Note that the orientation of the standard coordinates, as seen from the outside of the celestial sphere, is opposite to the orientation of the spherical polar coordinates. The relations are

$$x = \tan r \sin \theta,$$
$$y = \tan r \cos \theta.$$

The relation between (r, θ) and $(\Delta\alpha, \Delta\delta)$ is obtained using formulae (3.45) and (3.46) in the spherical triangle PAB for r,

$$\cos r = \sin \delta_0 \sin(\delta_0 + \Delta\delta) + \cos \delta_0 \cos(\delta_0 + \Delta\delta) \cos \Delta\alpha, \tag{3.50}$$
$$\sin r \sin \theta = \cos(\delta_0 + \Delta\delta) \sin \Delta\alpha. \tag{3.51}$$

Dividing (3.51) by (3.50), one obtains x. Applying now formula (3.47), one gets,

$$\sin r \cos \theta = \sin(\delta_0 + \Delta\delta) \cos \Delta\alpha,$$

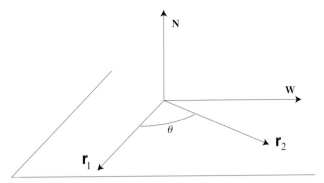

Fig. 3.7. Geometry in the $\mathbf{r}_1$–$\mathbf{r}_2$ plane.

which, divided by (3.50), gives y. Finally, replacing $\delta + \Delta\delta$ by δ, one finally obtains,

$$
\begin{aligned}
x &= \frac{\cos\delta\sin\Delta\alpha}{\sin\delta_0\sin\delta + \cos\delta_0\cos\delta\cos\Delta\alpha}, \\
y &= \frac{\cos\delta_0\sin\delta - \sin\delta_0\cos\delta\cos\Delta\alpha}{\sin\delta_0\sin\delta + \cos\delta_0\cos\delta\cos\Delta\alpha}.
\end{aligned}
\tag{3.52}
$$

3.9.2 Displacement of a body

A problem encountered in many instances is the determination of the position of a body after it has moved by an angle θ. Let $\mathbf{r}_1$ be the unit vector of the initial position and $\mathbf{r}_2$ of the new direction. The transformation from $\mathbf{r}_1$ to $\mathbf{r}_2$ is a rotation by the angle θ about the normal unit vector, $\mathbf{N}$, to the $\mathbf{r}_1$–$\mathbf{r}_2$ plane (see Fig. 3.7):

$$\mathbf{N} = \mathbf{r}_1 \times \mathbf{r}_2.$$

In this plane let us consider a system of rectangular coordinates defined by $\mathbf{r}_1$ and $\mathbf{W}$ directly perpendicular

$$\mathbf{W} = \mathbf{N} \times \mathbf{r}_1.$$

The projection of $\mathbf{r}_2$ on $\mathbf{W}$ is $\sin\theta$ and as a result

$$\mathbf{r}_2 = \mathbf{r}_1 \cos\theta + (\mathbf{N} \times \mathbf{r}_1)\sin\theta. \tag{3.53}$$

3.9.3 Space standard coordinates

It is useful, for instance, in determining the space velocity of a star, to compute the rectangular equatorial coordinates (x, y, z) of a vector $\mathbf{V}$, whose coordinates in the standard system are (p, q, r) with $r \neq 0$ (Fig. 3.8). Two rotations are necessary to

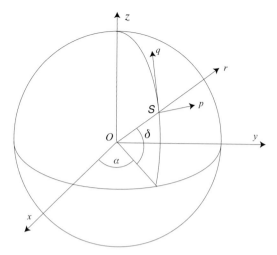

Fig. 3.8. Space standard coordinates.

transform space standard coordinates into the equatorial system:

(i) a rotation by an angle $(\delta - \pi/2)$ about the Sp axis, to align Sr with Oz,
(ii) then a rotation of an angle $(-\alpha - \pi/2)$ about Sr (now Oz) to align Sp and Sq
with Ox and Oy.

One gets,

$$
\begin{pmatrix} x \\ y \\ z \end{pmatrix} = \begin{pmatrix} -\sin\alpha & -\cos\alpha & 0 \\ \cos\alpha & -\sin\alpha & 0 \\ 0 & 0 & 1 \end{pmatrix} \times \begin{pmatrix} 1 & 0 & 0 \\ 0 & \sin\delta & -\cos\delta \\ 0 & \cos\delta & \sin\delta \end{pmatrix} \times \begin{pmatrix} p \\ q \\ r \end{pmatrix}
$$

$$
= \mathcal{R} \times \begin{pmatrix} p \\ q \\ r \end{pmatrix}
$$

with

$$
\mathcal{R} = \begin{pmatrix} -\sin\alpha & -\sin\delta\cos\alpha & \cos\delta\cos\alpha \\ \cos\alpha & -\sin\delta\sin\alpha & \cos\delta\sin\alpha \\ 0 & \cos\delta & \sin\delta \end{pmatrix}. \tag{3.54}
$$

3.10 Reference coordinates

There are a number of natural reference frames to use in astronomy. Origins for
the reference frames can include the barycenter of the Solar System, the geocenter,
and the observer. Similarly, the equator, ecliptic, and horizon are natural reference

planes. Thus, reference frames are developed based on these natural origins and planes, but there is the complication that everything is in motion. Also, there are necessary transformations for the appropriate relativistic time scales for the different reference frames.

3.10.1 Equatorial celestial coordinates

Using a fixed reference frame of some epoch, the barycenter as the origin, and the equator as a reference plane, from a fiducial point in the equatorial plane, a right-handed coordinate frame is specified. In this reference frame, right ascensions can be measured in the equatorial plane, usually in time measurement from 0 h to 24 h, but also from 0° to 360°. Declination is measured perpendicular to the equatorial plane, positive to the North from 0° to 90°. For a complete specification, the distance is needed along with the motions of the object in all three directions. Such a fixed barycentric reference is natural for a reference frame based on extragalactic sources, which is now called the International Celestial Reference Frame (ICRF), (see Section 7.2). Celestial reference frames can also be geocentric, with the origin at the geocenter, or topocentric, with the origin at the observer.

3.10.2 Terrestrial geocentric and geodetic coordinates

Geocentric terrestrial coordinates are Earth fixed and rotating with the Earth. The prime reference plane of such geocentric coordinates is the plane of the equator. The fiducial point is defined by the arbitrary prime meridian called the International Meridian, close to the Greenwich Meridian (see Fig. 3.9). This frame defines a system of rectangular coordinates (x, y, z) or of spherical coordinates (longitude, λ', latitude, ϕ', and radial distance, ρ'). Longitude is measured positively to the East from the International Meridian and usually designated to $\pm180°$. Latitude is measured from the equator to $\pm90°$, positive to the North. As in the case of the sky, there is also an International Terrestrial Reference Frame (ITRF) described in Section 9.3.

The Earth is not a rigid body, nor is it a sphere. Owing to the shape of the Earth there is a need for geodetic coordinates (longitude, latitude, and height) that refer to an adopted spheroid, which best represents the shape of the Earth, and whose equatorial radius and flattening are specified. Geodetic longitude (λ) is the same as the geocentric longitude. Geodetic latitude (ϕ) is equal to the inclination to the equatorial plane of the normal to the spheroid (see Fig. 3.9). Geodetic height (h) is the distance above the spheroid measured along the normal to the spheroid (see also Section 9.4). One must note that this normal does not coincide with the vertical,

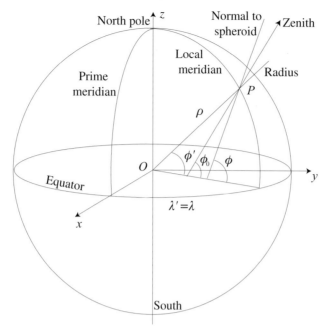

Fig. 3.9. Geocentric and geodetic coordinates. The direction of the zenith is different from the radius and the normal to the spheroid.

pointing towards the zenith, and is perpendicular to the local equipotential surface. The corresponding latitude (ϕ_0) and longitude (λ_0) are therefore different. The angle between the two directions is the *deflection of the vertical* (see Section 9.7.2).

3.10.3 Topocentric coordinates

Topocentric coordinates are celestial coooordinates, which can be defined in terms of the local zenith and the resulting horizon plane, or in terms of the Earth's axis of rotation and the plane of the celestial equator. The local meridian, which contains the directions of the local zenith and the axis of rotation North–South, is common to both frames as shown also on Fig. 3.10.

   With respect to the horizon and the local zenith, the altitude (a) is measured positively from the horizon toward the zenith, while, alternatively, the zenith distance (z) is measured from the zenith toward the horizon ($z = 90° - a$). Azimuth (A) is measured from 0° to 360° from North in the direction of East. However, sometimes it is measured from the South, so it is best to specify the convention being used (see Fig. 3.11). In the topocentric equatorial frame, the angular coordinates are known as local hour angle (LHA) and declination (δ). LHA is measured from 0° to 360°, or from 0 h to 24 h, from South in the direction of West, along the equator. The

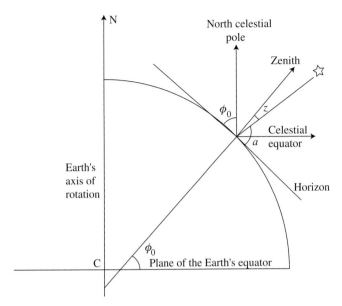

Fig. 3.10. Topocentric coordinates.

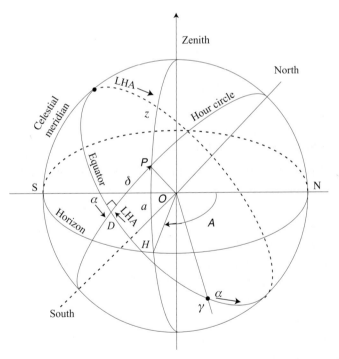

Fig. 3.11. The observer O and an arbitrary point P shown with respect to the equatorial and hour circles, with the reference directions of zenith and North celestial pole.

LHA of a celestial object increases by about 15°, or 1 h, for every hour of sidereal time. Declination is measured from the equator to the celestial pole, positive toward the North.

3.11 Time

For each origin there is an appropriate coordinate time (see Chapter 5). The time scale on the geoid based on atomic time standards and the international System of Units (Système international d'unités, SI) unit of time (second) is the International Atomic Time (TAI). This is the measurable time scale and the basis for all other time scales, including the Coordinated Universal Time (UTC), and standard times used around the world.

The idealized time on the geoid is Terrestrial Time (TT), which is a continuation of an earlier time scale (Ephemeris Time), and is currently defined by its origin in 1977 with respect to TAI (International Atomic Time) and the SI second. Realizations of TT are designated as TT(x). For instance,

$$TT(TAI) = TAI + 32.184 \, s.$$

By a transformation, Geocentric Coordinate Time (TCG), the time scale appropriate for geocentric coordinates, can be determined from TT (Section 5.5.1). Another transformation from TCG will give Barycentric Coordinate Time (TCB), the coordinate time appropriate for barycentric coordinates centered at the barycenter of the Solar System (Section 5.5.2). TT, TCG, and TCB are defined to satisfy the relativistic transformations between the different origins, such that units of measurements are consistent between the different coordinate systems.

There were previously defined time scales, Terrestrial Dynamical Time (TDT), which is equivalent to Terrestrial Time (TT) and Barycentric Dynamical Time (TDB), which was a time scale for the barycenter of the Solar System, such that there were only periodic differences between TDT and TDB. Unfortunately, this led to ambiguities in the determination of TDB, and differences in the values of units, depending on the time scale being used. The time scales will be further studied in Sections 5.5.1, 5.5.2, and 10.4.4.

3.12 Extragalactic reference frame

The extragalactic, or fundamental, reference frame provides a reference frame to which all observations can be referred. Owing to the number, wavelengths, and magnitudes of the reference sources, direct observation on the extragalactic reference frame is not always possible. Thus, observations must be made in the coordinate system available and then transformed into the extragalactic reference frame (see Section 7.2). The extragalactic reference frame defines a fixed reference frame, while most other coordinate systems are subject to motions, or rotations, so the epoch of the coordinate system must be considered. Also the motion of the observer's platform must be included. An observer on the surface of the Earth is subject to the kinematics of the Earth with respect to the celestial reference frame, referred to as precession and nutation, and the dynamics of the Earth itself, called polar motion and Earth rotation (see Section 10.3). There are also effects due to parallax (Section 6.2) and to aberration (Section 6.3).

Distinction must also be made between small- and large-angle observations. Small-angle observations can be made with respect to the reference stars whose positions are on the extragalactic reference frame. These observations are differential and require only differential corrections (see Section 3.9). Large-angle observations must be transformed from the coordinate system of the instrument and date to the extragalactic reference frame.

In the past, the vernal equinox, one of the intersections of the ecliptic with the celestial equator was used as the origin for measuring right ascensions (α) along the equator. The equinox has been replaced by an origin, defined to have no rotation with respect to the extragalactic reference sources (Section 8.4). Thus, in the ICRF, right ascensions are now measured from this fixed origin along a fixed conventional equator to the hour circle.

One should mention that a reference frame can also be defined in terms of our galaxy by the plane of the Milky Way, called the *galactic plane* (see Section 7.7). Thus, galactic latitude is measured from the plane in the North and South directions and galactic longitude is measured in the galactic plane with a specified origin. The details concerning all the reference systems will be given in Chapters 7, 8, 9 and 10.

4

Treatment of astronomical data

As a rule, an astrometric observation does not directly provide the quantity sought, but rather some function of it and of a number of other parameters. The procedure that is used to obtain it is called *reduction of observations*. Generally, one needs to use several observations taken at different times and conditions, and treat them together. Astrometric data reduction includes taking into account a variety of effects that are described in the following chapters. But it also consists of a mathematical treatment of the data, for which one generally assumes some statistical properties. Among other things, one must be aware of the fact that a numerical result has no physical significance, and is not useful, unless there is an indication of how good the numbers are. The results must be supplemented by another number, called *uncertainty*, which provides an interval (sometimes called error bars) within which there is a stated probability that it includes the true value. Sometimes, when several quantities are computed simultaneously, and are mutually dependent or correlated, other numbers called *correlation coefficients* must be determined.

This chapter presents succinctly the most commonly used statistical tools to perform the data reduction. There are many textbooks in statistics or data treatment to which one should refer in order to go deeper in the understanding of the procedures or get more information. Let us mention, for instance, Jaech (1985), Liebelt (1967) and Silvey (1991). For more mathematical approaches, see Mendenhall *et al.* (1986) or Stark and Woods (1986).

4.1 Introduction to data reduction

Let us first give an example to illustrate how data reduction occurs in a classical problem of astrometry.

4.1.1 Example

Suppose that a minor planet is photographed, or observed with a CCD, together with N stars, whose positions on the sky, A_i, are known. We then wish to determine the equatorial coordinates, α and δ, that we denote by the vector A. The positions are determined on the plane of the plate, or of the CCD, and one obtains the coordinates $X(x, y)$ of the image of the planet and $X_i(x_i, y_i)$ of the N stars.

The reduction of the plate consists of determining the relation

$$F(X, A) = 0$$

that links the measured quantities to the unknowns. This function is not arbitrary, but its form can be deduced from the physics of the transformation between points on the celestial sphere and points on the photographic plate, namely the path of the light rays through space and the atmosphere, then through the telescope. This includes the effect of the differential atmospheric refraction, then the geometric transformation from a sphere onto a plane (gnomonic projection, see Section 3.9.1), the irregularities of the plate or of the CCD, and the aberrations of the optics. The refraction depends upon zenith distance, color of the object, temperature, pressure and the water content of the atmosphere (see Section 6.1). The gnomonic projection is a well-defined transformation, which is a function of X and A, as well as of the focal length of the telescope. The optical aberrations produce an additional modification of the formulae for gnomonic projection that are usually represented by a power series of x and y.

One may compute the leading part of F from the refraction formulae and observed values of temperature, pressure and hygrometry, and the gnomonic transformation from a nominal value of the focal length and the position of the optical axis of the telescope. The remaining effects are small, so that, together with the optical aberrations, they can be represented by power series of x and y. In addition, one must take into account measurement uncertainties ε_j on X and A. So the relation will become

$$F(X, A, p_1, \ldots p_j, \ldots p_k, \varepsilon_X, \ldots \varepsilon_A) = 0, \tag{4.1}$$

where the parameters p_j are the unknown coefficients of the development in power series.

4.1.2 Principle of data reduction

Prior to obtaining (4.1), we have actually performed two preliminary steps of the reduction, that are present in all problems. The first step has three parts.

(i) *Modeling*. This consists of writing the theoretical equations that link the observed and the unknown quantities.

(ii) *Linearization*. The model is applied to the observed quantities so that only a small part is not taken into account. This small part is assumed to be in general increments to roughly known quantities or neglected small effects. It depends upon k parameters p_j that appear in the equations as first degree quantities, so that $\mathbf{F}$ is a linear equation in p_j.

(iii) *Solving the linearized equations*. To determine these parameters, one uses the N measurements of reference stars, each giving rise to *equations of condition*:

$$\mathbf{F}_n\left(\mathbf{X}_i, \mathbf{A}_i, p_1, \ldots p_j, \ldots p_k, \varepsilon(\mathbf{X}_i), \ldots \varepsilon(\mathbf{A}_i)\right) = 0, \tag{4.2}$$

with $1 < n < N$.

The next step is to solve these N vectorial Equations (4.2) with k linear unknowns. This is usually, but not necessarily, done by a least-squares adjustment, which provides the values of the parameters, their uncertainties, and correlations that depend on the uncertainties with which the measurements are performed.

Finally, when these parameters p_i are determined, one again writes Equation (4.1), $\mathbf{A}$ being unknown and $\mathbf{X}$ being known. Now, A, the minor planet position, can be determined.

In the following sections, we shall describe the theoretical background and practical procedures of the steps of the reduction process.

Note. It may be that the modeled equations do not represent the reality well enough to justify linearization. One must then solve a set of nonlinear equations. There is no general method to do so, and one must search, case by case, for the best way to reach an adequate solution (see Section 4.4.7).

4.1.3 Basic concept of uncertainties

No observation is exact, there is always an error. The error is the difference between the value measured and the true value. Since the latter is not known, it is impossible to determine it. For this reason it is preferable to speak of *uncertainties*. Indeed, if one could evaluate part of the error, one would add this evaluation to the value measured and improve it. Therefore this procedure would be part of the evaluation of the unknown value. At some time, however, when all possible components of errors are analyzed and their effects added, the result is still not the true value, and there remains an unknown difference, or error. However, its order of magnitude may be evaluated with some associated probability. The result is the uncertainty, which characterizes the dispersion of values that can be assigned to a measured quantity. It is a number attached to a numerical result, which depends upon the probability of the true value being within some domain around the value determined. This probability quantifies the *level of confidence*. One may distinguish two types of error.

(i) **Random errors.** These presumably arise from unpredictable, or stochastic, variations of a parameter influencing the measurement of a quantity. This is the case, for example, for the rapid atmospheric fluctuations giving rise to an agitation of the images and scintillation. The origin of these effects is the presence in the atmosphere of small (5–20 cm) moving turbulent cells, each acting like a small lens modifying the mean direction of the rays, while focusing or de-focusing them. Another example is the residual of an ensemble of settings on the same star image by a measuring machine, or the centering error of an image spread over several CCD pixels. These errors are treated by various statistical methods, some of which are presented in Sections 4.2 and 4.4. Their evaluation provides the *precision* of the measurement.

(ii) **Systematic errors.** These correspond to effects which, if they were known, would provide a correction to the result. They may arise from an insufficient accuracy of the model, from the omission of a significant parameter in the reduction, or from some unknown instrumental bias. There are a few hints to estimate the order of magnitude of systematic errors (Section 4.3), but it is not an easy task, and it also involves a deep understanding of the problem and accumulated practice. The evaluated systematic errors provide the *accuracy* of the measurement.

The combination of estimated random and systematic errors is the *uncertainty*. Let $\hat{x}$ be the estimated value of a parameter x, and σ_x be a quantity such that the true value x_0 has a probability of 0.683 of being within the interval

$$\hat{x} - \sigma_x < x_0 < \hat{x} + \sigma_x.$$

Then, we can write

$$x_0 = \hat{x} \pm \sigma_x.$$

The quantity σ_x is called the *standard* uncertainty or deviation. Often it is also referred to as the root mean square uncertainty (r.m.s.). Its definition will be given in the next section.

4.2 Random errors

Generally, the best available estimate of the expected value of x that varies randomly, and for which n independent values, x_k, have been obtained, is the arithmetic mean or average, μ,

$$x = \mu = \frac{1}{n} \sum_{k=1}^{n} x_k. \tag{4.3}$$

The residual of each measurement is

$$r_k = x_k - \mu.$$

One introduces the *variance* of the measurements

$$\sigma^2(x_k) = \frac{1}{n-1} \sum_{k=1}^{n} r_k^2, \tag{4.4}$$

and the *standard deviation*, $\sigma(x_k)$, is the square root of the variance.

4.2.1 Combination of random quantities

Generally, the quantity, z, to be determined depends upon N different random parameters, x_i. Let us assume that there exists a model that links these different quantities, and that one has

$$z = f(x_1, x_2, \ldots x_N).$$

Let us assume that each of the parameters, x_i, is random with an expected value μ_i and a standard deviation σ_i. Then, for small deviations of these quantities from their expected values, the first-order Taylor expansion yields

$$z - \mu_z = \sum_{i=1}^{N} \frac{\partial f}{\partial x_i}(x_i - \mu_i). \tag{4.5}$$

Hence, the expected value of z is

$$\mu_z = f(\mu_1, \mu_2, \ldots \mu_N), \tag{4.6}$$

and

$$(z - \mu_z)^2 = \sum_{i=1}^{N} \left(\frac{\partial f}{\partial x_i}\right)^2 (x_i - \mu_i)^2$$
$$+ 2 \sum_{i=1}^{N} \sum_{j=i+1}^{N} \left(\frac{\partial f}{\partial x_i}\right)\left(\frac{\partial f}{\partial x_j}\right)(x_i - \mu_i)(x_j - \mu_j). \tag{4.7}$$

The expected value of $(z - \mu_z)^2$ is the variance σ_z^2 of z, and similarly, for each of the parameters, the expected values of $(x_i - \mu_i)^2$ are σ_i^2.

Now, in the second part of (4.7), the expected value of $(x_i - \mu_i)(x_j - \mu_j)$ is written as $\sigma_i \sigma_j \rho_{ij}$, where ρ_{ij} is, by definition, the *correlation coefficient* of x_i and x_j. The product, $\sigma_{ij} = \sigma_i \sigma_j \rho_{ij}$, is called the covariance of x_i and x_j. If x_i and x_j are independent variables, meaning that their variations do not depend one on the other nor on a common cause, then the correlation coefficient is null. With these notations, (4.7) becomes

$$\sigma_z^2 = \sum_{i=1}^{N} \left(\frac{\partial f}{\partial x_i} \right)^2 \sigma_{x_i}^2 + 2 \sum_{i=1}^{N} \sum_{j=i+1}^{N} \frac{\partial f}{\partial x_i} \frac{\partial f}{\partial x_j} \sigma_i \sigma_j \rho_{ij}. \tag{4.8}$$

Finally, one introduces the *variance–covariance matrix*, the symmetric matrix $|\sigma_{ij}|$, whose elements are $\sigma_{ii} = \sigma_i^2$ in row i, column i and σ_{ij} both in row i, column j and row j, column i.

Example. Coordinates α and δ of a star in a fixed reference system change with time proportional to its proper motion μ_α and μ_δ. Let α_0 and δ_0 be its position at some time origin; its values at time t are

$$\alpha = \alpha_0 + t\mu_\alpha,$$
$$\delta = \delta_0 + t\mu_\delta, \tag{4.9}$$

provided that t is not too large, and the star is not close to the poles. Following the notation of the *Hipparcos Catalogue*, let us set

$$\alpha^* = \alpha \cos \delta_0,$$
$$\mu_\alpha{}^* = \mu_\alpha \cos \delta_0.$$

In these notations, Equations (4.9) become :

$$\alpha^* = \alpha_0^* + t\mu_\alpha^*,$$
$$\delta = \delta_0 + t\mu_\delta.$$

Since the determination of proper motions is based upon observed values of the position, μ_α^* and α^* on one side, μ_δ and δ on the other side are correlated. The corresponding correlation coefficients are given in the catalogue. Applying (4.8), one gets

$$\sigma_{\alpha^*}^2(t) = \sigma_{\alpha^*}^2(0) + 2t\rho_{\alpha^*\mu\alpha^*}(0)\sigma_{\mu\alpha^*}(0)\sigma_{\alpha^*}(0) + t^2\sigma_{\mu\alpha^*}^2(0),$$
$$\sigma_\delta^2(t) = \sigma_\delta^2(0) + 2t\rho_{\delta\mu\delta}(0)\sigma_\delta(0)\sigma_{\mu\delta}(0) + t^2\sigma_{\mu\alpha}^2(0).$$

The degradation of the standard errors of positions depends on the correlation coefficients, as well as on individual variances.

4.2.2 Generalization to functions of N variables

Let us generalize Equation (4.5), assuming a transformation of an N-vector $\mathbf{x}$ into an N-vector $\mathbf{z}$, and let us call, respectively, $\Delta \mathbf{x}$ and $\Delta \mathbf{z}$ the differences from their expected values. The equivalent to (4.5) now is

$$\Delta \mathbf{z} = \mathcal{J}_f(\mathbf{x}) \Delta \mathbf{x},$$

where $\mathcal{J}_f$ is the $N \times N$ Jacobian of the transformation,

$$\mathcal{J}_f = \left(\frac{\partial f_j}{\partial x_i}\right)_{1 \le i,j \le N},$$

as evaluated at the expected values. Let $\mathcal{C}$ be the variance–covariance matrix of $\Delta\mathbf{x}$. It can be computed as

$$\mathcal{C} = \Delta\mathbf{x} \times \Delta\mathbf{x}^\top,$$

where $\Delta\mathbf{x}$ is a column vector and $\Delta\mathbf{x}^\top$ is the corresponding transposed (row) vector. Then, the theory of transformation of functions provides the new variance–covariance matrix

$$\mathcal{S} = \mathcal{J}_f \times \mathcal{C} \times \mathcal{J}'_f,$$

where $\mathcal{J}'_f$ is the transposed matrix of $\mathcal{J}_f$. This result is to be applied in particular to coordinate transformations for which the number of dimensions is not modified by the transformation.

4.2.3 Linear combination of random quantities

Let us consider a function z that has the following form

$$z = \sum_{i=1}^{N} A_i x_i,$$

where A_i are constants. In this case, the variance of z is

$$\sigma_z^2 = \sum_{i=1}^{N} A_i^2 \sigma_i^2 + 2 \sum_{i=1}^{N} \sum_{j=i+1}^{N} A_i A_j \sigma_{ij}. \tag{4.10}$$

A particular case is when all the parameters x_i are independent. Then, the variance of z is:

$$\sigma_z^2 = \sum_{i=1}^{N} A_i^2 \sigma_i^2. \tag{4.11}$$

The variance of z is the same linear combination of variances as z.

Example. Let us compute the variance of the uncertainty in measuring a plate. Let us assume that it is because of the random atmospheric error, whose variance is σ_A^2, the random error of the measuring machine, whose variance is σ_M^2, and the random error due to the sensitivity distribution in the plate, whose variance is σ_S^2. Clearly these three phenomena are independent and have no influence on one another. Then, the variance of the measurement is

$$\sigma_z^2 = \sigma_A^2 + \sigma_M^2 + \sigma_S^2,$$

provided that the three quantities are expressed in the same units. The standard uncertainty of z is therefore:

$$\sigma_z = \sqrt{\sigma_A^2 + \sigma_M^2 + \sigma_S^2}.$$

This formulation is very commonly used to evaluate the standard uncertainties of measurements.

4.2.4 Normal probability density function

One may extend the notion of successive values of a random parameter to a continuous function, that takes for each possible value of the random parameter – called hereafter the *random variable* – the associated probability. Such a function is called the *probability density function* (pdf). Among the many possible pdfs, one has a particular status. It results from the *central limit theorem*, which states that the sum of a large number of mutually independent random variables tends to a probability density function called the Gaussian, or *normal*, density function given by

$$f(x) = \frac{1}{\sqrt{2\pi}\sigma} \exp\left[\frac{-(x-\mu)^2}{2\sigma^2}\right], \tag{4.12}$$

where μ is the mean of the probability distribution, and σ is the standard deviation. It is represented in Fig. 4.1. Table 4.1 gives the level of confidence, C, for different intervals around the mean, that is the probability for a value to be within the interval

$$\mu - k_p\sigma < x < \mu + k_p\sigma,$$

where k_p is called the *coverage factor*. It is given by

$$C = \int_{\mu-k_p\sigma}^{\mu+k_p\sigma} f(x)\mathrm{d}x.$$

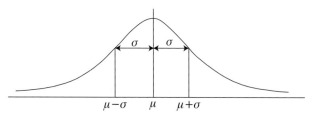

Fig. 4.1. The normal probability density function.

Table 4.1. *Levels of confidence for*
different intervals of x symmetric with
respect to the mean

Coverage factor k_p	Level of confidence C
1σ	68.27%
1.645σ	90.00%
2σ	95.45%
2.5σ	98.70%
3σ	99.73%

An important property of this pdf is that any random variable, formed by taking a linear combination of independent normally distributed random variables, is itself also normally distributed. This normal pdf is commonly assumed to represent the probability density function of random quantities contributing to measurements. It is indeed a fact that uncertainties of many physical measurements are well represented by the normal law. This justifies the importance of this pdf in astrometry.

4.2.5 Other probability density functions

The Gaussian pdf can be generalized to as many independent random variables as needed. For instance, let X_1 and X_2 be two random variables, and x_1 and x_2 the measurements with mean values μ_1 and μ_2, and the corresponding variances σ_1^2 and σ_2^2. To be general, one should also allow for a correlation coefficient ρ. Then the bivariant pdf is

$$f(x_1, x_2) = \frac{1}{\pi\sigma_1\sigma_2\sqrt{1-\rho^2}} \exp\left[-\frac{1}{2(1-\rho^2)}\right.$$
$$\left. \times \left(\frac{(x_1-\mu_1)^2}{\sigma_1^2} + \frac{2\rho(x_1-\mu_1)(x_2-\mu_2)}{\sigma_1\sigma_2} + \frac{(x_2-\mu_2)^2}{\sigma_2^2}\right)\right].$$

If X_1 and X_2 are independent, then $\rho = 0$.

This can be extended to N random variables $X_1, \ldots X_N(x_1 \ldots x_N)$ in which we introduce all the quantities of the variance–covariance matrix $|\sigma_{ij}|$ defined in Section 4.2.1:

$$f(x_1, x_2 \ldots x_N) = \left(\frac{1}{2\pi}\right)^{0.5N} \sqrt{|\sigma_{ij}|} \exp\left[-\frac{1}{2}\sum_{i=1}^{N}\sum_{j=1}^{N}\sigma_{ij}(x_i-\mu_i)(x_j-\mu_j)\right],$$

where $\mu_1, \ldots \mu_N$ are the expected values of $x_1, \ldots x_N$.

There are many other probability density functions that may be useful in some particular situations. Let us present the three most frequently used.

(i) **Chi-square pdf.** This is closely related to the Gaussian distribution. It describes the statistics of a sample of N observations of a quantity that obeys a Gaussian pdf with a variance σ^2. One is led to compute the average

$$\bar{x} = \sum_{i=1}^{N} \frac{x_i}{N},$$

and the sample variance

$$s^2 = \frac{1}{N-1} \left(\sum_{i=1}^{N} x_i^2 - N\bar{x}^2 \right).$$

One then considers the random variable

$$y = \frac{\nu s^2}{\sigma^2},$$

where $\nu = N - 1$ is the *degree of freedom*. This random variable has a pdf called the *chi-square* pdf, $\chi_\nu^2(y)$, which has the following form,

$$\chi_\nu^2(y) = K y^{(\nu-2)/2} \exp(y/2),$$

where the constant K is chosen in such a way that the integral from $-\infty$ to $+\infty$ of $\chi^2(\nu)$ is equal to 1. One has

$$K = \frac{1}{2^{\nu/2} \Gamma(\nu/2)},$$

in which Γ is the gamma function:

$$\Gamma(x) = \int_0^\infty \exp(-t) t^{x-1} dt.$$

The chi-square pdf is usually tabulated as a function of the degree of freedom, and is used to find the probability that s^2 exceeds some given number, showing how well the sample is representative of the distribution. Figure 4.2 gives graphs of some chi-square pdfs.

(ii) **Gamma-type pdfs.** These constitute a large family of probability density functions,

$$G(y, a, b) = \frac{1}{b\Gamma(a)} y^{a-1} \exp(-y/b)$$

with a, b and y positive. The chi-square pdf is a particular case of gamma pdfs. Another particular case is the exponential pdf

$$E(y, b) = \frac{1}{b} \exp(-y/b),$$

corresponding to $a = 1$.

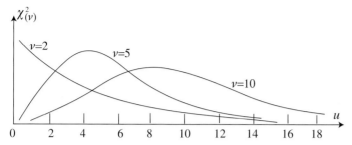

Fig. 4.2. Examples of chi-square pdfs for various values of the degrees of freedom ν.

(iii) **Uniform pdf**. This is a useful probability density function for evaluating the effect of rounding errors, which may, in some cases, be non-negligible in an error budget. The pdf is constant in a finite interval of the variable

$$f(y) = \frac{1}{b-a}, \quad \text{with} \quad a \leq y \leq b,$$
$$f(y) = 0, \text{ elsewhere.}$$

The corresponding variance is

$$\sigma^2 = \frac{(b-a)^2}{12}.$$

4.3 Systematic errors

In contrast to random errors, there is no rigorous definition of systematic errors. They are the types of error that cannot be characterized by an uncertainty computed by using the methods described in the preceding section. The uncertainty of a given source of errors can in some cases be treated by random methods (e.g. described by a normal pdf), and in some cases it has to be considered as giving rise to an offset depending upon the level of accuracy required.

A simple way to distinguish the two types of error is to see what happens if the variance is multiplied by some number larger than one and all the parameters increased accordingly. If the random error treatment is justified, the mean value should not change, and only its standard deviation changes. If the mean value is modified, then one says that parameters produce a *bias*. In this case, one should model the effect involved, and consider the parameters that describe this effect as part of the unknowns, either directly in the equations, or by adding further conditions from parallel measurements, or theoretical considerations. However, in all cases, there exist some remaining unidentified biases, and there is no way to recognize their existence, so that in practice, one is led to treat the left over errors as an uncertainty.

The vocabulary to be used in stating the uncertainty of a measurement depends upon the type of errors that are present in the process. If all the parameters introduced in the model are random, then the standard uncertainty represents what is called *precision*. If there are, in addition, biases, one should evaluate them, and add them to the precision and obtain what is called the *accuracy* of the determination. In other terms, one may say that the precision describes the random distribution of variations in the measurements, whereas the accuracy is an estimation of the difference between the measurement and the true value, which is inaccessible. For this reason, it is preferable to use the word *uncertainty*, which represents a combination of both types of errors.

The significant quantity is accuracy, but it cannot be treated rigorously as a random quantity when using the result of the measurement to obtain some other quantity. Some effects of the bias can be computed by using the model, while the effect of the precision may be extended by using formula (4.7). It can also be evaluated from comparisons between observations performed independently, and under different conditions. The uncertainty of the result will be the quadratic sum of the precision and of the assumed effects of the biases.

The only possible way to reduce biases is to introduce their effects in the model, and consider the parameters governing them as unknown parameters, whose variances will be determined together with the variances of other unknowns.

4.4 Method of least squares

The method that we are going to present is not the only possible method capable of estimating the unknown parameters of a measurement model. It has, however, many important positive properties which justify its very wide use, provided that some conditions are satisfied; in particular, that the model can be linearized with respect to the unknown parameters (next section).

Let $X(x_1, x_2 \ldots x_n)$ be the vector of unknowns, A the vector of measurements $(a_1, a_2 \ldots a_m)$, and $P(p_1, p_2 \ldots p_q)$ the vector of the parameters entering the model which relates A to X. One has

$$A + E = F(X, P), \tag{4.13}$$

with $n + q$ unknowns and $E(\varepsilon_1, \varepsilon_2 \ldots \varepsilon_m)$ is an unknown error vector of the measurements A.

4.4.1 Linearization of the equations

The method of least squares is designed to find the most probable estimation, $\hat{a}_i$, of the components a_i of A. The assumptions are:

- the pdf of all the unknowns is the normal probability density function;
- the unknowns are sufficiently small, so that one may neglect their squares.

The first assumption is met if all the parameters, which may cause a bias, are taken into account by (4.13), so that their uncertainties may be considered as random. The second assumption implies that, unless the unknowns are indeed small, one knows an approximate value $\mathbf{X}_0(x_1^0, x_2^0 \ldots x_n^0)$ and $\mathbf{P}_0(p_1^0, p_2^0 \ldots p_q^0)$, so that one can develop (4.13) as follows,

$$
\begin{aligned}
\mathbf{A} = \mathbf{F}(\mathbf{X}_0, \mathbf{P}_0) + \sum_{i=1}^{n} \left(\frac{\partial \mathbf{F}(\mathbf{X}_0, \mathbf{P}_0)}{\partial x_i} \right) \delta x_i \\
+ \sum_{j=1}^{q} \left(\frac{\partial \mathbf{F}(\mathbf{X}_0, \mathbf{P}_0)}{\partial p_j} \right) \delta p_j - \mathbf{E},
\end{aligned}
\tag{4.14}
$$

where δx_i and δp_j are the new unknowns and, following our assumption, the second order of the development is negligible, and can be added to $\mathbf{E}$ without changing its randomness. This equation does not formally distinguish the actual unknowns and the parameters. So we shall not individualize the second part further, and shall include in $\mathbf{X}$ all unknowns, whether they are of astronomical interest or are additional parameters. Writing (4.14) for each component of $\mathbf{A}$, one gets a set of m equations

$$
a_j - F_j(x_1, \ldots x_k) = \sum_{i=1}^{k} \left(\frac{\partial F_j(x_1, \ldots x_k)}{\partial x_i} \right) \delta x_i - \varepsilon_j,
\tag{4.15}
$$

where $1 \leq j \leq m$ and $k = n + q$ is the number of unknown quantities, while ε_j describes the statistical properties of the uncertainties of a_j. This set of linear equations in δx_i cannot be solved if $k < j$. Therefore, it is necessary to increase the number of equations by performing a series of observations obeying the same model. In the example of a plate reduction given in Section 4.1, one measures a sufficient number of reference stars, each providing two equations. The beauty of the method of least squares is that it allows any number m of Equations (4.15), provided that m is strictly larger than the number k of unknowns. A general experience is that it is desirable that m be of the order of two or three times k. The quantity $m - k = \nu$ is the degree of freedom as introduced in Section 4.2.5.

The ensemble of Equations (4.15) is called the *equations of condition*, or *design equations*, the coefficients of the second member forming the design matrix $\mathcal{D}$.

Equations (4.15) can then be written as

$$
\mathbf{C} = \mathcal{D}\mathbf{Y} + \mathbf{E},
\tag{4.16}
$$

where $\mathbf{Y}$ is the vector of the new unknowns $(\delta x_1 \dots \delta x_k)$, and $\mathbf{C}$ is the vector of the left-hand members of (4.15)

$$\mathbf{C} = (a_j - F_j(x_1 \dots x_k)), \quad j = 1, m.$$

4.4.2 Principle of the method of least squares

If the conditions stated in the beginning of Section 4.4.1 are met, the expected value of $\mathbf{E}$ is zero, and its variance is

$$\frac{1}{m-1} \sum_{j=1}^{m} \varepsilon_j^2.$$

In reality, of course, this is not the case. The objective of the method of least squares is to determine the vector $\mathbf{Y}$ that minimizes the sum of the squares of the components of $\mathbf{E}$,

$$\sum_{j=1}^{m} \varepsilon_j^2 = \mathbf{E}^\mathsf{T} \cdot \mathbf{E} = (\mathbf{C} - \mathcal{D}\mathbf{Y})^\mathsf{T} \cdot (\mathbf{C} - \mathcal{D}\mathbf{Y}), \tag{4.17}$$

where the superscript T denotes that the vector, or the matrix, is transposed. The quantity in (4.17) is minimum, if its derivative with respect to $\mathbf{Y}$ is zero. The right-hand member of (4.17) can also be written as

$$\mathbf{C}^\mathsf{T} \cdot \mathbf{C} - (\mathcal{D}\mathbf{Y})^\mathsf{T} \cdot \mathbf{C} - \mathbf{C} \cdot \mathcal{D}\mathbf{Y} + (\mathcal{D}\mathbf{Y})^\mathsf{T} \cdot \mathcal{D}\mathbf{Y}.$$

Its derivative with respect to $\mathbf{Y}$ is

$$-\mathcal{D}\mathbf{C} + \mathcal{D}^\mathsf{T}\mathcal{D}\mathbf{X} + (\mathcal{D}\mathbf{Y})^\mathsf{T}\mathcal{D} - \mathbf{C}^\mathsf{T}\mathcal{D} = 0,$$

which is equivalent to

$$\mathcal{D}^\mathsf{T}\mathcal{D}\mathbf{Y} = \mathcal{D}^\mathsf{T}\mathbf{C}. \tag{4.18}$$

The product of the transposed design matrix by itself is a square matrix of dimension k. One can, hence, solve (4.18), and obtain the solution

$$\hat{\mathbf{Y}} = (\mathcal{D}^\mathsf{T}\mathcal{D})^{-1}\mathcal{D}^\mathsf{T}\mathbf{C}, \tag{4.19}$$

which is the estimate of $\mathbf{Y}$ by the least-squares procedure.

4.4.3 Weighted least-squares solution

An important result concerning this solution is the Gauss–Markov theorem, which states that the least-squares estimate, $\hat{\mathbf{Y}}$, given by (4.19) is a better estimation than a solution given by any other possible linear estimator, provided that all the components of $\mathbf{E}$ have the same variance and are uncorrelated. Now, to estimate the variance of ε_j, which is the variance of an observation, is not a straightforward task, and it often presents a non-negligible qualitative aspect. We shall assume that this is done.

The absence of correlation normally implies that the unknowns are well chosen, in the sense that they correspond to the minimum needed to describe the model. For example, a rotation around a given point in a plane should be represented by a single angle θ, and not by $\sin\theta$ and $\cos\theta$.

A simple way to equalize the variances in Equation (4.13) is to multiply each equation by the inverse of ε_j. Let $\mathcal{G}$ be a square matrix of order m, whose elements are all zeroes except for the main diagonal components, which are $g_j = \varepsilon_j^{-1}$. This operation is performed on (4.16) by multiplying both sides of the equation by $\mathcal{G}$

$$\mathcal{G}D\mathbf{Y} + \mathcal{G}\mathbf{E} = \mathcal{G}\mathbf{C},$$

so that the right-hand member of (4.17) becomes

$$(\mathcal{G}\mathbf{C} - \mathcal{G}D\mathbf{Y})^\top (\mathcal{G}\mathbf{C} - \mathcal{G}D\mathbf{Y}).$$

Let $\mathcal{W}$ be the square matrix of order k

$$\mathcal{W} = \mathcal{G}^\top \mathcal{G}.$$

After some calculations, (4.19) becomes

$$\hat{\mathbf{Y}} = (\mathcal{D}^\top \mathcal{W} D)^{-1} \mathcal{D}^\top \mathcal{W}\mathbf{C}. \tag{4.20}$$

The matrix $\mathcal{W}$ is the weight matrix. All its terms are zeroes except on the main diagonal where they are $w_{jj} = \varepsilon_j^{-2}$.

The weight matrix may be generalized in the case when there are correlations between the unknowns. Then, the non-diagonal terms are not all null. We shall not deal with this case.

The procedure described by Equation (4.20) is the weighted least-squares procedure to which the Gauss–Markov theorem applies.

Remark. The definition of $\mathcal{G}$, as adopted here, implies that the variances and, hence, the standard deviations are unity. One could use $g_{ii} = s\varepsilon_j^{-1}$, where s is any appropriate value of the common standard deviation.

It is possible to verify whether the assumption concerning the variances are significant by computing the variance of the residuals, by applying (4.17) to the estimated unknowns, and extending (4.4) to $m - k$ degrees of freedom. One gets

$$s_0^2 = \frac{1}{m - k} (\mathbf{C} - \mathcal{D}\hat{\mathbf{Y}})^\top \mathcal{W}(\mathbf{C} - \mathcal{D}\mathbf{Y}).$$

Or, if we call $\mathbf{R}(r_i)$ the vector of residuals,

$$\mathbf{R} = \mathbf{C} - \mathcal{D}\hat{\mathbf{Y}},$$

$$s_0^2 = \frac{1}{m-k}\mathbf{R}^\top \mathcal{W}\mathbf{R}. \tag{4.21}$$

One can show that s_0^2 is an unbiased estimate of the *unit weight variance*.

4.4.4 Variance–covariance matrix of the estimation

An important result is that the variance-covariance matrix for the estimation $\hat{\mathbf{Y}}$ is

$$\mathcal{V} = (\mathcal{D}^\top \mathcal{W}\mathcal{D})^{-1}.$$

The proof involves complicated algebra, and we shall not give it. Let us note that (4.20) can also be written as

$$\hat{\mathbf{Y}} = \mathcal{V}\mathcal{D}^\top \mathcal{W}\mathbf{C}. \tag{4.22}$$

The diagonal terms of the matrix $\mathcal{V}$, v_{ii} are the *formal* variances of $\hat{y}_i$. They are computed as though the model described by (4.15) is exact. They are actually underestimations of the true values, and in order to get more realistic estimations, one should multiply them by the unit-weight variance, s_0^2. So, finally, the best estimate of the standard deviation of $\hat{y}_i$ is

$$\hat{\sigma}_i = s_0 \sqrt{v_{ii}}.$$

The best estimates of correlation coefficients are not affected by s_0, and one has

$$\hat{\rho}_{ij} = \frac{\hat{\sigma}_{ij}}{\hat{\sigma}_i \hat{\sigma}_j} = \frac{v_{ij}}{\sqrt{v_{ii}}\sqrt{v_{jj}}}.$$

4.4.5 Chi-square test

An important problem is to check whether the model used is indeed representative of the reality as described by the observations ($\mathbf{C}$). Many tests exist that help in answering this question. Let us describe the simplest one, which is particularly well fit to a least-square solution: the chi-square test.

Let us introduce the random variable u

$$u = \sum_{i=1}^{n} \left(\frac{r_i}{s_0}\right)^2,$$

and apply the probability distribution of u, as given in Section 4.2.5,

$$\chi^2(v, u) = \left(2^{v/2}\Gamma\left(\frac{v}{2}\right)\right)^{-1} u^{(v-2)/2} \exp(-u/2),$$

Table 4.2. *Limiting values χ_0^2 of χ^2 for given probability p*

ν	$p = 0.1$	$p = 0.05$	$p = 0.01$	$p = 0.001$
5	9.2	11.2	15.1	20.5
10	16.0	18.3	23.2	29.6
20	28.4	31.4	37.6	45.3
30	40.3	43.8	56.9	59.7
40	51.8	55.8	63.7	73.4
60	74.4	79.1	88.4	99.6
80	96.6	101.9	112.3	124.8
100	118.5	124.3	135.8	149.5

where ν is the degree of freedom. When ν increases, the chi-square pdf tends towards a normal distribution, whose mean and variance are ν and 2ν, respectively. These functions, or their integrals, giving the probability of having $u < u_0$ are found in most books on probability, or current statistical software. One can associate the level of confidence, c, given by

$$c = \int_{\chi_0^2}^{\infty} \chi^2 d\chi^2,$$

which is the probability for χ^2 to exceed the observed value of χ_0^2. Table 4.2 gives, for a few values of the degree of freedom ν, the probability p of getting values of χ^2 larger than χ_0^2.

4.4.6 Goodness of fit

The computation of χ^2, corresponding to the residuals of a least-squares solution, and the determination of the probability that it exceeds the computed value, gives an indication of the quality of the fit as regards the adopted model. If the value obtained is such that the probability of exceeding it is small, this is a clear indication that some systematic effects remain in the residuals, and that there is a good probability that the observations do not fit the model. If the values obtained are larger than those given in, or interpolated from, Table 4.2, one must very seriously consider the adequacy of the model.

However, it might be that only a very small number of large residuals increases χ^2. If this is the case, it is possible that the corresponding observations were spoiled by some spurious event. For instance, in the case of the reduction of a photographic plate, it could be that a reference position of a star was incorrectly copied, or even that a star was mistaken for another. These cases should first be investigated by considering the distribution of residuals, which should obey the normal law. If there is a *small* number of large residuals, one should re-do the least-square solution omitting these outliers. If this is sufficient to significantly increase the probability of getting the new χ^2, the new solution is acceptable. But, if that is not the case, and the χ^2 is still too high, then the linearized model used to describe the observations is not adequate, and there remains at least one cause for systematic errors.

A drawback of the chi-square test is that the values used to compare χ^2 with χ_0^2 are dependent upon the degree of freedom. If ν is large enough ($\nu > 20$ or 30), there exists another test which has the advantage of being independent of ν. It is derived from what is called the $F2$ statistics, which is a transformation of χ^2 statistics of the least-square fit,

$$F2 = \left(\frac{9\nu}{2}\right)^{1/2} \left(\left(\frac{\chi^2}{\nu}\right)^{1/3} + \frac{2}{9\nu} - 1\right). \tag{4.23}$$

If χ^2 follows the chi-square distribution with a degree of freedom equal to ν, then $F2$ is approximately normal with a mean value of zero and a unit standard deviation. The value of $F2$ is the *goodness of fit*. A value larger than 3 is a clear indication of a modeling error (probability smaller than 0.0027), especially after removing outliers.

4.4.7 Problems in linearization

There exist cases for which no linearization is possible. This happens when no evident approximate solution exists and, in particular, when there exists a choice among several sets of parameters that could fit the observations and still differ by quantities of order zero. Such a case occurred, for instance, in the reduction of double-star data from the Hipparcos mission. Because of the periodicity of $s = 1''.208$ in the grid, the projected separation, ρ, of a double star would give the same modulation if it were $\rho \pm ks$, k being an integer (see Fig. 2.7). If there is no *a priori* knowledge of the actual separation, for instance from ground-based observations, k is an additional unknown quantity. However, since the observations were made at different angles relative to the two stars, k is not always the same, so it cannot enter explicitly in the equations of condition. The solution was to consider in the space of the unknowns ($X = \rho \sin\theta$, and $Y = \rho \cos\theta$, where θ is the angle of position) the sum of the square of the residuals as a function

$$F(X, Y, k)$$

and to search for the deepest among the minimums of F, which are distributed in a regular, two-dimensional grid. Then, starting from such a minimum, one may linearize the equations. In other cases, different methods have to be found by deep analysis of the problem, and no general rule can be suggested.

Now, if one is rather close to the solution, but still too far for a linearization to be justified, there exist, in mathematical software packages, nonlinear least-square programs that may be applied. However, caution should be taken that the convergence of the iterations is correct by monitoring the evolution of the sum of the squares of the residuals. Actually, it is good practice to apply an iterative procedure for the normal least-squares. One adds the solution **S** obtained in a first approximation to **X**, and iterate the procedure, thus controlling the behavior of the residuals. In all the cases, it is advisable to check the results with the chi-square test and compute the goodness of fit.

4.5 Additional aspects

The Gaussian distribution is not the only distribution useful in astronomy. In this section, we shall present the Poisson distribution, which has specific applications when observations have discrete values, for instance when they are photon counts. Another often-encountered problem concerns sparse data, which requires a treatment that is different from the ones presented above. Finally, statistical treatment is not limited to stationary cases; one often needs to evaluate the stability in time of some parameter. We shall consider these three special problems in this section.

4.5.1 Poisson distribution

A Poisson distribution is one of discrete probability expressed as

$$P(r) = \frac{\mu^r \exp(-\mu)}{r!}, \tag{4.24}$$

where r is an integer and μ is a positive number. There is a recurrence relation between successive probabilities,

$$P(r+1) = \frac{\mu P(r)}{r+1}.$$

This probability distribution corresponds to a random physical mechanism, called a *Poisson process*. It occurs, for example, when an event happens in the mean μ times in some interval of time Δt. The probability of having r such events during a given interval Δt is $P(r)$.

The variance of a Poisson distribution can be obtained, after some algebra, and is,

$$\sigma^2 = \sum_{r=0}^{\infty} (r - \mu)^2 P(r) = \mu.$$

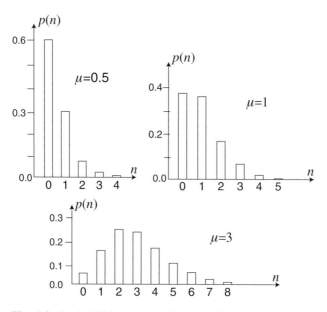

Fig. 4.3. Probabilities $p(n)$ of Poisson distributions for different μ.

Figure 4.3 gives the shape of Poisson distribution for different values for μ. The probability of observing at least N events is:

$$P(r \geq N) = \sum_{r=N}^{\infty} \frac{\mu^r \exp(-\mu)}{r!}.$$

4.5.2 Robust estimations

A major problem in estimation is what to do with outliers. Suppressing one or two observations modifies not only the variance – a result that is generally sought – but also the estimation itself. An estimation is *robust* if the removal of outliers has the least possible effect on the value found.

For instance, a least-square fit of points on a straight line (Fig. 4.4) may be biased by the presence of several outliers. In this particular case, the removal of outliers, as described in the preceding section, might affect good points rather than points actually widely distributed in the plane, because, as shown in Fig. 4.4, some correct points may be more distant from a wrong solution than false ones. Here a graphic representation of the situation allows one to recognize the real situation, and to find the best procedure to obtain the correct solution.

As an example, the median of a distribution is a robust estimation of the central value. For N values, $x_i, i = \Delta \ldots N$, sorted into ascending order, the median is

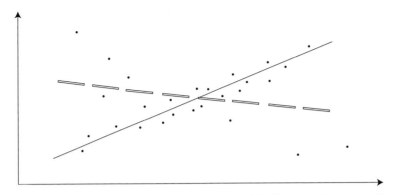

Fig. 4.4. Faulty linear least-square fit to a straight line sensitive to outliers.

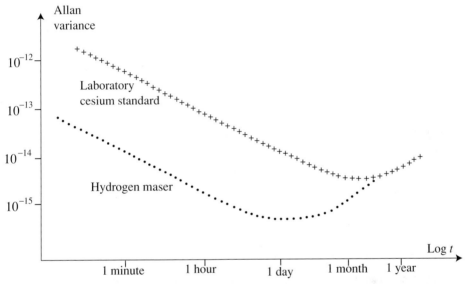

Fig. 4.5. Example of Allan variance for a cesium frequency standard and a hydrogen maser.

defined by

$$x_m = x_{N-1}/2 \quad \text{for } N \text{ odd,}$$

$$x_m = \frac{1}{2}(x_{N/2} + x_{N/2-1}) \quad \text{for } N \text{ even.}$$

If the distribution has a strong central clustering, the suppression of x_1 or x_N does not modify the estimation of the median by more than the mean distance between clustered values, while the mean is modified by x_1/N or x_N/N.

4.5.3 Allan variance

Let us consider a parameter that may not be stable in time; one wishes to quantify its stability, and see how much it changes with time. This can be the case of the frequency of an atomic clock, the rate of rotation of a pulsar, or the period of a variable star.

Let T be the repetition rate of measurements, x_k, such that their times are $t_k = t_0 + kT$. Let us consider three successive samples, x_k, x_{k+1}, and x_{k+2}, and consider the deviation of the central sample from the mean of the adjacent ones:

$$y_k = x_{k+1} - (x_{k+2} + x_k)/2.$$

The Allan variance for a period $\tau = mT$ is proportional to the average of the square of the deviation for this period. An estimation is

$$\sigma_a = \frac{K}{2(m-1)} \sum_{k=1}^{m-1} y_k^2, \tag{4.25}$$

where K is a constant that depends on the characteristics of the parameter, in particular its power spectrum. Computing the Allan variance for increasing values of τ, one can evaluate, for instance, for how long a clock is stable, and at what point various perturbations become predominant. Figure 4.5 shows, as an example, the Allan variances of a laboratory cesium standard and a hydrogen maser. The latter is very stable, but only for one day, then instabilities appear. This is particularly useful in VLBI. On the contrary, the stability of cesium standards remains comparable over several years, a fundamental quality for an accurate clock.

5

Principles of relativity

For many years, the theory of relativity was ignored for astrometry because the effects were much smaller than the accuracies being achieved. For the motions of bodies of the Solar System, Newtonian theory was adequate, provided that some relativistic corrections were introduced. Actually, there was a long period of questioning whether the theory of relativity was correct or not, and thus relativistic corrections were introduced in a manner to determine whether observational data could then be better represented.

However, over the past 30 years, there has been a very significant improvement in the accuracies of observations, all confirming the conclusions of general relativity. So its introduction became not only acceptable, but necessary. In 1976, the International Astronomical Union introduced relativistic concepts of time and the transformations between various time scales and reference systems. In 1991, it extended them to reference frames and to astrometric quantities. Now, with plans for microarcsecond astrometry and with time standards approaching accuraries of one part in 10^{-16}, and better in the future, it is necessary to base all astrometry, reference systems, ephemerides, and observational reduction procedures on consistent relativistic grounds. This means that relativity must be accepted in its entirety, and that concepts, as well as practical problems, must be approached from a relativistic point of view. In 2000, the IAU enforced this approach further by extending the 1991 models for future requirements in such a way that they become valid to accuracies several orders better than those currently achieved in 2000.

The aim of this chapter is to provide the physical background, the definitions, and the basic expressions of the Theory of General Relativity that are useful for astrometric applications in regions of weak gravitational fields, such as the one that characterizes the Solar System. Whenever it is necessary, they will be applied in the next chapters of the book. For more details on the concepts developed in this chapter, see Mould (1994), Fock (1976), Ohanian (1980), or Rindler (1977).

5.1 Relativity principles

The general concept of relativity is as old as the advent of mechanics. It was based on what was called, and still is in a more particular sense, an inertial frame. A reference frame is called inertial if a free particle, that is not subject to any external force, moves without acceleration with respect to it. In the case of Newtonian mechanics, such free particles move with constant speed along straight lines. The Galilean principle of relativity states that if a reference frame is inertial, all reference frames that move with a constant linear speed with respect to it are also inertial. This Newtonian mechanical relativity principle was generalized by Einstein, who stated that all inertial frames are totally equivalent for the performance of all physical experiments. In Einstein's Theory of Relativity, the inertial frame is generalized by the notion of a *free falling* isolated local frame, in an electrically and magnetically shielded environment, sufficiently small that inhomogeneities in the external fields can be ignored throughout the volume, and in which self-gravitating effects are negligible. Under these conditions, any local non-gravitational test experiment performed in a free falling frame is independent of where and when in the Universe it is performed and of the velocity of the frame.

Two theories of relativity, respectively called special and general, are modifications of the classical physical laws in order to enter into the framework of this principle. To take just one example, the invariance of the speed of light, whatever the speed of the observer, was demonstrated for the first time in 1887 by Michelson and Morley with an interferometer that could rotate. They attempted to show that there was a difference in the speed of light as measured in the direction of the Earth orbital motion, compared with the speed at right angles. The result was negative and led to the principle of relativity, as far as the independence on the velocity of the frame is concerned. This will be developed in the following sections.

5.2 Special relativity

Special relativity, proposed by Einstein in 1906, was the outcome of an effort to build a coherent theory that would explain, without introducing an hypothetical ether, several effects that were found to contradict the Newtonian principles upon which physics – and especially mechanics – were based until that time.

5.2.1 The Newtonian mechanical concepts

Newtonian mechanics postulates the existence of an absolute, Euclidean space and an independent, absolute time. The points in space are referred to three coordinates in some system of rectangular axes. The Galilean relativity concept applies, which is that all systems of coordinates are equivalent (or interchangeable) provided they move with respect to one another with a constant velocity; the laws of mechanics are everywhere the same. Let us consider a fixed, reference coordinate system S ($oxyz$ and a time scale t) and a second system S' ($o'x'y'z'$ and a time scale t'), which moves with respect to S with a velocity $\mathbf{v}$ and is identical to S at time $t = 0$. Assuming that $\mathbf{v}$ is along the ox axis with a speed equal to v, the relations between the coordinates are

$$x = x' + vt,$$
$$y = y',$$
$$z = z',$$
$$t = t'. \tag{5.1}$$

The last equation expresses the fact that the time is absolute, that is, the same in all equivalent reference frames.

Let P be a point at x_0, y_0, z_0 at time $t = 0$, which moves in S' along the $o'x'$ axis with a velocity v'.

$$x' = x_0 + v't,$$
$$y' = y_0,$$
$$z' = z_0,$$
$$t' = t. \tag{5.2}$$

The position of P in system S is therefore

$$x = x_0 + (v + v')t,$$
$$y = y_0,$$
$$z = z_0. \tag{5.3}$$

So, we have the following result: the velocity with respect to the fixed system is equal to the sum of the velocity v of the moving system and the velocity v' with respect to the moving system

$$V = v + v'. \tag{5.4}$$

This can of course be generalized to the three coordinates.

5.2.2 *Invariance of the speed of light*

As already mentioned, the law of addition of velocities was tested by Michelson and Morley in 1887, assuming that P is moving at the speed of light, c. At that time, it was believed that the light was a property of ether, and an experiment

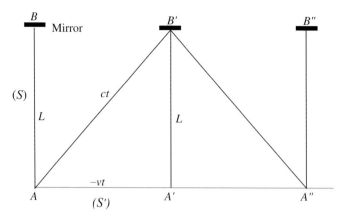

Fig. 5.1. A light pulse bouncing between the mirrors A and B as seen by an observer moving with a velocity v.

was devised to determine the speed of the Earth's orbital velocity with respect to the ether, measuring it in the direction of the Earth's motion, or opposite to it, or perpendicular to it. In all the cases, the speed was found to be the same. Equation (5.4) does not match this experimental result if $v = c$. One has

$$c = c \pm v.$$

The interpretations of this result are that there is no ether and that no velocity larger than the speed of light, c, can exist. The light is seen travelling with the speed c by any observer, whether the source, or the observer, moves or not. This has many consequences. Let us describe one of them.

Consider a pair of parallel mirrors, AB, separated by a rod of length, L, and a light pulse bouncing from one to the other. This constitutes a periodic system, a clock, with a period $P = 2t = 2L/c$, which can be measured by an observer fixed with respect to this experimental equipment (system S).

A second observer in a system, S', moves with a speed v perpendicular to the rod (Fig. 5.1). After time t, the rod will appear to him at the location $A'B'$. For him, the light pulse will have followed the segment AB', but at the same speed c during the same time t. Let us apply the theorem of Pythagorus to the triangle $AA'B'$. One gets:

$$AB'^2 = c^2 t^2 = v^2 t^2 + L^2.$$

A similar conclusion will be obtained for the time when the light pulse bounces back to A seen as at A'' by the moving observer. Solving for t, one obtains

$$t = \frac{L}{c\sqrt{\frac{1-v^2}{c^2}}}.$$

The apparent period is $P' = P\gamma$, with $\gamma = (1 - v^2/c^2)^{-1/2}$. The moving clock ticks slower by the amount $1/\gamma$. The general result, a consequence of the constancy of the speed of light, is that a time interval measured in a moving frame is longer than the same one measured in a frame fixed with respect to the events. This property is known as the *time dilatation*. One may similarly describe another fundamental result, which is the contraction of moving objects. A rod in motion relative to an observer will have its length contracted along the direction of the motion as compared with an identical rod at rest. If L is the length at rest, the apparent length is $L' = L/\gamma$ when moving at velocity v along L.

For more details on these properties, together with some others, see Mould (1994).

5.2.3 The Lorentz transformation

The preceding properties were formalized by Lorentz, who proposed the following transformation between coordinates x, y, z, $(\mathbf{r})$, and time t in a given reference frame S, and x', y', z', $(\mathbf{r}')$, and t' in another frame S' moving with a constant speed $\mathbf{V}$ with respect to S.

The generalized vectorial form of Lorentz transformation (Fock, 1976) is

$$\mathbf{r}' = \mathbf{r} - \gamma \mathbf{V}t + (\gamma - 1)\mathbf{V}(\mathbf{V} \cdot \mathbf{r})/V^2$$
$$t' = \gamma[t - (\mathbf{V} \cdot \mathbf{r})/c^2], \tag{5.5}$$

where γ is the quantity already mentioned, $\gamma = (1 - V^2/c^2)^{-1/2}$. The inverse formulae are

$$\mathbf{r} = \mathbf{r}' + \gamma \mathbf{V}t' + (\gamma - 1)\mathbf{V}(\mathbf{V} \cdot \mathbf{r}')/V^2$$
$$t = \gamma[t' + (\mathbf{V} \cdot \mathbf{r}')/c^2]. \tag{5.6}$$

Let us now assume that the velocity $\mathbf{V}$ is along the x-axis, $\mathbf{V} = (v_x, 0, 0)$. The y and z components reduce to an identity. The x component becomes:

$$x' = x - \gamma v_x t + (\gamma - 1)\left(\frac{v_x}{v_x}\right)^2 x.$$

The terms independent of γ vanish and Equations (5.5) become:

$$x' = \gamma(x - v_x t),$$
$$y' = y,$$
$$z' = z,$$
$$t' = \gamma(t - v_x x/c^2). \tag{5.7}$$

The inverse formulae are similar except that the sign of v_x changes, since it represents the velocity of S with respect to S'.

These simplified formulae are in the form in which the Lorentz transformation is usually presented in reference books. But in astrometry, the situation is generally such that one wishes to choose axes that are not constrained by the relative motion of reference planes.

Let us now derive the Lorentzian equivalent to the Newtonian law of addition of velocities. Let V be the constant velocity of a reference system S' with respect to the system S. In S, a point r has a velocity $U = dr/dt$, and in S', its coordinates are r', t' and the velocity is $U' = dr'/dt'$.

Let us differentiate (5.6) with respect to t,

$$\frac{dr}{dt} = \frac{dr'}{dt'}\frac{dt'}{dt} + \gamma V \frac{dt'}{dt} + (\gamma - 1)\left(V \cdot \frac{dr'}{dt'}\frac{dt'}{dt}\right) V/V^2$$

$$dt = \gamma \left[dt' + \left(V \cdot \frac{dr'}{dt'}dt'\right)/c^2\right].$$

Or, with the new notation,

$$\frac{dt'}{dt} = \frac{1}{\gamma(1 + V \cdot U'/c^2)},$$

from which one derives,

$$U = \frac{U' + \gamma V + (\gamma - 1)(V \cdot U')V/V^2}{\gamma(1 + V \cdot U'/c^2)}. \tag{5.8}$$

The inverse formula is

$$U' = \frac{U - \gamma V + (\gamma - 1)(V \cdot U)V/V^2}{\gamma(1 - V \cdot U/c^2)}. \tag{5.9}$$

If (5.8) is applied to the simplified case with $V = (v_x, 0, 0)$, one gets

$$u_x = \frac{u'_x + \gamma v_x + (\gamma - 1)(v_x u'_x)v_x/v_x^2}{\gamma(1 + v_x u_x'/c^2)},$$

the term u'_x cancels out and γ is a factor of both terms of the quotient, so that one gets

$$u_x = \frac{v_x + u'_x}{1 + v_x u'_x/c^2},$$

which is the formula that is usually given in reference books together with

$$\frac{dt}{dt'} = \gamma(1 + v_x u'_x/c^2).$$

5.2.4 The metric of special relativity

Let us consider a light pulse emitted by O at time $t = 0$. At time t, in the system S, it reaches a point P of coordinates x, y, z, and one has

$$c^2t^2 = x^2 + y^2 + z^2,$$

or

$$x^2 + y^2 + z^2 - c^2t^2 = 0.$$

In the system S' the coordinates of P are x', y', z', but because of the invariance of the speed of light, which is also c in the system S', one has similarly

$$x'^2 + y'^2 + z'^2 - c^2t'^2 = 0. \tag{5.10}$$

The quantity (5.10) is an invariant interval. Written in a differential form, considering an infinitesimal distance OP, one has

$$dx'^2 + dy'^2 + dz'^2 - c^2dt'^2 = 0. \tag{5.11}$$

Now, let us generalize this by considering an event characterized by its three coordinates x, y, z and the time t in the system S. Let us compute the quantity $x'^2 + y'^2 + z'^2 - c^2t'^2$ by applying the simplified Lorentz transformation (5.8). Let us consider the inverse equations:

$$x = \gamma(x' + vt)$$
$$y = y'$$
$$z = z'$$
$$t = \gamma(t' + vx'/c^2)$$

and compute

$$x^2 + y^2 + z^2 - c^2t^2 = \gamma(x' + vt')^2 + y'^2 + z'^2 - c^2\gamma^2(t' + vx'/c^2)^2.$$

The terms multiplying γ are

$$G = x'^2 + 2vx't' + v^2t'^2 - c^2t'^2 - 2vx't' - v^2x'^2/c^2,$$
$$G = x'^2(1 - v^2/c^2) - c^2t'^2(1 - v^2/c^2) = (x'^2 - c^2t'^2)/\gamma.$$

Therefore, one finally gets:

$$x^2 + y^2 + z^2 - c^2t^2 = x'^2 + y'^2 + z'^2 - c^2t'^2 = s^2.$$

The quantity is an invariant for all transformations between reference systems in uniform motion, one with respect to others. This quantity, s^2, is a generalized distance that is also called the invariant interval or metric of special relativity. It is usually presented for an infinitesimal displacement and written

$$ds^2 = dx^2 + dy^2 + dz^2 - c^2 dt^2. \tag{5.12}$$

By analogy with the space coordinates x, y, z, and because ct has the dimension of a length, t is called coordinate time. It results that in special relativity (and actually also in general relativity), an event is specified by four coordinates: three space coordinates and the time coordinate.

All the properties already described, and those that will be presented in the next subsections, are a consequence of the metric (5.12). This fully describes the properties of space-time, invariant from the coordinate system that could be chosen, provided that it is inertial, meaning that it is not accelerated in any fashion.

5.2.5 Coordinate time and proper time

Let us consider two events that occur after a time dt at the same point of the reference frame, so that dx = dy = dz = 0. Then the metric reduces to

$$ds^2 = -c^2 dt^2.$$

The time t is linked to the object. This new time is called *proper time* τ and is the one that rules all the local physics of the point considered. The metric (5.12) can therefore be written in the general case as

$$ds^2 = -c^2 d\tau^2 = dx^2 + dy^2 + dz^2 - c^2 dt^2.$$

In an integrated form, this relation means that for any interval s^2 between two events, the quantity

$$\sqrt{-s^2}/c$$

equals the reading of a clock that travels at a constant speed between the two events.

In order to obtain the relation between the proper and the coordinate times, let us consider the components of the velocities as measured with respect to the coordinate time t. Changing the signs, one gets

$$c^2 \left(\frac{d\tau}{dt}\right)^2 = c^2 - \left[\left(\frac{dx}{dt}\right)^2 + \left(\frac{dy}{dt}\right)^2 + \left(\frac{dz}{dt}\right)^2\right],$$

$$\frac{d\tau}{dt} = \sqrt{1 - \frac{v^2}{c^2}} = 1/\gamma. \tag{5.13}$$

This is exactly the expression obtained in presenting the time dilatation in Section 5.2.3. Coming back to it, one sees that the moving observer measures

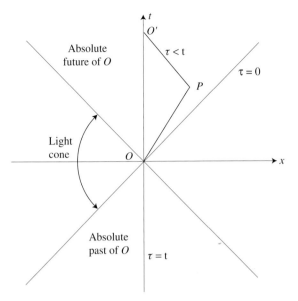

Fig. 5.2. Minkowsky's *x*–*t* diagram. Events in the lower part of the cone have been seen by *O* in the past. Those in the upper part will be seen sometime in the future. Outside the cone, are events inaccessible to the observer until some time when the light issued will reach the world line of the observer.

the period of the clock in coordinate time, while the experimenter, fixed with respect to it, measures it in proper time.

For an observer at rest within his or her reference frame – for instance at the origin – the coordinate time and the proper time are identical, but this is not the case as soon as he or she observes a moving object and assigns time to events on this object. This can be visualized in Minkowsky's diagram (Fig. 5.2). A line on the diagram represents the path of an object and is called a *world line*. The world lines of light reaching *O* are straight lines defined by $x = ct$. One can see, using Equation (5.13), that $\tau = 0$ along the line for a photon; the proper time between emission and reception does not change. The vertical line corresponds to a fixed point in space and one has $t = \tau$. Inside the cone, one has $|x| < |ct|$. In the lower part (past), events occur before the origin event. Outside the cone no causal connection with the origin event is possible, because the velocity of any information linking the two events would be larger than the speed of light.

If we consider the path OPO' of a traveler, *A*, leaving *O* for *P*, and then coming back to the origin *O'* after some time t_0. Since during all the path, the proper time of *A* is always smaller than *t*, the proper life time of *A* between the event *O* and *O'* is smaller than for an observer who remained fixed. This is the famous paradox of Langevin's twins. However, since accelerations must exist to reverse the path, the problem is not as simple, and one has to introduce the acceleration expressions in general relativity in order to describe completely the behavior of the traveling twin, so that one cannot consider that there is a strict symmetry between the twins. We shall not dwell more on this *paradox* (see for instance Mould, 1994), but we have taken this opportunity to illustrate the basics of mechanics in special relativity.

5.2.6 Mass and acceleration

We have already seen that the expression of velocity is not the same if reckoned in the local environment with proper time, or in a general reference frame with coordinate time. In any reference frame, the components of the velocity are:

$$\left(\frac{dx}{dt}, \frac{dy}{dt}, \frac{dz}{dt}\right) = \gamma(v_x, v_y, v_z) = \gamma \mathbf{v}, \tag{5.14}$$

where v_x, v_y and v_z are the components of the velocity $\mathbf{v}$. In local coordinates, with $t = \tau$ (proper time), we have

$$v = \sqrt{v_x^2 + v_y^2 + v_z^2}$$

and

$$\gamma = 1.$$

Now, if we consider the Newtonian law of conservation of linear momentum in this inertial frame, we can still write it in the local frame using proper time

$$\mathbf{p} = m\mathbf{v},$$

where m is the inertial mass.

But if we transcribe it in the general coordinate system, using coordinate time t instead of τ, $\mathbf{p}$ and $\mathbf{v}$ become $\mathbf{P}$ and $\mathbf{V}$, and, we have

$$\mathbf{P} = m\mathbf{V}.$$

In this case the modulus can be written as a function of proper quantities

$$P^2 = m^2 v^2 \left(\frac{d\tau}{dt}\right)^2 = m^2 v^2 \left(1 - \frac{v^2}{c^2}\right).$$

The law of conservation of momentum implies here that m is not constant and is velocity dependent. If m_0 is the mass at rest (that is $v = 0$ in a local frame), one has

$$m = m_0 \gamma = m_0 \left(1 - \frac{v^2}{c^2}\right)^{-1/2}. \tag{5.15}$$

This result implies that the mass increases toward infinity when the velocity approaches c, a direct consequence of the limitation of velocities by the speed of light. It expresses the impossibility to accelerate a particle to this limiting velocity.

Let us note, at this point, that the acceleration $\mathbf{A}$ of a particle in general has to be computed taking into account that γ depends upon the velocity v. The derivative of (5.14) is hence,

$$\mathbf{A} = \frac{d\mathbf{V}}{d\tau} = \gamma \frac{d\mathbf{V}}{dt} = \gamma \frac{d}{dt}(\gamma \mathbf{v}) = \gamma \left(\frac{d\gamma}{dt} \mathbf{v} + \gamma \frac{d\mathbf{v}}{dt} \right)$$

and, if $\mathbf{a}$ is the acceleration with respect to the coordinate time, that is $\mathbf{a} = d\mathbf{V}/dt$, one gets

$$\mathbf{A} = \gamma \frac{d\gamma}{dt} \mathbf{v} + \gamma^2 \mathbf{a}. \tag{5.16}$$

5.3 General relativity

General relativity is an extension of the theory of special relativity in the sense that it keeps its results locally, but supplements them with a theory of gravity to which it gives a geometrical form. Let us remember that classical Newtonian celestial mechanics implicitly postulates that the gravitational attraction is a force that propagates at an infinite speed, a concept that contradicts the basic assumption of special relativity, in which no information can be transferred with a velocity higher than the speed of light. General relativity postulates that the presence of masses curves the Euclidian four-dimensional space-time of special relativity, and that the motions are guided by the deformed shape of the space-time, which is described by a metric analogous to, but more complex than, the special relativity metric. Finally, this geometrization of the gravitational field is a formulation of a more general basic principle of general relativity, the equivalence principle.

5.3.1 The equivalence principles

In classical mechanics, there are two distinct properties of matter that are involved in mechanics: the gravitational mass, m, which links the gravitational force, $\mathbf{F}$, to the acceleration, $\mathbf{a}$, by $\mathbf{F} = m\mathbf{a}$, and the inertial mass, m', which is a factor that links any inertial type of force, like the centrifugal force, to the acceleration, by $\mathbf{F} = m'\mathbf{a}$. It has been recognized since Galileo that to the accuracy of observations, one has $m' = m$, and classical mechanics uses this weak equivalence principle. In the generalization of this weak equivalence principle, Einstein postulated the generalized relativity principle as already described in Section 5.1, and, in particular, that a freely falling laboratory, even near a strongly gravitating mass, is strictly equivalent to a laboratory in free motion on an artificial satellite, or a space-probe outside the Solar System. This is just a particular situation; the equivalence principle states more generally that there is no experiment that would allow one to distinguish whether a room is in a uniformly accelerating elevator, or if it is a fixed room in a gravitational field having a uniform strength over the dimensions of the room.

The consequence is that there is no preferred standard of rest. All are equivalent, and in each of them all non-gravitational laws of nature take the same form as in special relativity. This means that locally, in this standard of rest, there exists a metric that has the form and the properties of the special relativity metric (5.12) describing a flat four-dimensional space-time. The extension to full space-time is not possible by straight translation, because of the presence of gravitating masses. General relativity states that there exists an extended metric, which depends on the distribution and motion of masses in the Universe; and that at each point, there exists a flat special relativity space-time, tangent to the general relativity space-time in the sense that the four coordinates and their derivatives on the two space-times can be set to correspond. In this general relativity space-time, a test particle moves on a geodesic of this generalized metric. In particular, light bends in a gravitational field, because the geodesic is no longer a straight line. It also implies that light traveling down a gravitational field suffers a blueshift, which corresponds to a gravitational time dilatation and adds to the time dilatation due to velocity as described in Section 5.2.5.

The dependence of the metric of general relativity on the distribution and motion of masses is realized essentially through the gravitational potential U and, with a much smaller intensity, through the Lense–Thirring acceleration, **W**, due to the gravito-magnetic field of a rotating body. Then,

$$ds^2 = f(x, y, z, t, U, \mathbf{W}).$$

This will be presented in the following section.

5.3.2 The potentials in general relativity

In Newtonian mechanics, the masses produce at a point, P, of coordinates x, y, z, a potential $U(x, y, z)$, such that the components of forces along the Px, Py and Pz axes are respectively

$$\frac{\partial U}{\partial x}, \frac{\partial U}{\partial y}, \frac{\partial U}{\partial z},$$

or, in other terms, that the force **f** at a point P is related to U by

$$\mathbf{f} = \nabla U.$$

We have adopted here the sign convention that is in common use in celestial mechanics.

Let $Q_i(m_i)$ be the N masses acting on the point $P(X, Y, Z)$. The expression of the Newtonian potential at P is

$$U(P) = G \sum_{i=1}^{N} \frac{m_i}{|PQ_i|},$$ (5.17)

where G is the universal constant of gravitation. For a finite gravitating body, one has to integrate with respect to infinitesimal volumes with density ρ over the volume V of the body

$$U(P) = G \int \int \int_{(V)} \frac{\rho(x, y, z)}{\sqrt{(X - x)^2 + (Y - y)^2 + (Z - z)^2}} dx dy dz.$$ (5.18)

If P is outside a sphere including the volume V, this expression can be expanded in spherical harmonics. If the origin of axes is taken at the center of mass O of the body, and r, ϕ (latitude), and λ (longitude) are the spherical coordinates of P, one has

$$U = \frac{GM}{r} \left[1 + \sum_{n=2}^{\infty} \frac{J_n a_e^n}{r^n} P_n (\sin \phi) + \sum_{n=2}^{\infty} \sum_{k=1}^{\infty} \frac{J_{nk} a_e^n}{r^n} P_{nk} (\sin \phi) \cos k (\lambda - \lambda_{nk}) \right],$$

(5.19)

where M is the mass of the body, and a_e is a scaling factor. In the case of the Earth, a_e is the equatorial radius of the Earth, M is the mass of the Earth, r, ϕ, λ are the spherical equatorial coordinates of P, J_n, J_{nk} are dimensionless coefficients, respectively called zonal and tesseral harmonics, λ_{nk} are the phases of the tesseral harmonics, P_n are Legendre polynomials of degree n, and P_{nk} are associated Legendre polynomials of degree n and order k.

In the case of general relativity, an additional term of second order in c^{-2} must be added (Soffel, 2000):

$$\Delta U = \frac{1}{2c^2} G \frac{\partial^2}{\partial t^2} \int \int \int_{(V)} \rho(x, y, z) \sqrt{(X - x)^2 + (Y - y)^2 + (Z - z)^2} \, dx dy dz,$$

where $\rho(x, y, z)$ is the gravitational mass density. Actually, this expression can also be developed in spherical harmonics, so that both terms can be added and an expansion identical to (5.19) can be used in a coordinate system with its origin at O. The coefficients have slightly different values, but the form remains the same.

In addition, if the body is rotating around an axis crossing the origin O, in analogy to the situation in electro-magnetism (see Section 3.5.2), matter currents produce gravito-magnetic fields that lead to a magnetic type gravitational vector potential (see Soffel, 1989, or Brumberg, 1991a), whose value, outside the body is given by

$$\mathbf{W} = G \int \int \int_{(V)} \frac{\rho'(x, y, z)\mathbf{V}(x, y, z)}{(X - x)^2 + (Y - y)^2 + (Z - z)^2} dx\,dy\,dz, \tag{5.20}$$

where $\rho'(x, y, z)$ is the mass-flux density and $\mathbf{V}(x, y, z)$ is the velocity of a point of the body. In what follows, we shall call w_i with $i = 1, 2, 3$ the projections of $\mathbf{W}$ on the three axes.

Let us now turn to the potential at the surface of the Earth. Contrary to the situation outside the Earth, in which we can use a non-rotating system of coordinates, on the Earth, one must take into account the additional acceleration due to the rotation of the Earth. One uses a local reference that is linked to the site, and hence is rotating with respect to the outside world.

Following the principle of equivalence, the gravitational acceleration and the rotational inertial acceleration combine into a single one, which depends on a potential that is the sum of the gravitational potential of the Earth, U_G, and the rotational potential, U_R. The latter is equal to

$$U_R = \frac{1}{2}\omega^2(x^2 + y^2),$$

where ω is the angular velocity of the Earth, and x and y are the coordinates of the point P in a plane perpendicular to the axis of rotation. The equipotential surfaces are defined by $U_G + U_R = K$. The gravity vector at some point P is perpendicular to the surface passing by P. For a certain value of K, the equipotential surface matches the mean sea level. It is called the geoid. It plays an important role in the realization of the Earth time scales, as well as in geodesy (see Section 9.5.2).

5.3.3 The metric in general relativity

The most general form of a four-dimensional metric is

$$ds^2 = \sum_{i=0}^{3} \sum_{j=0}^{3} g_{ij} dx_i dx_j, \tag{5.21}$$

with $g_{ij} = g_{ji}$ (the index 0 corresponds to the time-like coordinate). In the development of the theory, this is often represented by a four-dimensional symmetric tensor $\mathcal{T} = \|g_{ij}\|$. In the case of general relativity, if the potential vanishes, the ds^2 reduces, in any space-time point, to a tangental special relativity metric,

$$ds^2 = -c^2 dx_0^2 + dx_1^2 + dx_2^2 + dx_3^2.$$

This form with one negative sign is a characterization of pseudo-Euclidean geometry. The geometry of general relativity is therefore pseudo-Euclidean, but, in this presentation, we shall not need to use specific results of this geometry. Furthermore, the vanishing terms have to be small and the small parameter happens to be proportional to $1/c^2$. The coefficients, g_{ij}, can be expanded as a function of small parameters. As recommended by the International Astronomical Union (IAU) in 2000, the only terms that are, or could be, significant in astrometry at the microarcsecond accuracy, or in time transfer at a 10^{-17}s precision, in a weak field like the case of the Solar System, are the following:

$$g_{00} = -1 + \frac{2U}{c^2} - \frac{2U^2}{c^4},$$

$$g_{ii} = 1 + \frac{2}{c^2}U,$$

$$g_{0i} = -\frac{4}{c^3}w_i. \tag{5.22}$$

All other terms, in particular the cross space terms g_{ij}, are null. In tensor notation, this gives:

$$\begin{pmatrix} -1 + \frac{2U}{c^2} - \frac{2U^2}{c^4} & -\frac{4}{c^3}w_1 & -\frac{4}{c^3}w_2 & -\frac{4}{c^3}w_3 \\ -\frac{4}{c^3}w_1 & 1 + \frac{2}{c^2}U & 0 & 0 \\ -\frac{4}{c^3}w_2 & 0 & 1 + \frac{2}{c^2}U & 0 \\ -\frac{4}{c^3}w_3 & 0 & 0 & 1 + \frac{2}{c^2}U \end{pmatrix}. \tag{5.23}$$

It is usually assumed that U vanishes at infinity. As in special relativity, one distinguishes the proper time, τ, and the coordinate time, t. The metric can be written as, in the usual notation:

$$ds^2 = -c^2 d\tau^2 = \left(-1 + \frac{2U}{c^2} - \frac{2U^2}{c^4} \right) c^2 dt^2$$
$$+ \left(1 + \frac{2U}{c^2} \right) (dx^2 + dy^2 + dz^2)$$
$$- \frac{4}{c^3} (w_1 dx + w_2 dy + w_3 dz) dt. \tag{5.24}$$

The relation between τ and t is an extension of the relation (5.13). Neglecting terms of order 3 and more in $1/c$, one gets

$$\frac{d\tau}{dt} = \sqrt{1 - \frac{2U}{c^2} - \frac{v^2}{c^2}}. \tag{5.25}$$

The presence of the potential describes quantitatively the gravitational time dilatation mentioned in Section 5.3.1.

The motion of a point mass – or of a photon – is fully described by the metric which depends on time, since masses are moving and consequently the potentials vary with time. It is a geodesic of the space-time manifold. If (x_0, x_1, x_2, x_3) are the coordinates of a free-falling particle, its motion – that is the variation of its coordinates with time – can be expressed in terms of some parameter p by solving the four geodesic differential equations

$$\frac{\mathrm{d}}{\mathrm{d}p}\left(\sum_{i=0}^{3} g_{ki}\frac{\mathrm{d}x_i}{\mathrm{d}p}\right) - \frac{1}{2}\sum_{i=0}^{3}\sum_{j=0}^{3}\frac{\partial g_{ij}}{\partial x_k}\frac{\mathrm{d}x_i}{\mathrm{d}p}\frac{\mathrm{d}x_j}{\mathrm{d}p} = 0, \tag{5.26}$$

where the coefficients g_{ij} are given as functions of $U(x_0, x_1, x_2, x_3)$ by the expressions (5.22).

Remark. In almost all the practical cases of astrometry, in particular in the Solar System, it is sufficient to keep only the terms in $1/c^2$ in (5.22), or use any other simplified expression like the Schwarzschild space-time (see Section 5.3.5). Even in a number of instances, as will be indicated in the following chapters, the Newtonian or the special relativity models may be quite sufficient to describe the observations, but this is no longer true when observed objects are out of the Solar System (e.g. VLBI observations, see Kopeikin and Schäffer, 1999). A more sophisticated approach to relativistic astrometry is necessary in many other cases listed by Kopeikin and Gwinn (2000). Finally, when the objects are surrounded by a very strong gravitational field, one cannot limit the relativistic equations to the truncated expressions given above. This is, for instance, the case of gravitational waves, pulsar, quasar, or black hole theories. We shall not consider these cases, except millisecond pulsar timing observations, which are presented in a simplified form in Section 14.7.

5.3.4 Transformation between reference frames

In practice, two different space-time coordinate systems are currently used:

(i) a Geocentric Celestial Reference System (GCRS), whose origin is at the center of mass of the Earth, realized by a terrestrial reference frame, bounded somewhat outside the orbit of the Moon, and

(ii) a Barycentric Celestial Reference System (BCRS), whose origin is the barycenter of the Solar System, realized by a barycentric reference frame and extending to the whole Solar System.

In many problems, it is necessary to use successively, or simultaneously, these two (and possibly others like topocentric or satellite) reference frames. The transformation from one to another is not trivial as in the case of Newtonian mechanics. In the most general case, the metric of one of the space-time systems (x_0, x_1, x_2, x_3) is:

$$\mathrm{d}s^2 = \sum_{\alpha=0}^{3}\sum_{\beta=0}^{3} g_{\alpha\beta}\mathrm{d}x_\alpha \mathrm{d}x_\beta \quad (\text{with } g_{\alpha\beta} = g_{\beta\alpha})$$

and the metric of the second (x_0', x_1', x_2', x_3') is

$$ds'^2 = \sum_{i=0}^{3} \sum_{j=0}^{3} g_{ij}' dx_i' dx_j' \quad \text{(with } g_{ij}' = g_{ji}'),$$

where the quantities g and g' are respectively functions of the corresponding coordinates. They are related by the fundamental tensor relation describing the change of reference

$$g_{\alpha\beta}(x_0, x_1, x_2, x_3) = \sum_{i=0}^{3} \sum_{j=0}^{3} g_{ij}'(x_0', x_1', x_2', x_3') \frac{\partial x_i'}{\partial x_\alpha} \frac{\partial x_j'}{\partial x_\beta}. \tag{5.27}$$

Solving these ten differential equations permits one to express the new coordinates with respect to the old ones, or *vice versa*. The full general solution is very complex and can be found in Brumberg and Kopeikin (1990). Even if we apply this transformation for reduced metrics (5.22), the calculation is lengthy and will not be presented here. The results for the transformation between GCRS and BCRS are given in Recommendation B1-3 of the General Assembly of the IAU in 2000 (Andersen, 2001) for the metric defined in (5.24). This is the geodesic precession and nutation (Section 7.5).

5.3.5 The Schwarzschild approximation

An often-used simplified form of the metric (5.22), reduced to its second term, is the Schwarzschild space-time, which is spherically symmetric around a single particle of mass M, so that the potential reduces to $U = GM/r$. It is convenient to use the transformation defining polar coordinates (radius vector r, latitude ϕ, and longitude λ), and the transformed metric is

$$ds^2 = -c^2 d\tau^2 = \left(\frac{dr^2}{1 - \frac{2GM}{rc^2}} + r^2 d\phi^2 + r^2 \cos^2 \phi d\lambda^2 \right) - \left(1 - \frac{2GM}{rc^2} \right) c^2 dt^2. \tag{5.28}$$

It can be extended to the fourth order in $1/c$, assuming that the bodies are not rotating, that is setting $\mathbf{W} = 0$. It is sufficient to replace in (5.28) the term in $c^2 dt^2$ by

$$\left(1 - \frac{2GM}{rc^2} + \frac{2G^2 M^2}{r^2 c^4} \right) c^2 dt^2. \tag{5.29}$$

These expressions can conveniently be used whenever one may neglect all the masses except the central non-rotating mass.

5.3.6 The parametrized post-Newtonian formulation

Another simplified and useful tool for the evaluation of astrometric consequences of general relativity is the parametrized, post-Newtonian (PPN) formalism introduced in its earlier form by Eddington (1922), but essentially developed by Nordtvedt (1970) and used by Will (1981). It has been widely used in describing various relativistic effects, and in comparing them with observations. It is actually a generalization of the expressions (5.28) and (5.29) with the same assumptions.

The idea is to consider the most general form (5.21) of the metric, accepting that the main terms are those given to the second order in $1/c$. This takes into account a large set of possible metrics of alternative theories of gravity that include, for special values of the parameters, the Einsteinian Theory of General Relativity. Will has introduced ten parameters related to the ten terms of the general metric (5.21), with particular physical meanings that can, in principle, be determined from observations. However, in practice, most of them are not, at present, accessible to observation, so we shall keep only the two that are generally used: γ, which describes the space curvature, and β, which gives the amount of nonlinearity of the gravitational field. The corresponding metric is given by

$$g_{00} = -1 + \frac{2U}{c^2} - \beta \frac{U^2}{c^4}$$

$$g_{ii} = 1 + 2\gamma \frac{U}{c^2}$$

$$g_{0i} = -\frac{2(1+\gamma)}{c^3} w_i. \tag{5.30}$$

Comparing this with (5.22), one sees that in general relativity, $\beta = \gamma = 1$. Most of the tests of general relativity use this representation and, up to now, the two parameters were found to be equal to 1 with uncertainties of the order of 10^{-3} to 10^{-4}.

5.3.7 Harmonic coordinates

In the most general approach in general relativity, the motion and the distribution of masses are mutually dependent. This is expressed by field equations which permit the determination of the ten elements, g_{ij}, of the metric (5.21), usually expressed as a four-dimensional tensor. There are ten covariant field equations, but they admit four arbitrary functions and hence satisfy four identities, which can be used to simplify the equations by a convenient choice of coordinate systems. In practice, the specific choice that is used to simplify the field equations, and consequently the solution, corresponds to what are called harmonics coordinates (Fock, 1976, and Brumberg, 1991a). In 2000, the IAU recommended that they should systematically be used, at least when terms of order larger than $1/c^2$ are to be kept.

Let g be the determinant of the $\|g_{ij}\|$ tensor. The four ($j = 0$ to 4) conditions that set the four identities and define the harmonic coordinates are (Brumberg, 1991a)

$$\frac{\partial \left(\sum_{i=0}^{3} g_{ij} \sqrt{-g} \right)}{\partial x_j} = 0. \tag{5.31}$$

In the case of the tensor of the general metric (5.23), and neglecting the terms of higher order than $1/c^4$, the effect of **W** disappears, and one is left with

$$g = -1 - 4\frac{U}{c^2} - \frac{2U^2}{c^4},$$

and

$$\sqrt{-g} = 1 + \frac{2U}{c^2} - \frac{U^2}{c^4}.$$

Applying the definition (5.31), one finally gets the harmonic conditions up to order $1/c^4$:

- for the time-like argument

$$\frac{\partial}{\partial x_0} \left(-1 + \frac{3U^2}{c^4} - \frac{4}{c^3}(w_1 + w_2 + w_3) \right) = 0,$$

- and for the three space-like arguments

$$\frac{\partial}{\partial x_j} \left(1 + \frac{4U}{c^2} + \frac{3U^2}{c^4} - 4\frac{w_j}{c^3} \right) = 0.$$

However, in this book, which is not devoted to relativistic celestial mechanics, we shall not use the harmonic conditions. To get more acquainted with their use, refer to Fock (1976) and Brumberg (1991a).

5.4 Space-time coordinate systems

Locally, a space-time coordinate system is defined by the flat, special relativity space-time tangent to the general relativity space, as is postulated by the equivalence principle (Section 5.3.1). In a small environment, like a laboratory or an astronomical instrument, one can use it in conjunction with the laws of physics of special relativity. To what extent this is adequate depends on the accuracy with which one needs to represent the physical phenomena. This is discussed in detail by Guinot (1997). For instance, if one wishes to synchronize two clocks to 10^{-17}s, they should remain at a distance less than 1 m. The problem is how to extend a local reference system to the astronomical universe.

5.4.1 Practical reference frames

Our knowledge of positions in space is dependent on the light received from the celestial objects. What is measured, in the local tangent, flat frame, is the direction from which the light reaches the instrument. Between the source and the observer, the light travels in a space where the gravitational potentials do not vanish. It follows a geodesic as does any test particle, but at speed c. So, one is led to determine the geodesic line by solving Equations (5.26) using the reduced form of the metric (5.22) or, if necessary, the full metric (5.24).

Let us, however, remark that, in reality, the Universe is not empty and has a certain mass distribution and a certain density. In constructing a cosmic reference frame, or discussing any cosmological observation, one should take into account the mean density of the Universe (Kopeikin, 2001). In modeling pulsar observations, the gravity field surrounding the object is a major parameter to be taken into account.

But in constructing an astrometric reference frame for observations from the Solar System, it is not necessary to know the shape of the light path, provided that it does not change in an observable manner during the few decades of validity of a reference frame. So, in practice, except within the Solar System itself, one ignores the effects of the outside gravitational fields. It is conventionally assumed that the reference system is such that directions correspond to the directions from which the light arrives, when it reaches the outskirts of the Solar System. In other terms, they correspond to what they would be, if all the masses outside the Solar System vanished.

The background of this procedure is that we model the environment by an asymptotically flat, or Minkowskian, space-time. The details can be found in Soffel (1989). The procedure is based on the notion of Weyl parallelism in such a space. Provided that the axes are defined far away from the Sun, the space-time is practically flat, and can, therefore, be used as if it was Euclidean. So, what is left is to determine the path of a light beam through the Solar System. At this level, we shall ignore the various effects that bend the light in the vicinity of the observer (refraction), or produce an apparent deviation (aberration). At this point, we shall assume that the local frame is free falling, which means that we are neglecting the effects of Earth rotation and gravity. All these additional effects have to be taken into account and will be described in Chapter 6.

5.4.2 Light paths

Light crosses the Solar System on a geodesic line defined by the distribution of potentials due to the Sun and the planets. One should, therefore, know the positions of the bodies of the Solar System and their motions during the travel of the light beam.

As a first approximation, sufficient in practice, let us limit ourselves by considering independently the principal effect due to the solar potential, and then add to it the effects due to individual planets which are smaller by at least a factor of 10^{-3}, the maximum in terms of solar mass (Jupiter). In the latter case, it is sufficient to assume that they are fixed at the position they have at the moment of closest approach of the light beam. We shall describe the path of a light beam in the potential of a point-like mass P, which is, again, a sufficiently good approximation of the potential of the Sun and the planets.

Let us use the extended Schwarzschild metric, (5.28) and (5.29), but simplifying the notation by setting $m = GM/c^2$. The four non-zero metric coefficients are:

$$g_{00} = -1 + \frac{2m}{r} - \frac{2m^2}{r^2},$$

$$g_{11} = \frac{1}{c^2(1 - 2m/r)},$$

$$g_{22} = \frac{r^2}{c^2},$$

$$g_{33} = \frac{r^2 \cos^2 \phi}{c^2}.$$

The path of the light is described by the coordinate time t, the three spherical coordinates, r, ϕ and λ, and the solution of Equations (5.26).

Since all the g_{ij} with i not equal to j are zero, Equations (5.26) reduce to

$$\frac{d}{dp}\left(g_{kk}\frac{dx_k}{dp}\right) - \frac{1}{2}\sum_{i=0}^{3}\frac{\partial g_{ii}}{\partial x_k}\left(\frac{\partial x_i}{\partial p}\right)^2 = 0, \tag{5.32}$$

with $k = 0$ to 3.

Now, the potential has a spherical symmetry, so that the plane comprising P and the initial velocity vector of the light is a plane of symmetry of the problem, so that the path is in this plane. Let us chose it as the equatorial plane of the system of coordinates, and we have

$$\phi \equiv 0$$

and, of course,

$$\frac{dx_2}{dt} = \frac{d\phi}{dt} = 0.$$

If we turn now to other equations, we note that $\lambda = x_3$ is absent from all four coefficients. Hence, the second term of (5.32) disappears, and one has, with $\cos \phi = 1$,

$$\frac{d}{dp}\left(\frac{r^2}{c^2}\frac{d\lambda}{dp}\right) = 0,$$

or

$$r^2 \frac{d\lambda}{dp} = h. \tag{5.33}$$

Similarly, the time $x_0 = t$ does not appear in (5.32) and the equation for x_0 becomes

$$\frac{d}{dp}\left(\left(-1 + \frac{2m}{r} - \frac{2m^2}{r^2}\right)\frac{dt}{dp}\right) = 0,$$

or

$$\left(1 - \frac{2m}{r} + \frac{2m^2}{r^2}\right)\frac{dt}{dp} = k. \tag{5.34}$$

Let us remark, that at this level, we have not yet assumed anything about ds^2. This means that all this is valid for the one body problem in general relativity. The results obtained mean that:

(1) the motion is plane,
(2) the constant h is the generalization of the law of conservation of the angular momentum,
(3) the constant k is the generalization of the law of conservation of energy.

At this point, let us return to the ds^2 which is a constant on a geodesic and, in the case of the light, $ds^2 = 0$. Substituting the equation $\theta = 0$ into the metric (5.29), one gets

$$c^2 \left(\frac{ds}{dp}\right)^2 = 0 = \frac{1}{1 - \frac{2m}{r}}\left(\frac{dr}{dp}\right)^2 + r^2\left(\frac{d\lambda}{dp}\right)^2 - \left(1 - \frac{2m}{r} + \frac{2m^2}{r^2}\right)c^2\left(\frac{dt}{dp}\right)^2,$$

or, introducing the integrals (5.33) and (5.34), and leaving only the first- and the second-order terms in m,

$$\left(\frac{dr}{dp}\right)^2 = \left(1 - \frac{2m^2}{r^2}\right)c^2 k^2 - \left(1 - \frac{2m}{r}\right)\frac{h^2}{r^2} \quad \text{and} \quad \frac{d\lambda}{dp} = \frac{h}{r^2}. \tag{5.35}$$

These equations give the parametrized path of the light in its plane of propagation. They can be integrated from any departure position and the path followed is a function of the parameter p, which, in a first approximation, neglecting m and setting $k = 1$, is the coordinate time. In any case, at least in the Solar System, where m is very small, it is a valid approximation.

In summary, these equations allow the computation of the path of light throughout the Solar System, assuming that the bodies are symmetric, non-rotating, and non-moving with respect to the observer, and that the planetary effects are additive.

The application of these equations to the apparent shift of the direction from which the light reaches the observer is given in Chapter 6. They are to be applied in order to determine the coordinate axes and the directions of extra-solar objects.

5.5 Time scales

Coordinate time is one of the coordinates of general relativity, so there is a specific time scale with each coordinate frame.

5.5.1 Terrestrial time scales

Since we are on the Earth, terrestrial reference frames and time scales have a particular importance, and the compatibility with general relativity concepts is imperative. Having defined a Geocentric Celestial Reference Frame (GCRS) on the basis of the preceding section, let us find the corresponding time coordinate scale.

The origin of the GCRS is the center of mass of the Earth, E, at which point the potential U is set to zero. The corresponding time scale is called Geocentric Coordinate Time (TCG, Temps-coordonnée géocentrique). The problem is to build a practical time scale on the surface of the Earth that coincides with TCG.

If we come back to Equation (5.25), we can determine the coordinate time, t, at any place in which the potential is U, and the rotational potential is

$$U_R = \frac{1}{2}\omega^2(x^2 + y^2) = v^2/2.$$

It results that, neglecting the terms in c^{-4},

$$\frac{dt}{d\tau} = 1 - \frac{U_G}{c^2}, \tag{5.36}$$

where U_G is the potential at the geoid (see Section 9.5.2),

$$U_G = U + U_R.$$

This coordinate time is called Terrestrial Time (TT) and differs from TCG uniquely by a constant rate. The unit of time of TT is the second of the International System of Units (SI) as realized on the geoid and defined as follows:

The second is the duration of 9 192 631 770 periods of the radiation corresponding to the transition between the two hyperfine levels of the ground state of the caesium 133 atom.

This definition refers to the atom in its ground state, undisturbed, and at a temperature of 0 K.

Atomic cesium clocks realize the SI second as a proper time which, at the geoid, is equal to the coordinate time TT. The ratio defined by the equation (5.36) provides the relationship between TT and TCG:

$$\text{TCG} - \text{TT} = L_G \times (JD - 2\,443\,144.5) \times 86\,400\,\text{s}, \tag{5.37}$$

with $L_G = U_G/c^2 = 6.969\,290\,134 \times 10^{-10}$, and JD is the Julian day. The bracket vanishes on January 1, 1977, 0h.TT. The uncertainty of this expression is 10^{-17}. It was decided at the General Assembly of the IAU in 2000, that this value of L_G will become a defining constant of TT. This overcomes the difficulties due to the temporal changes of the geoid and the intricacy of its definition.

The practical construction of TT is obtained by comparing the readings of many atomic clocks. Because they are at different altitudes, the local potentials are different, and the rates of their coordinate times are different. It is, therefore, necessary to reduce them to the geoid. If h_1 and h_2 are the heights of the clocks, and g is the mean gravity at $(h_1 + h_2)/2$, then the difference of rates is obtained by applying Equation (5.25). To a sufficient accuracy, this gives

$$\Delta\left(\frac{d\tau}{dt}\right) = \frac{g(h_1 - h_2)}{c^2}.$$

The time scale that is derived by the BIPM (Bureau International des Poids et Mesures) is actually the TAI (temps atomique international). One representation of TT is

$$\text{TT(TAI)} = \text{TAI} + 32.184\,\text{s}.$$

5.5.2 Barycentric time

In addition to the geocentric reference system, the other important reference system used in astronomy is centered at the barycenter of the Solar System. This point, x_B, is defined by the condition that the following integral over a volume $V(x)$ including all the masses of the Solar System is null,

$$\int_V \left[\rho(x) + \frac{1}{c^2}\left(\frac{1}{2}\rho(x)v(x)^2\right)\right](x - x_B)\,dV = 0. \tag{5.38}$$

This point is such that its motion in the Galaxy is linear, if one neglects the gravitational potential of the Galaxy. One of the major problems of astrometry is to refer the observations to a Barycentric Celestial Reference Frame. This is treated in the following chapters. Such a reference frame has a coordinate time called TCB (temps-coordonnée barycentrique = Barycentric Coordinate Time). The difficulty in computing TCB for any point of the Solar System is that the potentials vary with time because of the motion of planets.

Let us call t(TCB) and $\mathbf{x}$ the barycentric coordinates of an event in barycentric reference coordinates. The potential is described by the summation of particular potentials produced by N significant Solar System bodies with masses M_i and positions $\mathbf{x}_i$ that we describe as point-like masses

$$U_0 = G \sum_{i=1}^{N} M_i / |\mathbf{x} - \mathbf{x}_i|.$$

Let us consider an event at the center of mass of the Earth. Then, one can obtain the difference between TCB and TCG. In this case, U_0 should not contain the Earth potential, and let us denote it as U_0^{ext}. Let $\mathbf{v}_e$ be the coordinate velocity of the Earth's center of mass and $\mathbf{x}_e$ its coordinates; the relation between TCB and TCG can be obtained by computing

$$\text{TCB} - \text{TCG} = \frac{1}{c^2} \left[\int_0^t \left(U_0^{\text{ext}} + \frac{1}{2} v_e^2 \right) dt + \mathbf{v}_e \cdot (\mathbf{x} - \mathbf{x}_e) \right] + O(c^{-4}).$$

$$(5.39)$$

Additional terms in c^{-4} can be found in recommendation B1-4 of the IAU (Andersen, 2001). This expression has a linear term which represents the difference of their mean rates:

$$(\text{TCB} - \text{TCG})_{\text{secular}} = L_C (JD - 2\,443\,144.5) \times 86\,400\,\text{s},$$

$$(5.40)$$

where the best value at present of L_C is

$$L_C = 1.480\,826\,867\,41 \times 10^{-8},$$

with an uncertainty of 10^{-17}. In addition, the expression (5.38) has a nonlinear variation described by a number of periodic terms depending on the various periods present in the motions of planets. They are discussed in Fukushima (1995), who finds that there are 515 terms that are greater in amplitude than 0.1 ns. The most important terms in TCB–TCG are, in seconds,

$$0.001\,658 \sin g + 0.000\,014 \sin 2g,$$

$$(5.41)$$

where $g = 357°53 + 0°985\,003(JD - 2\,451\,545.0)$ essentially represents the mean anomaly of the Earth's orbit.

5.6 Astrometric effects

Whenever the accuracy of observations is such that a relativistic metric produces a non-negligible effect, one may compute the difference between what is obtained, assuming classical Newtonian Euclidean space and absolute time, and the solution by using the relativistic metric. Erroneously this is called a relativistic astrometric effect, because the complete effect results from the Theory of General Relativity. So, it is preferable to describe, as in the following sections, those phenomena observed by astrometric means, making use directly of the Theory of General (or sometimes Special) Relativity.

5.6.1 Relativistic reduction of astrometric measurements

In general, astrometric observations are related to some local frame of reference (telescope, artificial satellite, ranging system). But in order to compare measurements performed at different times and/or different locations, it is necessary to refer them to some conventional point, for instance the barycenter of the Solar System, or the center of mass of the Earth. In classical reduction procedures, one proceeds by taking into account the position and the velocity of the observer and the timing in some absolute time-scale. Examples of effects to be taken into account are aberration, parallactic displacements, precession and nutation, polar motion, refraction, corrections to reduce to some definite reference frame, etc. These will be described in the following chapters. But in the case of relativity, these reductions must take into account the potential in the field, and its dependence on time.

In special or general relativity, the transformation from local time and positions into some unique system of space-time coordinates implies working in a given metric, involving not only positions and velocities, but the Lorentz transformation in special relativity, and the effects of potentials in general relativity. A discussion of the problems in the most general context can be found in Brumberg (1991a). We shall apply them in the particular cases in the following chapters.

5.6.2 Main relativistic effects

Let us list the cases when it is essential to use a relativistic approach, and give a hint as to the treatment of the most important such phenomena, in particular those that have been – and still are – the classical tests of general relativity. There are three *classical*, relativistic effects that were originally used to verify the Einstein theory. Others have been added since.

(i) *The advance of perihelia of planets.* The part of the motion of perihelia that could not be explained by the classical Newtonian celestial mechanics amounts to 42″7 per century for Mercury and is fully explained by the theory of general relativity. The rotation is

$$\frac{6\pi m}{a(1 - e^2)}$$

per revolution for a planet of semi-major axis a, eccentricity e, and $m = GM/c^2$, where M is the mass of the central body.

(ii) *The bending of light*. The principle is given in Section 5.4.2 and the evaluation of the effect is presented in Section 6.4.5. For a ray grazing the Sun, the effect is $1''.75$. At $90°$ from the Sun, the effect is 4.0717 mas (see Section 6.4.3). It is the cause of the gravitational lensing by stars or quasars.

(iii) *Gravitational time dilatation*. This effect is described in Section 5.3.3 by Equation (5.24) and is now very well determined with accurate atomic clocks on board satellites.

(iv) *Geodesic precession and nutation*. This effect represents the rotation of a reference coordinate system in presence of forces. This is the case of the geocentric axes with respect to the barycentric ones, even if they are both defined by the apparent directions of the same fixed extragalactic objects (see Section 7.5).

(v) *Light time delays*. The travel time for light as measured by radar ranging depends on the potential in the path. This will be presented in Section 6.5. This is equivalent to the dependence of length with respect to the potential.

(vi) *Gravitational red shift*. The difference in gravitational potential between the place of emission and the observer has an effect that cannot be distinguished from the Doppler shift due to radial velocity. It has, however, to be taken into account and will be touched upon in Section 11.3.2.

(vii) *Gravitational wave emissions*. In analyzing the motion of a binary pulsar, Taylor and Weisberg (1989) have shown that there was an energy loss, interpreted in the framework of general relativity, as an emission of gravitational waves. However, until now, no direct detection of such waves has been achieved.

(viii) *Pulsar timing*. This problem is touched upon in Section 14.7.

Some other astrometric phenomena are not relativistic effects (aberration, parallax), but are modified by the Theory of General Relativity, and are observationally mixed with them. So, one has to keep them in mind, especially at the level of microarcsecond accuracies. For this reason, in the chapters that follow, and generally speaking in modern astrometry, one has to think in terms of general relativity, and return to more classical approaches, only if it is clear that this is consistent with the accuracy of the observations.

6

Apparent displacements of celestial objects

The apparent direction in the sky at which a celestial object appears is not the actual direction from which the light was emitted. What is observed is the tangent direction of the light when it reaches the observer. For reasons that will be discussed in this chapter, the light path is not rectilinear and several corrections describing the effects of bending, or shifts in direction, are to be applied to the direction from which the light is observed to determine the actual direction of the emission. We shall not deal here with the various transformations undergone by the light within the observing instrument; they are particular to each case. Some examples are given in Chapter 14. We shall consider only the direction from which the light came when it entered into the instrument. One has to consider the atmospheric refraction, the shift in direction due to the combination of the speed of light with the motion of the observer, called *aberration*, and the bending of light in the presence of gravitational fields. The latter has been already presented in Section 5.4.2, but will be revisited in Section 6.4. Similarly, the geodesic precession and nutation are to be considered when relating the positions from a moving reference frame of fixed orientation to a fixed reference frame of the same orientation (see Section 7.5). In addition, one often wants to get the direction of the source as seen from a point other than the observatory, for example the center of the Earth or the barycenter of the Solar System. These are the *parallactic corrections*. Sometimes, particularly for objects in the Solar System, one wishes to know the position of an object at the time of the observation, or the exact time at which the observed light was emitted. This is called the *light-time* correction. All these effects are discussed in this chapter.

6.1 Atmospheric refraction

When light crosses a surface which separates two gas layers with refractive indices n and $n + dn$, it is deviated – refracted – in a manner described by Snell's law. Let ξ be the angle of incidence (angle of the ray with the normal of the separation surface) in the medium characterized by the refractive index n. The angle of refraction in the medium of refractive index $n + dn$ is $\xi + d\xi$ and obeys the relation

$$(n + dn) \sin(\xi + d\xi) = n \sin \xi. \tag{6.1}$$

The total bending of the light is the quantity

$$R = z - z_0,$$

where z_0 is the observed zenith distance of the object, and z is the zenith distance it would have had without the atmosphere and is referred to the vertical direction (zenith) $O\Omega$ of the observer,

$$z = z_0 + R. \tag{6.2}$$

In theory, the determination of R should imply the knowledge of n at all points of the atmosphere crossed by the ray, and the angle of the local normals to iso-index layers with the vertical V at the origin. It would be obtained by integrating (6.1) from the ground with a measured refractive index n_0 to space – at a height of the order of 50 km – from which one may consider that the vacuum is sufficient to assume $n = 1$. The theory was made by Garfinkel (1944) and applied using lidar measurements at various altitudes. For practical applications in astrometry, no lidar observations are available, and one is led to represent the atmosphere by a model that allows computation of R as a function only of conditions (temperature, pressure, etc.) at the observatory.

6.1.1 Planar stratified layers

The simplest model is to consider that the atmosphere is described by plane horizontal layers. In this case, the normals to the layers are all parallel to the vertical, so that ξ can be replaced by z, and formula (6.1) simply reduces to consider that the quantity $n \sin z$ is an invariant, so that at the two limits of the integration, one has

$$n_0 \sin z_0 = \sin z.$$

Replacing z by its expression (6.2) and neglecting R^2, one gets the first-order refraction formula,

$$R = (n_0 - 1) \tan z_0.$$

This is a very poor representation of the actual refraction, except very close to the zenith.

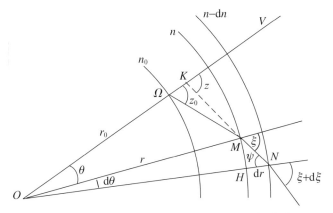

Fig. 6.1. Refraction through a spherical atmospheric layer.

6.1.2 Spherical atmosphere model

The next step is to assume that the atmosphere is radially symmetric around the local vertical. The iso-refractive index curves are spheres centered at the center of the Earth, O. Let r_0 be the radius vector at the observer, Ω, and z_0 the apparent zenith distance of the ray, with respect to the vertical $V(O\Omega)$. Let us consider a small layer with a width dr and a refractive index n. Let NM (Fig. 6.1) be the linear path of the light within this layer. At N, one may apply Snell's law to the light coming from the outer medium with the refractive index equal to $n - dn$. Calling ψ and $\xi + d\xi$, respectively, the refractive and incidence angles in N, one has

$$n \sin \psi = (n - dn) \sin(\xi + d\xi).$$

Let θ be the angle between the verticals at Ω and M, $d\theta$ its increment from M to N, and r the radius vector at M. The sine relation in OMN gives

$$r \sin \xi = (r + dr) \sin \psi.$$

Multiplying both equations, member by member, one gets, after simplification by $\sin \psi$:

$$nr \sin \xi = (r + dr)(n - dn) \sin(\xi + d\xi). \tag{6.3}$$

This relation proves that the quantity $nr \sin \xi$ is an invariant, so that

$$nr \sin \psi = n_0 r_0 \sin z_0. \tag{6.4}$$

Now, since $z = \theta + \psi$ in the triangle KOM, and since in MHN, $\tan \psi = r d\theta/dr$, one gets

$$dz = d\psi + \frac{\tan \psi}{r} dr. \tag{6.5}$$

Differentiating (6.4), one gets

$$nr \cos \psi \, d\psi + r \sin \psi \, dn + n \sin \phi \, dr = 0, \tag{6.6}$$

or

$$rn \cos \psi \left(d\psi + \frac{dr}{r} \tan \psi \right) = -r \sin \psi \, dn,$$

which, with (6.5), gives

$$dz = -\frac{dn}{n} \tan \psi. \tag{6.7}$$

Another way of writing (6.4) is

$$\tan \psi = \frac{r_0 n_0 \sin z_0}{\left(r^2 n^2 - r_0^2 n_0^2 \sin^2 z_0 \right)^{1/2}},$$

and substituting it into Equation (6.7), one finally gets, after integration,

$$R = r_0 n_0 \sin z_0 \int_1^{n_0} \frac{dn}{n \left(r^2 n^2 - r_0^2 n_0^2 \sin^2 z_0 \right)^{1/2}}, \tag{6.8}$$

which is an exact equation, but assumes that one knows the variation of n with r or $r = r(n)$. This is not directly available from observations, and one would need to use a model. So, in practice, what is useful is to reduce (6.8) to a simpler formulation and later consider some integrals as parameters to be determined.

6.1.3 The Laplace formula

Let us re-write (6.8) as

$$R = \sin z_0 \int_1^{n_0} \frac{dn}{n((rn/r_0 n_0)^2 - \sin^2 z_0)^{1/2}},$$

and introduce two small parameters. The first is the refractivity on the ground, $\alpha = n_0 - 1$. Applying the Gladstone–Dale law, which states that for small densities the refractivity is proportional to the density of the gas, and calling ρ the relative density with respect to the density on the ground, one has

$$n - 1 = (n_0 - 1)\rho = \alpha\rho. \tag{6.9}$$

Setting h, the height with respect to the ground, as

$$r = r_0 + h,$$

one gets

$$\left(\frac{nr}{n_0 r_0}\right)^2 = \left(\frac{1+\alpha\rho}{1+\alpha}\right)^2 \left(1+\frac{h}{r_0}\right)^2$$

$$= 1 + \frac{2h}{r_0} + 2\alpha\,(\rho-1) + \cdots$$

$$= 1 + 2\eta$$

in which 2η stands for the increment to 1. With these parameters, one has

$$R = \sin z_0 \int_0^1 \frac{\alpha\,\mathrm{d}\rho}{(1+\alpha\rho)\sqrt{1-\sin^2 z_0 + 2\eta}},$$

or, after replacing $\cos^2 z_0$ by $1/(1+\tan^2 z_0)$ and developing R with respect to small quantities α and η, one obtains

$$R = \alpha \tan z_0 \int_0^1 (1+\alpha\rho)\,(1-\eta\,(1+\tan^2 z_0))\,\mathrm{d}\rho.$$

Let us remark that

$$\int_0^1 \eta\,\mathrm{d}\rho = \int_0^1 \frac{h}{r_0}\,\mathrm{d}\rho + \alpha \int_0^1 (\rho-1)\,\mathrm{d}\rho.$$

The first integral can be integrated in parts

$$\int_0^1 \frac{h}{r_0}\,\mathrm{d}\rho = \left[\frac{h\rho}{r_0}\right]_0^1 + \frac{1}{r_0} \int_0^\infty \rho\,\mathrm{d}h.$$

Since $h = 0$ when $\rho = 1$, the term within brackets is null. The second integral is the scale factor L of the atmosphere. Let us now call $\beta = L/r_0 = p/\delta_0 r_0$, p being the pressure and δ_0 the air density at the site. Then we obtain after some simplifications

$$R = \alpha(1-\beta)\tan z_0 - \alpha\left(\beta - \frac{\alpha}{2}\right)\tan^3 z_0. \tag{6.10}$$

This is the Laplace formula, usually written

$$R = A \tan z_0 - B \tan^3 z_0, \tag{6.11}$$

where $A = \alpha(1-\beta)$ and $B = \alpha(\beta - \alpha/2)$.

6.1.4 Dependence on physical parameters

A simple look at the meaning of A and B shows that they depend upon the density of the air at the site, that is on temperature and pressure. Furthermore the refractive indices depend on the wavelength. We shall examine how these parameters affect R, as well as some other smaller effects. Several approaches have been followed, and the resulting refraction tables were published. The most complete were published by the Pulkovo Observatory (1985), and then discussed by Guseva (1987), who found that these tables closely correspond to the mean atmospheric structure under a variety of observational conditions, and significantly reduce systematic errors of refraction determinations. These were put in the form of analytical expressions with a few short tables by the Bureau des Longitudes (Simon *et al.*, 1996), and the following discussion is based upon this last work (note, however, that misprints for $f(z_0)$ and $g(z_0)$ have been corrected here).

6.1.4.1 Temperature and pressure

The well-known Boyle's law of perfect gases is only a first approximation to the variation of density, and consequently of α. One starts with the Laplace formula for *normal conditions*, that is:

- temperature t: $t_0 = 15\,°C$
- pressure p: $p_0 = 101\,325$ Pa (pascals)
- wavelength λ: $\lambda_0 = 0.590$ nm
- partial water vapor pressure f: $f_0 = 0$ Pa.

Then,

$$\alpha_0 = 0.000\,277\,117\ (= 57''\!.1595)$$
$$\beta_0 = 0.001\,3037$$

and the Laplace formula is:

$$R_0 = 60''\!.236 \tan z_0 - 0''\!.0675 \tan^3 z_0.$$

The corrected formulae for R at temperature t (in degrees Celsius) and pressure p (in pascals) as deduced from the Pulkovo tables is

$$R(t,\,p) = R_0\,\frac{p}{p_0}\,\frac{1.055\,2126}{1 + 0.003\,680\,84t}\,F\,G, \tag{6.12}$$

with

$$F = [1 - 0.003\,592(t - 15) - 0.000\,0055(t - 7.5)^2][1 + f(z_0)],$$
$$G = [1 + 0.9430 \times 10^{-5}(p - p_0) - 0.78 \times 10^{-10}(p - p_0)^2][1 + g(z_0)],$$

where the correcting factors depending upon the zenith distance z_0 are tabulated in Table 6.1. They increase with z_0 and remain close to 1, reaching a maximum respectively of 1.01 and 1.0013 for $z_0 = 70°$.

Table 6.1. *Correction factors as*
functions of zenith distance

z_0	$10^{-4}f(z)$	$10^{-4}g(z)$
0°	0	4
10°	0	4
20°	0	4
30°	0	5
35°	0	5
40°	2	5
45°	6	6
50°	12	6
55°	21	7
60°	34	8
65°	56	10
70°	97	13

6.1.4.2 Chromatic effects

The actual dependence of n, and hence of α, is a rather complicated function of the inverse square of the wavelength. An approximate expression for the chromatic refraction at wavelength λ in accordance with the Pulkovo tables is

$$R(t, p, \lambda) = R(t, p)(0.982\,82 + 0.005\,981/\lambda^2). \tag{6.13}$$

However since, in general, the observations are made in a finite interval of wavelengths using a filter whose transparency is also a function of λ, the above formula is sufficient when applied to the mean wavelength of the ensemble star-filter.

For this reason, it is often preferable to make two observations at two widely separated mean wavelengths. Let z_1 and z_2 be the two observed zenith distances for $\lambda = \lambda_1$ and λ_2, and C be the wavelength-dependent coefficient in the Laplace formula which takes the form

$$R = A_0 \tan z + \frac{C}{\lambda^2} \tan z - B_0 \tan^3 z,$$

so that the two observations give two equations

$$z = z_1 + A_0 \tan z_1 + \frac{C \tan z_1}{\lambda_1^2} - B_0 \tan^3 z_1,$$

$$z = z_2 + A_0 \tan z_2 + \frac{C \tan z_2}{\lambda_2^2} - B_0 \tan^3 z_2,$$

with two unknowns z and C. It is, of course, assumed that the other terms of the Laplace formula are known, that is, determined, from measurements of temperature, pressure, etc.

6.1.4.3 Water vapor

The content of water vapor in air varies considerably with meteorological conditions and the proportion of water vapor defined by its partial pressure f modifies n and α. The Pulkovo refraction tables are well represented by the following correction to $R(t, p, \lambda)$:

$$R(t, p, \lambda, f) = R(t, p, \lambda)(1 - 0.152 \times 10^{-5} f - 0.55 \times 10^{-9} f^2).$$

$$(6.14)$$

6.1.4.4 Refraction at a finite distance

If the observed object, although outside the atmosphere, is close to the Earth, it is not possible to assume that the true direction is parallel to the asymptote of the light ray.

On Fig. 6.1, if the object is in N the actual point from which the true zenith distance is given is K. If the origin of the ray is not at infinity, there is an apparent parallactic displacement, so that in order to get the true direction, one has to correct the position of Ω by $x = \Omega K$ along the vertical.

In the triangle OMK of Fig. 6.1, one has

$$OK \sin z = r \sin \psi.$$

Applying the invariant (6.4), one has

$$n_0 r_0 \sin z_0 = nr \sin \psi.$$

The combination of these two equations gives, assuming, in addition, that the source, M, is outside the atmosphere so that $n = 1$:

$$OK = x + r_0 = \frac{n_0 r_0 \sin z_0}{\sin z}.$$

By construction, if R is the refraction, one has, neglecting the third-order small quantities,

$$\sin z = \sin(z_0 + R)$$

$$= \sin z_0 \left(1 - \frac{R^2}{2} + R \cot z_0\right).$$

Then,

$$\frac{x}{r_0} = n_0 \left(1 + \frac{R^2}{2} - R \cot z_0\right) - 1,$$

$$\frac{x}{r_0} = n_0 \alpha_0 - n_0 R \cot z_0 + n_0 \frac{R^2}{2}.$$

$$(6.15)$$

6.1.5 *Accuracy of refraction formulae*

The formulae given in Section 6.1.4 are, at best, good to a few milliarcseconds (mas) for small zenith distances. They degrade for larger values of z_0. One can estimate the errors to be of the order of $0''.01$ around $55°$, and $0''.05$ for $z_0 = 70°$. At lower altitudes, additional multiplicative coefficients are to be applied for chromatic and water vapor corrections. For A and B, special tables exist in the literature (Simon *et al.*, 1996), and additional corrections must be introduced for latitude in order to take into account the non-sphericity of the Earth and of its atmosphere, as well as for the height of the observer – a purely geometrical effect due to the curvature of the Earth, independent of the effects of atmospheric pressure. The precision of these improved formulae is rather poor, and becomes worse for increasing zenith distances. The maximum uncertainty of the refraction occurs at the horizon where it reaches a fraction of a degree (see Section 13.3.3).

Even at small zenith distances the errors of refraction formulae, or tables, forbid sub-milliarcsecond astrometry. It is necessary to determine simultaneously the refraction using two or more color observations. But even then, the uncertainty in the actual mean wavelength of the observation, as well as the short period variations of refraction due to atmospheric turbulence, do not permit one to expect anything better than a few tenths of a mas. To obtain higher accuracies, one must get rid of refraction completely by observing from space.

6.1.6 *Refraction in distance measurements*

The speed of light in a gas is c/n, where n is the refractive index of the medium. Let us consider a distance between the ground (index 0) and a point, P, determined using c as the speed of light. The actual distance is smaller, and the difference is given by the integral

$$\Delta D = \int_0^P (n(s) - 1)\,ds,$$

taken along the ray path. Neglecting the curvature of the light and assuming that the atmosphere is composed of plane layers, if h is the height of the layer, one has

$$dh = ds \cos z,$$

and

$$\Delta D = \frac{1}{\cos z} \int_0^A (n(h) - 1)\,dh.$$

Using the Gladstone–Dale law (formula (6.9)) and introducing the coefficients α

and β used in (6.10), or the scale factor of the atmosphere, one gets

$$\Delta D = \frac{\alpha \beta r_0}{\cos z} = \frac{\alpha L}{\cos z}. \tag{6.16}$$

This has to be corrected for chromatic refraction and the particular effect of water vapor, as in the case of the optical refraction, using the same corrective terms for α and β.

This theoretical approach is not ideal when high accuracy is sought. So, in practice, it is better to use an empirical formula deduced from the analysis of a large number of observations. The Marini and Murray (1973) formula, adopted by the IERS standards, is the following, expressed in meters:

$$\Delta D = \frac{\Phi(\lambda)}{\Psi(\phi, h)} \times \frac{A + B}{\cos z + B/((A + B)(\cos z + 0.01))}, \tag{6.17}$$

where

$$A = 0.000\,023\,57\,p + 0.000\,001\,41\,f,$$

$$B = (1.084 \times 10^{-10})pTK + \frac{(9.468 \times 10^{-12})p^2}{T(3 - 1/K)},$$

where $T = t + 273.16$ is the absolute temperature, p and f are the pressure and the partial water vapor pressure expressed in pascals, λ is the wavelength in micrometers, h is the height in kilometers, and ϕ is the latitude of the observer. In addition,

$$K = 1.163 - 0.009\,68 \cos 2\phi - 0.000\,04T + 0.000\,000\,1435\,p,$$

$$\Phi(\lambda) = 0.9650 + 0.0164/\lambda^2 + 0.000\,228/\lambda^4,$$

$$\Psi(\phi, h) = 1 - 0.0026 \cos 2\phi - 0.000\,31h.$$

6.1.7 Refraction in radio waves

The refractive index in radio waves is very different, and a special formula, due to Smith and Weintraub (1953), is generally adopted,

$$\alpha_R = \frac{0.2840 \times 10^{-8}\,p}{1 + 0.003\,66t} - \frac{0.469 \times 10^{-9}\,f}{1 + 0.003\,66t} + \frac{0.508 \times 10^{-7}\,f}{(1 + 0.003\,66t)^2},$$

where the same notations and units as above are used. The refraction is

$$R = \alpha_R \tan z_0 - 0''\!.067 \tan^3 z_0. \tag{6.18}$$

However, in general, the radio observations for astrometry (VLBI) reduce to time measurements, so that the main effect to take into account is the refraction in distance, which is dealt with in a way similar to that in Section 6.1.6.

In addition, radio waves undergo an important delay due to the ionosphere. The speed of the waves is

$$V = cn = c\sqrt{1 - \frac{v_p^2}{v^2}},$$

where v is the frequency of the wave and v_p is the proper frequency of the plasma given, in hertz, by

$$v_p = \frac{N_e^2}{\varepsilon_0 m_e},$$

where N is the electronic density of the plasma per cubic meter, e and m_e are the charge and the mass of the electron, respectively, and ε_0 is the constant of permittivity of free space, (or electric constant) equal to

$$8.854\,187\,817 \times 10^{-12} \text{ farads per meter } (f/m).$$

The distance correction, after introducing the various numerical values in the formula, is

$$\Delta D_R = c\Delta t = 40.308\frac{N}{v^2}. \tag{6.19}$$

It is determined by making observations at two frequencies and solving the two equations (6.19) after eliminating N.

Then if ΔD_R^0 is the distance correction at the zenith, the actual effect of the ionospheric refraction is

$$\Delta D_R = \frac{\Delta D_R^0}{\cos\left(\sin^{-1}\left(\frac{r_0 \sin z}{r_0 + h}\right)\right)},$$

where h is the mean height of the ionosphere. This formula takes into account the sphericity of the ionosphere, but assumes that the latter is homogeneous. A more detailed analysis can be found in Thomson *et al.*, 1986.

6.2 Parallactic corrections

As a rule, observations made at different times from different locations should be referred to a single common system. For determining star positions, the preference is the celestial barycentric reference frame (Section 7.2). If the observed objects were at infinity, the translation from the observer to the barycenter of the Solar System would not affect their apparent direction. But this is not the case, since stars are at some finite distance. The result is that the motion of the observer produces an apparent displacement of the star with respect to the fixed reference system, called

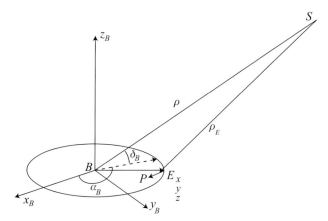

Fig. 6.2. Parallactic correction of the star position.

the parallactic displacement. The motion of the observer is conveniently split into three components.

- A linear uniform motion, representing the motion of the Solar System as a whole. It combines with analogous linear motions of stars in the Galaxy and will be considered in Section 11.3 dealing with proper motions and radial velocities.
- The motion of the center of mass of the Earth around the barycenter. It is the combination of the motion of the barycenter of the Earth–Moon system about the barycenter of the Solar System, and of the motion of the center of mass of the Earth about the barycenter of the Earth–Moon system. The result is a yearly periodic displacement called *annual parallax* and a small monthly component. Actually, the yearly periodicity is only an approximation that is valid at the level of a milliarcsecond. It is not adequate at higher precisions, and the actual motion of the center of mass of the Earth must be used.
- The motion of the observer around the center of mass of the Earth. If the observer is on the ground, the effect produced is called the *diurnal parallax*. The same phenomenon exists for a spacecraft, and one can speak of *orbital parallax*.

6.2.1 Correction for annual parallax

Let us assume that the barycentric spherical coordinates of the star S are α and δ and that the distance of the star is ρ (Fig. 6.2). The components of the three-dimensional vector **BS** are

$$\left\{ \begin{array}{l} \rho \cos \delta \cos \alpha \\ \rho \cos \delta \sin \alpha \\ \rho \sin \delta. \end{array} \right.$$

Let E be the center of the Earth. The components x, y, z of the vector $\mathbf{BE}$ are given by the ephemerides of the Earth's motion and are functions of the time t. The actual geocentric coordinates α_E, δ_E and the geocentric distance ρ_E are given, as a function of the barycentric coordinates and distance, α_B, δ_B, and ρ by

$$\rho_E \cos \delta_E \cos \alpha_E = \rho \cos \delta_B \cos \alpha_B - x,$$
$$\rho_E \cos \delta_E \sin \alpha_E = \rho \cos \delta_B \sin \alpha_B - y$$
$$\rho_E \sin \delta_E = \rho \sin \delta_B - z. \tag{6.20}$$

The corrections $\Delta\alpha = \alpha_B - \alpha_E$ and $\Delta\delta = \delta_B - \delta_E$ to the barycentric positions are readily obtained if one neglects the second-order effects, an assumption that is justified because, for almost all stars, the ratio BE/ρ is smaller than 10^{-6} radians, the largest value, for Proxima Centauri, being 3.7×10^{-6}. One then may differentiate (6.20) and obtain the following formulae expressed in radians as functions of $\Delta\alpha$, $\Delta\delta$, and $\Delta\rho = \rho - \rho_E$:

$$\Delta\alpha \cos \delta = \frac{x}{\rho} \sin \alpha - \frac{y}{\rho} \cos \alpha,$$

$$\Delta\delta = \left(\frac{x}{\rho} \cos \alpha + \frac{y}{\rho} \sin \alpha \right) \sin \delta - \frac{z}{\rho} \cos \delta, \tag{6.21}$$

and the variation of the distance of the star

$$\Delta\rho = -x \cos \alpha \cos \delta - y \sin \alpha \cos \delta + z \sin \delta. \tag{6.22}$$

It is to be noted that, although these formulae are valid to better than a microarcsecond, one cannot, as was classically done, replace the Earth's orbit by a conventional ellipse. It is necessary to derive x/ρ, y/ρ, and z/ρ from precise barycentric ephemerides of the Earth's center of mass, with a relative accuracy of 10^{-6}, so as to avoid any erroneous effect.

6.2.2 Stellar parallax

Formulae (6.21) are expressed in radians, which is an inconvenient unit for astrometry in general, and especially for small angles. So, it is customary to express small angles in seconds of arc (and submultiples mas and μas). This leads to defining the *stellar parallax* ϖ as the ratio $1/\rho$ where ρ is expressed in astronomical units, transformed into arcseconds,

$$\varpi = k/\rho,$$

with $k = \pi/648\,000$. It is also the angle in arcseconds with which one sees one astronomical unit from the star.

However, one generally expresses the distance ρ in units of $648\,000/\pi = 206\,264.81$ astronomical units, a unit called *parsec*, then one has simply,

$$\varpi = 1/\rho,$$

The parsec is the distance corresponding to a parallax of $1''$, and is also equal to $3.261\,5637$ light-years. If, in a first approximation, we describe the motion of the Earth as an ellipse, with a semi-major axis equal to one astronomical unit, the apparent geocentric motion of the star around its barycentric position is an ellipse, whose semi-major axis is equal to its parallax.

Using the stellar parallax and the parsec, the formulae (6.21) are transformed into:

$$\Delta\alpha\cos\delta = \varpi\,(x\sin\alpha - y\cos\alpha)\,,$$
$$\Delta\delta = \varpi\,[(x\cos\alpha + y\sin\alpha)\sin\delta - z\cos\delta]\,, \tag{6.23}$$

where the coordinates x, y, z of the Earth are expressed in astronomical units and the corrections to the position are in arcseconds.

Note. The interpretation of the parallax and the determination of its associated uncertainty in terms of the distance to the star is not straightforward. It will be discussed in Section 11.2.

6.2.3 Diurnal and orbital parallaxes

The diurnal parallax is the displacement of the geocentric position of the star **S** due to the vector **GP** where P is the observer and G is the geocenter. Let λ, ϕ and R be the geocentric coordinates of P, and T the sidereal time or the stellar angle (see Section 10.4.4). The components of **GP** are

$$x' = R\cos\phi\cos(T + \lambda),$$
$$y' = R\cos\phi\sin(T + \lambda),$$
$$z' = R\sin\phi.$$

To compute the parallactic correction, it is sufficient to apply Equations (6.21), replacing x, y, z by $x + x'$, $y + y'$, $z + z'$, or using only x', y', z' if only the diurnal parallax correction is needed.

If the observation is performed from a satellite, the geocentric coordinates of the satellite x', y', z' are obtained from the orbital ephemerides. In both cases, formulae (6.21) apply again, provided that the quantities x', y', z' are respectively added to x, y, z.

Generally, in the past, diurnal parallax corrections were not applied to the observations of stars. The effect, of the order of R/ρ, is smaller than the effect of the stellar parallax by a factor equal to the ratio between the Earth's radius and the astronomical unit (4.26×10^{-5}). Nevertheless, for a parallax equal to $0\rlap{.}''1$, this represents a correction of 4 μas. So it is not to be neglected for nearby stars. In the case of orbital parallax, which gives a larger correction (a factor of 6 for a geostationary satellite), the correction must be made, and was indeed performed in the reduction of Hipparcos data.

In the case of the Moon and the planets, the situation is different. It is important to get the position of the body at a well-determined time, and to take into account the fact that during the time the light traveled from the body to the observer, the Moon, or the planet, as well as the Earth and the observer moved. This will be presented in Section 6.3.6.

6.3 Aberration

Since the velocity of light is finite, the apparent direction to an object is affected by the motions of the object and of the observer. Let us first consider the case of very distant objects like stars. The correction for the motion of the observer is called stellar aberration. The motion of an observer on Earth can be divided into three parts: the diurnal rotation of the Earth, the orbital motion of the Earth about the barycenter of the Solar System, and the motion of the barycenter in space. Thus, stellar aberration is made up of the three components, which are referred to as diurnal, annual, and secular aberration. For a spacecraft in orbit around the Earth, the diurnal aberration is replaced by orbital aberration.

In the case of objects in the Solar System, the situation is different. One speaks of planetary aberration, but it is actually a different concept. In this case, aberration corresponds to the displacement of a planet between the moment the light was emitted and the time when it was received by the observer. Here, the light-time and the relative motions of the body and of the Earth can be accurately computed in order to determine the planetary aberration.

6.3.1 Secular aberration

The stars and barycenter of the Solar System can generally be considered to be in uniform rectilinear motion. The aberrational displacement due to the relative motion is equal to the proper motion of the star multiplied by the light-time, which results in a correction of the position of each star. These are seldom well known and, since the knowledge of the barycentric position of a star at the time of observation has no scientific interest, this displacement is in general ignored. However, one might wish to refer the stars to another reference frame than the barycentric. One may think of the *local standard of rest*, which is at rest with respect to close-by stars.

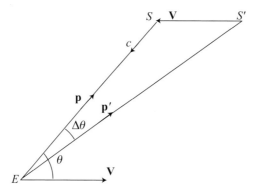

Fig. 6.3. Stellar aberration.

Then the secular aberration would be computed from the motion of the Sun with respect to it; the value is 19.5 km/s in the direction of the apex. The corresponding magnitude of the secular aberration is about 13 arcseconds. But the apex itself depends upon the classes and the population of stars used to determine it, so that the physical significance of the local standard of rest is questionable.

One may also wish to take a galactocentric reference frame. In that case, the Sun has a velocity of the order of 220 km/s, leading to a secular aberration of $150''$. Let us, however, note that, for very accurate astrometry, the curvature of the orbit of the Sun may have to be taken into account. This part of the secular aberration is proportional to the square of time. In one year, assuming a period of the Sun's galactic orbit of $P = 250\,000\,000$ years, the rotation of the instantaneous velocity vector is $2\,\pi/P$ radians. This represents a change of the velocity vector of 5.5 mm/s giving rise to a variation of the secular galactic aberration of about 4 μas. Stars in the vicinity of the Sun have similar motions around the center of the Galaxy, which cancel out most of this effect. But it remains to be considered for very distant stars and extragalactic objects. This aberration in proper motion may reach a fraction of a milliarcsecond close to the center of the galaxy, where stars have short revolution periods (Kovalevsky, 2003).

6.3.2 Stellar aberration

In general, the observer at E is moving with a velocity $\mathbf{V}$ relative to the fixed frame. The apparent change in the geometric direction of the celestial object at S, due to the orbital motion of the Earth about the barycenter, is called stellar aberration. In Fig. 6.3, $\mathbf{p}$ is a unit vector in the geometric direction ES, i.e. in the direction of the body at time t in a fixed frame. The velocity vector of the light is $-c\mathbf{p}$ in the same frame. At time t the observer will see the body at S', a direction defined by the vectorial sum of the two velocities. The unit vector of this direction is $\mathbf{p}'$. Here,

$(S'E, \ \mathbf{V}) = \theta$ is the angle between the direction of motion and $\mathbf{p}$ in the stationary frame, and $SES' = \Delta\theta$ is the displacement due to aberration in the moving frame, which is always toward the direction of motion.

The classical Newtonian expression for the direction of the source as seen by the moving observer is obtained by vector addition of velocities as follows:

$$\mathbf{p}' = \frac{\mathbf{p} + \mathbf{V}/c}{|\mathbf{p} + \mathbf{V}/c|.} \tag{6.24}$$

Taking the scalar part of the vector cross product of $\mathbf{p}$ with $\mathbf{p}'$, then

$$\sin \Delta\theta = \frac{(V/c)\sin\theta}{\sqrt{(1 + 2(V/c)\cos\theta + (V/c)^2)}}$$

$$= \frac{V}{c}\sin\theta - \frac{1}{2}\left(\frac{V}{c}\right)^2 \sin 2\theta + \cdots, \tag{6.25}$$

since $|\mathbf{p} \times \mathbf{p}'| = \sin \Delta\theta$, $|\mathbf{p} \times \mathbf{p}| = 0$, and $|\mathbf{p} \times \mathbf{V}/c| = \frac{V}{c}\sin\theta$.

The term of order V/c is about 0.0001, which, expressed in arcseconds, is of the order of $20''$. Similarly, the term in $(V/c)^2$ has a maximum value of about $0''.001$. The third-order terms are not significant, particularly because they are smaller than the effect of general relativity.

6.3.3 Aberration in relativity

In special relativity, the velocity of light is constant in the moving and stationary frame, and the Lorentz formula for the addition of velocities applies. This is also true in general relativity whose local kinematic properties are those of special relativity plus additional effects caused by the curvature of space-time. The latter effects are applied separately as light deflection (see Section 6.4).

Let us apply the velocity addition formula in special relativity (5.8). With the notations set above for the classical case, we have:

- the velocity of light in the geometric direction: $c\mathbf{p}$,
- the velocity of light in the apparent direction: $c\mathbf{p}'$,
- the velocity of the observer: $\mathbf{V}$.

Formula (5.8) becomes:

$$c\mathbf{p}' = \frac{c\mathbf{p} + \gamma\mathbf{V} + (\gamma - 1)(\mathbf{V} \cdot c\mathbf{p})\mathbf{V}/V^2}{\gamma(1 + c\mathbf{p} \cdot \mathbf{V}/c^2)},$$

with $\gamma = (1 - V^2/c^2)^{-1/2}$. Dividing by c and dividing both terms of the fraction by γ, one obtains:

$$\mathbf{p}' = \frac{\mathbf{p}\gamma^{-1} + \mathbf{V}/c + (1 - \gamma^{-1})(\mathbf{V} \cdot \mathbf{p})\mathbf{V}/V^2}{1 + \mathbf{p} \cdot \mathbf{V}/c}.$$

From the definition of γ, it is easy to see that

$$(1 - \gamma^{-1})(1 + \gamma^{-1}) = V^2/c^2.$$

Hence

$$\mathbf{p}' = \frac{\gamma^{-1}\mathbf{p} + (\mathbf{V}/c) + (\mathbf{p} \cdot \mathbf{V}/c)(\mathbf{V}/c)/(1 + \gamma^{-1})}{1 + \mathbf{p} \cdot \mathbf{V}/c}$$

$$= \frac{\gamma^{-1}\mathbf{p}}{1 + \mathbf{p} \cdot \mathbf{V}/c} + (1 + \gamma^{-1})\mathbf{V}/c. \tag{6.26}$$

Again taking the modulus of the vector cross products of Equation (6.26) with $\mathbf{p}$, then

$$\sin \Delta\theta = \frac{(V/c)\sin\theta + 1/2(V/c)^2 \sin 2\theta/(1 + \gamma^{-1})}{1 + (V/c)\cos\theta}$$

$$= \frac{V}{c}\sin\theta - \frac{1}{4}\left(\frac{V}{c}\right)^2 \sin 2\theta + \frac{1}{4}\left(\frac{V}{c}\right)^3 \sin 2\theta \cos\theta + \cdots \tag{6.27}$$

$$\Delta\theta = \frac{V}{c}\sin\theta - \frac{1}{4}\left(\frac{V}{c}\right)^2 \sin 2\theta + \frac{1}{6}\left(\frac{V}{c}\right)^3 \sin\theta(1 + 2\sin^2\theta) + \cdots . \tag{6.28}$$

This shows that special-relativistic aberration and classical Newtonian aberration agree only up to the first order in V/c (mas precision). The difference in the second order is $\frac{1}{4}(\frac{V}{c})^2 \sin 2\theta$, that can reach 0.5 mas. For this reason, it is recommended that relativistic aberration (Equations (6.26)–(6.28)) be systematically used, particularly when high precision is required.

6.3.4 Accurate computation of the annual aberration

In accordance with recommendations of the International Astronomical Union in 1952, the annual aberration is calculated beginning in 1960 from the actual motion of the Earth, referred to an inertial frame of reference and to the center of mass of the Solar System. The resulting aberrational displacement, $\Delta\theta$, may be resolved into corrections to the directional coordinates by standard methods. Let us call $\dot{X}$, $\dot{Y}$, and $\dot{Z}$ the components of the Earth's velocity, $\mathbf{V}$, in equatorial rect-angular axes, and $x = \dot{X}/c$, $y = \dot{Y}/c$ and $z = \dot{Z}/c$ the components of the reduced velocity vector $\mathbf{v} = \mathbf{V}/c$. The components of the unit vector $\mathbf{u}$ of the direction of the star are

$$u = \cos\delta\cos\alpha,$$
$$v = \cos\delta\sin\alpha,$$
$$w = \sin\delta.$$

The aberration displacement is a rotation of $\Delta\theta$ about the normal $\mathbf{N}$ to this plane:

$$\mathbf{N} = \mathbf{u} \times \mathbf{v}$$

whose components are:

$$L = z\cos\delta\sin\alpha - y\sin\delta,$$
$$M = x\sin\delta - z\cos\delta\cos\alpha,$$
$$N = y\cos\delta\cos\alpha - x\cos\delta\sin\alpha.$$

According to Equation (3.22), the components of the displaced (apparent) direction $\mathbf{u}'$ of the star are given by

$$\mathbf{u}' = \mathbf{u}\cos\Delta\theta + (\mathbf{N} \times \mathbf{u})\sin\Delta\theta. \tag{6.29}$$

The components of $\mathbf{u}'$ are therefore:

$$x' = \cos\delta'\sin\alpha' = \cos\delta\sin\alpha\cos\Delta\theta + \frac{V}{c}(M\sin\delta - N\cos\delta\sin\alpha)\sin\Delta\theta,$$

$$y' = \cos\delta'\cos\alpha' = \cos\delta\sin\alpha\cos\Delta\theta + \frac{V}{c}(N\cos\delta\cos\alpha - L\sin\delta)\sin\Delta\theta,$$

$$z' = \sin\delta' = \sin\delta\cos\Delta\theta + \frac{V}{c}(L\cos\delta\sin\alpha - M\cos\delta\cos\alpha)\sin\Delta\theta. \tag{6.30}$$

From these expressions, one can derive the components u', v', w' of $\mathbf{u}'$ and consequently the displaced coordinates α' and δ' and the corrections to right ascension and declination, referred to the same coordinate system in the sense *corrected place minus uncorrected place*.

$$\Delta\alpha = \alpha' - \alpha,$$
$$\Delta\delta = \delta' - \delta.$$

6.3.5 Approximate formulae for annual aberration

These exact formulae are to be used when a full third order accuracy (to a µas), is needed. For uncertainties of the order of a mas, the following approximate expressions (Seidelmann, 1992) are sufficient:

$$\Delta\alpha \cos\delta = -\frac{\dot{X}}{c}\sin\alpha + \frac{\dot{Y}}{c}\cos\alpha + \frac{1}{c^2}(\dot{X}\sin\alpha - \dot{Y}\cos\alpha)$$
$$\times (\dot{X}\cos\alpha + \dot{Y}\sin\alpha)\sec\delta + \cdots,$$

$$\Delta\delta = -\frac{\dot{X}}{c}\cos\alpha\sin\delta - \frac{\dot{Y}}{c}\sin\alpha\sin\delta + \frac{\dot{Z}}{c}\cos\delta$$
$$-\frac{1}{2c^2}(\dot{X}\sin\alpha - \dot{Y}\cos\alpha)^2\tan\delta$$
$$+\frac{1}{c^2}(\dot{X}\cos\delta\cos\alpha + \dot{Y}\cos\delta\sin\alpha + \dot{Z}\sin\delta)$$
$$\times (\dot{X}\sin\delta\cos\alpha + \dot{Y}\sin\delta\sin\alpha - \dot{Z}\cos\delta) + \cdots. \quad (6.31)$$

Also to first order in V/c, we obtain the aberration in right ascension and declination due to the unperturbed elliptic component of the orbital motion of the Earth with respect to the Sun. It means, as earlier, the quantity to be added to the uncorrected place α,δ in order to obtain the place corrected for aberration α',δ' is:

$$\alpha' - \alpha = -\kappa\sec\delta((\sin\Lambda + e\sin\Pi)\sin\alpha - \cos\epsilon(\cos\Lambda + e\cos\Pi)\cos\alpha)$$
$$\delta' - \delta = -\kappa(\sin\Lambda + e\sin\Pi)\cos\alpha\sin\delta$$
$$-\kappa\cos\epsilon(\cos\Lambda + e\cos\Pi)(\tan\epsilon\cos\delta - \sin\alpha\sin\delta), \quad (6.32)$$

where Λ is the longitude of the Sun, e and Π are the eccentricity and longitude of perigee of the solar orbit, ϵ is the mean obliquity of the ecliptic, and κ is the constant of aberration. The latter is the ratio of the mean orbital speed of the Earth to the speed of light, where perturbations and the motion of the Sun relative to the barycenter are neglected. It is equal to

$$\kappa = na/c\sqrt{1-e^2},$$

where c is the speed of light, a is the mean distance of the Earth from the Sun, and n is the mean motion of the Earth's orbit.

Its value for the standard epoch J2000.0 is:

$$\kappa = 20\rlap{.}{''}49\,552.$$

It was used as the multiplicative factor to analytical expressions of the annual aberration, the parameters being the elements of the Earth's orbit around the Sun (see, for instance, Woolard and Clemence, 1966, p. 110). This is now obsolete. The aberration must be computed using the actual velocity of the observer with respect to the barycenter of the Solar System as provided by ephemerides. It is, however, convenient to handle separately the *annual aberration* due to the motion of the center of the Earth around the barycenter, B, of the Solar System as presented above, and the *diurnal aberration* due to the motion of the observer as a consequence of the Earth's rotation.

Note. The second term in each factor in Equation (6.32) depends explicitly on the eccentricity and represents the components of the displacement due to the departure of the elliptic orbital motion from a circle. The component of the aberration that depends on e is known as elliptic aberration.

6.3.6 Diurnal and orbital aberration

The rotation of the Earth on its axis carries the observer toward the East with a velocity $\omega \cos \phi'$, where ω is the equatorial angular velocity of the Earth (the standard value of ω is given in Appendix B; if a is the equatorial radius, then $a\omega = 0.464$ km/s is the equatorial rotational velocity of the surface of the Earth), and ρ and ϕ' are the geocentric distance and latitude of the observer, respectively. The corresponding constant of diurnal aberration is

$$\frac{a}{c}\omega\frac{\rho}{a}\cos\phi' = 0\rlap{.}''3200\frac{\rho}{a}\cos\phi'. \tag{6.33}$$

The aberrational displacement may be resolved into corrections in right ascension and declination:

$$\Delta\alpha = 0\rlap{.}''3200\frac{\rho}{a}\cos\phi'\cos H \sec\delta,$$
$$\Delta\delta = 0\rlap{.}''3200\frac{\rho}{a}\cos\phi'\sin H \sin\delta, \tag{6.34}$$

where H is the hour angle. The effect is small but is of importance for accurate observations. For a star at transit, $h = 0°$ or $180°$, so $\Delta\delta$ is zero, but

$$\Delta\alpha = \pm 0\rlap{.}''3200\frac{\rho}{a}\cos\phi',$$

where the plus and minus signs are used for upper and lower transits, respectively; this may be regarded as a correction to the time of transit.

Alternatively, the effect may be computed in rectangular coordinates using the following expression for the geocentric velocity vector $d\mathbf{r}/dt$ of the observer with respect to the celestial equatorial reference frame of date

$$\frac{d\mathbf{r}}{dt} = \begin{pmatrix} -\omega\rho\cos\phi'\sin(\theta+\lambda) \\ \omega\rho\cos\phi'\cos(\theta+\lambda) \\ 0 \end{pmatrix} \tag{6.35}$$

where θ is the Greenwich sidereal time, or stellar angle (see Section 10.4), and λ is the longitude of the observer (East longitudes are positive).

The geocentric velocity vector of the observer is added to the barycentric velocity of the Earth's center, to obtain the corresponding barycentric velocity vector of the observer. This procedure is valid insofar as the accuracy required permits the use of the Newtonian law of addition of velocities. At a microarcsecond accuracy, the addition of vectors should be performed in the barycentric reference frame and the Lorentz vector addition formulae (Section 5.2.3) should be applied.

In the case of observations that are performed from an artificial satellite, the diurnal aberration must be replaced by the orbital aberration. Since in general, there is no accurate analytical representation of the motion of the satellite and the velocity $d\mathbf{r}/dt$ is provided in interpolatable numerical form, the latter procedure is the only one possible.

If observations are performed from a space probe, the velocity vector $\mathbf{V}$ of the probe is given in barycentric coordinates. The procedure is then the same as the one described for annual aberration, $\mathbf{V}$ replacing the velocity of the center of mass of the Earth.

6.3.7 Planetary aberration

Planetary aberration is the displacement of the observed position of a celestial body produced by both the motion of the body and the motion of the Earth.

Let us denote the barycentric position of the Earth, E, as a function of the barycentric time t by $\mathbf{E_B}(t)$, and the barycentric position of a planet, P, by $\mathbf{P_B}(t)$. If the observation of P takes place at the time t_0, P is seen at the place $\mathbf{P_B}(t_0 - \tau)$, where τ is the light-time, that is the time taken by the light to travel from P to E. The light-time, assuming a Newtonian space-time, is then given by

$$\tau = |\mathbf{P_B}(t_0 - \tau) - \mathbf{E_B}(t_0)|/c. \tag{6.36}$$

This equation can be solved assuming that one has an ephemeris of the variation of $\mathbf{P_B}$ with time. This is the case, except for newly discovered objects, for which an orbit must be first determined. It is then sufficient to assume that the first and second derivatives are known:

$$\mathbf{P_B}(t_0 - \tau) = \mathbf{P_B}(t_0) - \frac{d(\mathbf{P_B})}{dt}(\tau) + \frac{1}{2}\frac{d^2(\mathbf{P_B})}{dt^2}(\tau)^2. \tag{6.37}$$

Equations (6.36) and (6.37) can be solved by successive approximations. The planetary aberration is the geocentric angle between two vectors:

$$(\mathbf{E_B}(t_0)\mathbf{P_B}(t_0) , \ \mathbf{E_B}(t_0)\mathbf{P_B}(t_0 - \tau)). \tag{6.38}$$

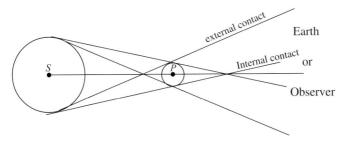

Fig. 6.4. Internal and external contacts for the transit of *P* in front of *S*.

This expression should be eventually corrected for light deflection and retardation described in the following sections, accounting for the actual relativistic space-time in the Solar System.

If ultimate accuracy is not required, one may assume, as in the Newtonian case, that the vector velocities of E and P can be added. The consequence would be that the aberration effects on the Earth and the planet can be computed independently and then added. This amounts to adding the correction for stellar aberration to the geometric position of the planet corrected for light-time. The error introduced in this simpler procedure can amount to a few milliarcseconds.

6.3.8 Differential planetary aberration

Aberration, because of its dependence on the relative motions and distances, sometimes has complex effects, where two or more bodies are involved, as, for instance, in eclipses, transits, and the phenomena of satellite systems; and on some past occasions, the determination of these effects has presented an intricate problem.

In a transit of a planet across the disk of the Sun, for example, the external contacts occur when the observer is on the conical surface that circumscribes the Sun and the planet and has its vertex between the Sun and the planet. The internal contacts occur when the observer is on the cone circumscribing the planet and the Sun (S) and has its vertex between the planet (P) and the Earth (Fig. 6.4). The observed contacts are at the instants when the apparent positions of a point on the limb of the planet and a point on the limb of the Sun are the same; i.e. the ray of light from the Sun, that reaches the geometric position of the observer at the instant T of contact, has grazed the planet on the way. Let τ_2 be the light-time between the limb of the Sun and the observer, and τ_1 the light-time between the limb of the Sun and the limb of the planet. This ray left the Sun at a previous time $T - \tau_2$ and reached the planet at time $(T - \tau_2) + \tau_1$. The circumscribing cones are formed by the grazing rays; hence, the points on the Earth and the planet that lie in the same straight line on one of the cones at the instant of a contact, are the geometric position of the observer at the time T, and the geometric position of the point on the planet at time $T - \tau_2 + \tau_1$. Therefore, in the formulas of the theory of transits, for any value of the time T, all quantities depending on the time must be derived from the values of the barycentric coordinates (r, l, b) of the planet at $T - \tau_2 + \tau_1$, and the barycentric coordinates (r', l', b') of the Earth at time T.

Similarly, in comparing observed positions of objects in the Solar System with one another, or with reference stars, in order to determine the coordinates of a body, great care is required in correcting the observations for aberration, according to the means of observation used, and the method of comparison.

In eclipsing binary systems, an apparent variation of the period may be produced by the variation in light-time with the changing distance from the observer, owing to an orbital motion of the eclipsing pair resulting from a distant third component (see Section 12.6).

6.4 Relativistic light deflection

As shown in Section 5.4.2, the light travels in space on a path that depends on the distribution of masses. To describe it, it is sufficient to treat bodies in the Solar System as point masses.

6.4.1 The light path

Let us compute the path of light under the influence of a body of mass M and in particular the Sun. It is described by the parametrized Equations (5.35). It was already shown that it follows a curve on a plane. So we shall study it in polar coordinates (r, λ) centered at the point mass. Let us repeat the equations:

$$\left(\frac{dr}{dp}\right)^2 = \left(1 - \frac{2m^2}{r^2}\right)c^2k^2 - \left(1 - \frac{2m}{r}\right)\frac{h^2}{r^2}, \tag{6.39}$$

$$\frac{d\lambda}{dp} = \frac{h}{r^2}, \tag{6.40}$$

where h and k are the generalized angular momentum and energy integrals, and $m = GM/c^2$. It is convenient to introduce a new variable $u = 1/r$ and eliminate dp between the two equations, so that

$$\left(\frac{du}{d\lambda}\right)^2 = \left(\frac{du}{dr}\frac{dr}{dp}\frac{dp}{d\lambda}\right)^2 = \frac{1}{r^4}\left(\frac{dr}{dp}\right)^2\frac{r^4}{h^2} = \frac{1}{h^2}\left(\frac{dr}{dp}\right)^2,$$

and, finally,

$$\left(\frac{du}{d\lambda}\right)^2 = (1 - 2m^2u^2)\frac{c^2k^2}{h^2} - (1 - 2mu)u^2. \tag{6.41}$$

It is important to evaluate the contribution of $2m^2u^2$ to the first term. In the case of the Sun,

$$GM = 1.327\,124 \times 10^{20}\,\text{m}^3/\text{s}^2,$$
$$c^2 = 8.987\,554 \times 10^{16}\,\text{m}^2/\text{s}^2,$$

hence, $m = 1476.62$ m.

No star can be observed at a distance from the center of the Sun that is less than the Sun's apparent radius, which is $R_\odot = 1/U_\odot = 6.96 \times 10^8$ m. So, $mU_\odot = 0.212 \times 10^{-5}$ and $m^2u^2 < 5 \times 10^{-12}$, and can be neglected with respect to 1. Hence, (6.41) reduces to

$$\left(\frac{du}{d\lambda}\right)^2 = K^2 - u^2 + 2mu^3,$$

where K^2 is a constant that is determined by the condition that, at the minimum distance R of the light ray to the central mass, one has $du/d\lambda = 0$. Finally, setting $V = 1/R$,

$$K^2 = 1/R^2 - 2m/R^3 = V^2 - 2mV^3.$$

Ultimately, the equation of the light path is:

$$\left(\frac{du}{d\lambda}\right)^2 = V^2 - 2mV^3 - u^2 + 2mu^3. \tag{6.42}$$

If $m = 0$, the path is given by the reduced Equation (6.42):

$$\left(\frac{du}{d\lambda}\right)^2 = V^2 - u^2. \tag{6.43}$$

Here, λ is made to vary from 0 to π when r varies from $-\infty$ to $+\infty$. It is easy to verify that the solution of (6.43) is a straight line defined, in polar cordinates centered at the Sun by

$$r = \frac{R}{\sin \lambda}.$$

Now, with regard to the full Equation (6.42), let us consider that the solution is a perturbed solution of (6.43), and write

$$Ru = \sin \lambda + mF(\lambda). \tag{6.44}$$

Let us call $F'(\lambda)$, the derivative with respect to λ. Differentiating (6.44), one gets

$$\frac{du}{d\lambda} = \frac{\cos \lambda}{R} + \frac{m}{R}F'(\lambda),$$

and

$$\left(\frac{du}{d\lambda}\right)^2 = \frac{\cos^2\lambda}{R^2} + \frac{2m}{R^2}F'(\lambda)\cos\lambda. \tag{6.45}$$

Let us now substitute the expression (6.44) into Equation (6.42) where we have replaced V and K^2 by functions of $1/R$. One obtains, neglecting terms in m^2,

$$\left(\frac{du}{d\lambda}\right)^2 = \left(\frac{1}{R^2} - \frac{2m}{R^3}\right) - \frac{\sin^2\lambda}{R^2} - \frac{2m}{R^2}\sin\lambda F(\lambda) + \frac{2m}{R^3}\sin^3\lambda. \tag{6.46}$$

Equating the two expressions (6.45) and (6.46), one gets, after some simplifications,

$$\frac{2m}{R^2}F'(\lambda)\cos\lambda = -\frac{2m}{R^3} - \frac{2m}{R^2}\sin\lambda F(\lambda) + \frac{2m}{R^3}\sin^3\lambda.$$

Finally, dividing by $2m/R^2\cos\lambda$, one obtains the differential equation giving $F(\lambda)$:

$$\frac{\cos\lambda F'(\lambda) + \sin\lambda F(\lambda)}{\cos^2\lambda} = \frac{d}{d\lambda}\left(\frac{F(\lambda)}{\cos\lambda}\right) = \frac{\sin^3\lambda - 1}{R\cos^2\lambda}. \tag{6.47}$$

This equation is easily integrable, and the result is

$$F(\lambda) = \frac{1}{R}(1 + \cos^2\lambda - \sin\lambda + C\cos\lambda),$$

where C is the constant of integration to be determined. Inserting this result into Equation (6.44), one gets

$$u = \frac{1}{r} = \frac{\sin\lambda}{R}(1 - m/R) + \frac{m}{R^2}(1 + \cos^2\lambda + C\cos\lambda). \tag{6.48}$$

The second part of the equation is small with respect to the first, so that one can put it as a factor, and develop the inverse of u to the first order of m:

$$r = \frac{R}{(1 - m/R)\sin\lambda} - m\left(\frac{1 + C\cos\lambda + \cos^2\lambda}{\sin^2\lambda}\right).$$

Let us first remark, that the reference straight path is slightly shifted towards the Sun (Fig. 6.5). In order to determine the value of the constant of integration C, let us note that we have assumed that the gravitational field of the Sun is isotropic. It results that the path is symmetrical with respect to the line of its closest approach to the Sun ($\lambda = 90°$) so that $F'(\pi/2)$ must be zero. This sets $C = 0$, and, in conclusion,

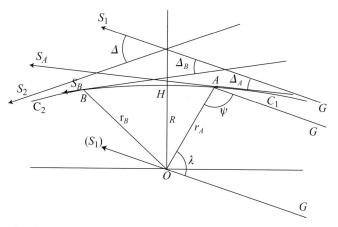

Fig. 6.5. The light path C_1C_2 and its closest approach H from the central mass O at a distance R. The apparent direction of a star G before bending is along the asymptote S_1 and after bending it is along the asymptote S_2, the deflection angle being Δ. If the observer is in A or B, the directions of rays emitted by the star are respectively S_A and S_B, and the deflections are Δ_A and Δ_B.

the distance from the Sun is

$$r = \frac{R}{(1 - m/R)\sin\lambda} - m\left(\frac{1 + \cos^2\lambda}{\sin^2\lambda}\right).$$

6.4.2 Deflection of light for a star

The curvature of the light path implies that the direction from which the light hits the observer is not the direction from which it was emitted. Let us first assume that the observer is at infinity. Then, the light path has two asymptotes (Fig. 6.5); one in the direction of the emitter, the other of the observer. They correspond to the values of λ for which $1/u \to \infty$. Following (4.48), this corresponds to

$$\sin\lambda(1 + m) + \frac{m}{R}(1 + \cos^2\lambda) = 0,$$

which, to the first order of m, gives $\lambda = 2m/R$ or $\pi - 2m/R$. The total deviation, returning to the definition of m is therefore

$$\Delta\lambda = \frac{4GM}{Rc^2}. \tag{6.49}$$

Taking the numerical values for $R_\odot$ of the Sun given above, the total deflection for a Sun grazing ray is $\Delta_\odot = 0.8486 \times 10^{-5}$ rad, or $1''.75$.

Let us now assume that the observer, A, is a point on the path defined by the angle λ between the undisturbed direction and the direction of A as seen from the Sun, O (Fig. 6.5) used in the previous derivations. We do not give the derivation

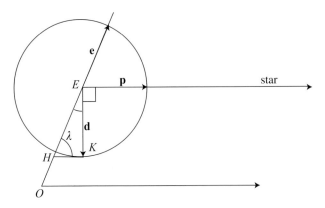

Fig. 6.6. Position of the unit vector **d** of the direction of the deviation in the Sun–Earth–star (*OES*) plane.

of the deflection, which is rather lengthy. It can be found in Will (1974) or Soffel (1989). The result is

$$\Delta_A(\lambda) = \frac{4GM}{Rc^2} \left(\frac{1 - \cos \lambda}{2} \right) = \frac{2GM}{rc^2} \left(\frac{1 - \cos \lambda}{\sin \lambda} \right)$$

$$= \frac{2GM}{rc^2} \left(\frac{\sin \lambda}{1 + \cos \lambda} \right) = \frac{2GM}{rc^2} \tan \lambda/2, \tag{6.50}$$

where r is the distance of the observer to the Sun ($R = r \sin \lambda$).

However, λ is not a directly observable quantity. It is preferable to use the angular elongation, ψ, of the star from the Sun. To an accuracy of the order of the square of the deflection, one has $\psi = \pi - \lambda$, and the deflection is

$$\Delta_A(\psi) = \frac{4GM}{Rc^2} \left(\frac{1 + \cos \psi}{2} \right) = \frac{2GM}{rc^2} \cot \psi/2. \tag{6.51}$$

This deviation must now be expressed as variations of the coordinates α and δ of the star. In the Sun–Earth–star (*OES*) plane already considered, the unit vector, **d** = **EK**, of the deviation is perpendicular to **ES** in the direction of the half-plane including the Sun (Fig. 6.6). Let **e** and **p** be the unit vectors of **OE** and **ES** respectively. Finally, let H be the point on OE, such that the projection of EH on **d** is K. Then, one can see from the figure that

$$\mathbf{d} = \frac{-\mathbf{e}}{\sin \lambda} + \mathbf{p} \cot \lambda.$$

The angular components of the deviation are obtained by multiplying this expression by the expression (6.50). One gets

$$\mathbf{D} = \frac{2GM}{rc^2} \frac{\sin\lambda}{1+\cos\lambda} \left(\frac{-\mathbf{e}}{\sin\lambda} + \mathbf{p}\cot\lambda \right) = \frac{2GM}{rc^2} \left(\frac{-\mathbf{e}+\mathbf{p}\cos\lambda}{1+\cos\lambda} \right).$$

(6.52)

Note that the minus sign corresponds to the fact that the deflection is always in the direction of the Sun, while **e** is in the opposite direction. In computing the actual deflection on the sky, the contribution of the **p** component, directed towards the star, is zero. One is left with

$$\mathbf{D} = \frac{2GM}{rc^2} \frac{-\mathbf{e}}{1+\cos\lambda} = \frac{2GM}{rc^2} \frac{-\mathbf{e}}{1-\cos\psi},$$

(6.53)

where ψ is the elongation of the star from the Sun $(\alpha_\odot, \delta_\odot)$, given by

$$\cos\psi = \sin\delta\sin\delta_\odot + \cos\delta\cos\delta_\odot\cos(\alpha-\alpha_\odot).$$

The modified coordinates of the star are then (Seidelmann, 1992),

$$\Delta\alpha = \frac{2GM}{rc^2} \frac{\cos\delta_\odot\sin(\alpha-\alpha_\odot)}{(1-\cos\psi)\cos\delta}$$

$$\Delta\delta = \frac{2GM}{rc^2} \frac{\sin\delta\cos\delta_\odot\cos(\alpha-\alpha_\odot)-\cos\delta\sin\delta_\odot}{1-\cos\psi}.$$

(6.54)

6.4.3 Deflection of light for planets

In the case of planets at a finite distance from the Earth, ψ and λ are no more equal, and Equations (6.50) and (6.51) are not equivalent. The formula representing the light deflection is (6.50), a function of λ. Figure 6.7 shows the geometry in the triangle Sun–Earth–planet (OEP), the angles ψ, λ and the third angle, ζ in P. Let K be the origin of the unit vector, $\mathbf{KO} = \mathbf{d}$, of the direction of the deviation and H the intersection of the parallel to OP drawn from K with OE. One has

$$\mathbf{d} = \mathbf{HO} + \mathbf{KH}.$$

(6.55)

In the triangle OHK, the angles can be expressed easily in λ, ψ, and ζ as shown in the figure. Let us apply the sine relations in this triangle:

$$\frac{KH}{\cos\psi} = \frac{OH}{\cos\zeta} = \frac{OK}{\sin\lambda},$$

and, since $OK = 1$, one has

$$KH = \frac{\cos\psi}{\sin\lambda} \quad \text{and} \quad OH = \frac{\cos\zeta}{\sin\lambda}.$$

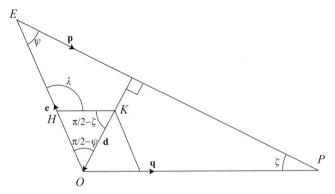

Fig. 6.7. Position of the unit vector **d** of the direction of the deviation in the Sun–Earth–planet (*OEP*) plane.

The lengths of the vectors in (6.55) being determined, the formula becomes

$$\mathbf{d} = -\frac{\mathbf{q}\cos\zeta + \mathbf{e}\cos\psi}{\sin\lambda}.$$

The angular components of the deviation are obtained by multiplying this expression by the magnitude of the deflection given by formula (6.50):

$$\mathbf{D} = -\frac{2GM}{rc^2}\left(\frac{\sin\lambda}{1+\cos\lambda}\right)\frac{\mathbf{q}\cos\zeta + \mathbf{e}\cos\psi}{\sin\lambda} = -\frac{2GM}{rc^2}\frac{\mathbf{q}\cos\zeta + \mathbf{e}\cos\psi}{1+\cos\lambda}.$$

$$(6.56)$$

The determination of the modified coordinates is to be performed in the same manner as in the case of stars.

6.4.4 Gravitational deflection by stars or quasars

The light emitted by some distant star and grazing another celestial body is deviated in the manner described above, irrespective of whether it is a star or another massive object. If the emitting object moves with a constant proper motion, the apparent path as well as its luminosity are modified by the massive object. For this reason, the deflector is also called a *gravitational lens*. Let us first consider a point-like star, *S*, exactly aligned with the deflector, *L*, of mass *M*, and the observer, *O*. The light is deviated by the same amount in all directions, and the image of *S*, as seen from *O*, is a circle, called an *Einstein circle or ring*. To describe the path of a ray in a plane passing by *S*, *L*, and *O*, it is legitimate to represent the nonlinear part by a point, *H*, since it is very short in comparison with distances between stars (upper part of Fig. 6.8). Let R_1 and R_2 be the distances *SL* and *OL* respectively,

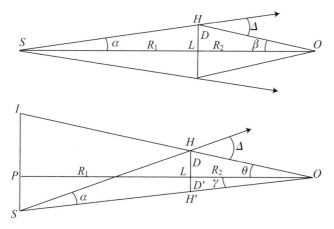

Fig. 6.8. The deflection of the light emitted by S by the lens L. The upper part of the figure represents the situation when the observer O is aligned with LS. In the lower part of the figure, S is shifted.

with $R = R_1 + R_2$. Let also $D = LH$ be the Einstein distance, that is the distance between L and the ray emitted by S and received by O.

The angle Δ between the rays is the total deviation given by (6.45). But it is also, in the triangle OHS, the sum of the small angles $\alpha = D/R_1$ and the $\beta = D/R_2$ under which LH is seen from S and O respectively:

$$\Delta = \frac{4GM}{c^2 D} = \frac{D}{R_1} + \frac{D}{R_2},$$

from which one obtains the distance D,

$$D = \left(\frac{4GM}{c^2} \frac{R_1 R_2}{R_1 + R_2} \right)^{1/2}. \tag{6.57}$$

The radius of the Einstein circle is, therefore,

$$\beta = \frac{D}{R_2} = \left(\frac{4GM}{c^2} \frac{R_1}{R R_2} \right)^{1/2}. \tag{6.58}$$

Suppose now that S is not aligned with OL as shown in the lower part of Fig. 6.8. The light is deflected in H. Let I be the point on the deviated ray at the distance of S from O. Let H' be the intersection of LH with OS. The deflection angle Δ is

$$\Delta = \frac{IS}{R_1}. \tag{6.59}$$

One sees on the figure that

$$\frac{IS}{HH'} = \frac{OP}{OL} = \frac{R}{R_2},$$

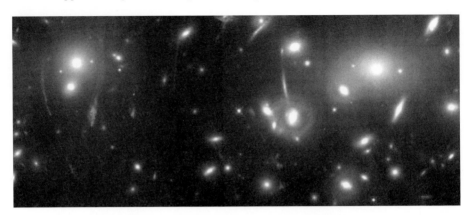

Fig. 6.9. Gravitational arcs in the cluster of galaxies Abell 2218 (Hubble Space Telescope). Material created with support from AURA/STScI under NASA contract NAS5-26555.

so that, introducing this in expression (6.59) for Δ, and setting $HH' = D + D'$, where $D' = LH'$, one gets

$$\frac{4GM}{Dc^2} = (D + D')\frac{R}{R_1 R_2}. \tag{6.60}$$

This is a second-order equation in D, which can be written as

$$D^2 + DD' = \frac{4GM}{c^2}\frac{R_1 R_2}{R}. \tag{6.61}$$

The right-hand term is precisely the square of the Einstein distance (6.57). Expressed as a function of the Einstein radius β, it becomes

$$D^2 + DD' = \beta^2 R_2^2. \tag{6.62}$$

There are two solutions, corresponding to each side of the deflector. As seen from the observer, calling γ the angle between the directions of L and S as seen from O, the angular directions ($\theta = D/R_2$) from L of the two images are given by

$$\theta = 0.5[\gamma \pm (\gamma^2 + 4\beta^2)^{1/2}]. \tag{6.63}$$

Note that one of the images is inside the Einstein circle and the other is outside it.

If S is not point-like, the shape is a more or less developed arc, a very common feature among clusters of galaxies (Fig. 6.9). If, in addition, S is an extended source with an irregular shape, the gravitational lensing produces complicated bright structures. Now, assume that S moves with an apparent proper motion, μ, with respect to L and parallel to an Lx axis. Taking the origin of time at the moment of the closest approach at the distance D along Ly (Fig. 6.8); the true positions of S are represented by

$$x = \mu t; \qquad y = D.$$

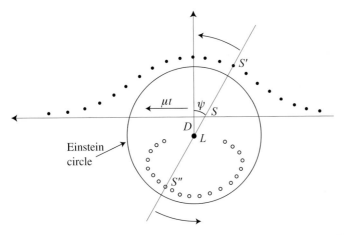

Fig. 6.10. Apparent paths S' and S'' of S when passing close behind the object L.

At some time t, the two apparent positions S' and S'' of S are given by (6.63) with the distance

$$\gamma = LS = \sqrt{D^2 + \mu^2 t^2} = D \sec \psi,$$

along the SL line. The shapes of the paths are drawn in Fig. 6.10. An analysis of the apparent motion of the outer component as a function of time shows that it is accelerated when it approaches the vicinity of L, and decelerated when it leaves it. A rough estimate of the time necessary to observe these paths is given by the *Einstein Time*, T_{E}, which is the time taken by S to travel the radius of the Einstein circle, in years

$$T_{\mathrm{E}} = \frac{390}{\mu} \frac{M R_1}{R R_2}, \tag{6.64}$$

where M is expressed in solar masses, the distances in kiloparsecs (kpc), and the proper motion μ in mas per year. With the same units, the radius of the Einstein circle β, in mas, is given by

$$\beta = 2.7 \frac{M R_1}{R R_2}. \tag{6.65}$$

Example. Taking $R_2 = 1\,\mathrm{kpc}$, $R = 3\,\mathrm{kpc}$, a one solar mass lens, and a relative proper motion of S with respect to L equal to 20 km/s, one has

$$\beta = 1.8\,\mathrm{mas}; \qquad T_{\mathrm{E}} = 1\,\mathrm{month}.$$

This shows that, if the detection of gravitational lensing in stars is still difficult, it will be within easy reach for microarcsecond astrometry.

Another property of the gravitational lenses is the magnitudes of the images. The lensing preserves the surface brightness, so that the magnification A is given, in a first approximation by the ratio of image to source areas:

$$A = \left| \frac{\gamma}{\theta} \frac{d\gamma}{d\theta} \right|.$$

Following Schneider *et al.* (1992), one has

$$A_{\pm} = \frac{u^2 + 2}{2u\sqrt{u^2 + 4}} \pm \frac{1}{2}, \tag{6.66}$$

where $\pm$ corresponds to one or the other solution of (6.63) and $u = \gamma/\beta$.

These characteristics of the disturbed proper motion of a star are used in attempts to recognize the presence of MACHOS (massive compact halo objects) that may exist in the outskirts of our Galaxy. They are also used very much in probing the distant Universe, particularly when there is a large magnification of the distant object. More detailed information can be found in Paszyński (1996) and Schneider *et al.* (1992).

6.5 Retardation of light

Deflection is not the only effect of the gravitational field on electro-magnetic radiation. Locally, measured with proper time, the speed of light is an invariant (c). But this is not the case if it is measured in coordinate time, which is a simple consequence of the gravitational time dilatation described by Equation (5.24). We shall consider here, as we did for light deflection, the gravitational effect of a central mass and neglect terms of order three or more in $1/c$. The light follows a geodesic defined by $ds^2 = 0$. Introducing again $m = GM/c^2$, we have

$$ds^2 = -d\tau^2 = 0 = -\left(1 - \frac{2m}{r}\right)c^2 dt^2 + \left(1 + \frac{2m}{r}\right)(dx^2 + dy^2 + dz^2). \tag{6.67}$$

We neglect the effect of the light deflection on the retardation (it is a second-order effect), and assume a rectilinear path along the line HE at the minimum distance $OH = R$ from the central object as shown in Fig. 6.6. The ds^2 becomes

$$0 = -\left(1 - \frac{2m}{r}\right)c^2 dt^2 + \left(1 + \frac{2m}{r}\right)dz^2,$$

from which one obtains the time of travel t between points of abscissæ z_1 and z_2,

$$t = \frac{1}{c} \int_{z_1}^{z_2} \sqrt{\frac{(1 + 2m/r)}{1 - 2m/r}} dz.$$

Developing this expression, neglecting terms in m^2, and replacing r by $\sqrt{R^2 + z^2}$, gives

$$t = \frac{1}{c}(z_2 - z_1) + \frac{1}{c}(z_2 - z_1) \int_{z_1}^{z_2} \frac{2m}{\sqrt{R^2 + z^2}} dz$$

The first term represents the travel time with a velocity c. What is left is the extra-time delay Δt. As in the case of the deflection, it is advisable to split the effect and compute it systematically from the point of closest approach corresponding to $z = 0$. Returning to the initial notations and performing the integration, one gets:

$$\Delta t(0, z) = \frac{2GM}{c^3} \ln \frac{|z| + \sqrt{z^2 + R^2}}{R}. \tag{6.68}$$

If now we assume that R/z is small and can be neglected, this reduces to

$$\Delta t(0, z) = \frac{2GM}{c^3} \ln \frac{2z}{R}. \tag{6.69}$$

The total retardation between two points z_1 and z_2 is, according to the respective positions of these points and H,

$$\Delta t(z_1, z_2) = |\Delta t(0, z_2) \pm \Delta t(0, z_1)|. \tag{6.70}$$

To give an example, the additional delay between the limb of the Sun and the Earth is 59.7 μs.

It is to be remarked, at this point, that formula (6.69) is coordinate dependent. It is important that the coordinate system be consistent with the one used to describe the light path. More details on the actual computation of delays can be found in McCarthy (2003), including the additional delays due to the bending of the light ray in space and at the arrival on the Earth. In particular, there is a delay correction ΔT to be introduced in the measurement of the return time in a laser ranging system. It amounts to

$$\Delta T = \frac{2GM}{c^3} \ln \left(\frac{R_1 + R_2 + \rho}{|R_1 + R_2 - \rho|} \right),$$

where R_1 is the distance from the Earth's center to the beginning of the light path, R_2 is the distance from the satellite center to the end of the light path, and ρ is the Euclidean distance between the beginning and the ending of the light path.

7

Extragalactic reference frame

In Section 1.4 we stated that astrometry must be developed within an extragalactic reference frame to microarcsecond accuracies. The objective of the present chapter is to provide the theoretical and practical background of this basic concept.

7.1 International Celestial Reference System (ICRS)

A reference system is the underlying theoretical concept for the construction of a reference frame. In an ideal kinematic reference system it is assumed that the Universe does not rotate. The theoretical background was presented in Section 5.4.1. The reference system requires the identification of a physical system and its characteristics, or parameters, which are determined from observations and that can be used to define the reference system. In 1991 the International Astronomical Union agreed, in principle, to change to a fundamental reference system based on distant extragalactic radio sources, in place of nearby bright optical stars (IAU, 1992; IAU, 1998; IAU, 2001). The distances of extragalactic radio sources are so large that motions of selected objects, and changes in their source structure, should not contribute to apparent temporal positional changes greater than a few microarcseconds. Thus, positions of these objects should be able to define a quasi-inertial reference frame that is purely kinematic. A Working Group was established to determine a catalog of sources to define this frame that is now called the ICRF. The XXIII-rd IAU General Assembly (IAU, 1998) adopted a resolution that:

(i) from 1 January 1998, the IAU celestial reference system shall be the International Celestial Reference System (ICRS) as defined by the International Earth Orientation Service (IERS) (Arias *et al.*, 1995),

(ii) the International Celestial Reference Frame (ICRF) shall be the fundamental reference frame as constructed by the IAU Working Group on Reference Frames (Ma and Feissel, 1997),

(iii) the *Hipparcos Catalogue*, with the exception of some problem stars (see Section 7.3) shall be the primary realization of the ICRF at optical wavelengths,

(iv) the ICRF shall be maintained by the IERS, and

(v) the ICRS shall have its origin at the barycenter of the Solar System, the directions of its axes are fixed with respect to the extragalactic sources and, for continuity with the past and with the FK5 system, aligned with the dynamical system of J2000.0 and the FK5 system of J2000.0, within the errors of the FK5 system.

Specifically, the origin of the right ascensions, the x-axis of the ICRS celestial system, was defined in the initial realization (Arias *et al.*, 1988) by adopting the mean right ascension of 23 radio sources from catalogs compiled by fixing the right ascension of the radio source 3C273B to the usual (Hazard *et al.*, 1971) FK5 value (12 h 29 m 6.6997 s at J2000.0) (Kaplan *et al.*, 1982). The definitions of the system and the reduction of observations are to be within the framework of general relativity (Chapters 5 and 6).

Actually, the IAU resolutions of the XXIV-th General Assembly introduced two *space fixed systems*: the Barycentric Celestial Reference System (BCRS) and the Geocentric Celestial Reference System (GCRS), which have been both defined in terms of the metric tensors given in Section 5.3.3 by Equation (5.22). The difference lies in the generalized potentials (Section 5.3.2) that are taken into account. The generalized Lorentz transformation between them, which contains the acceleration of the geocenter and the gravitational potential, is presented in Section 7.6. The ICRS is to be understood as defining the orientation of the axes of both these systems for each of the origins. These axes show no kinematical rotation between them, but they are related by a scale factor and have different time coordinates (TCB and TCG). For example, the metric of the BCRS with barycentric coordinates (t and $\mathbf{x}$) is

$$g_{00} = -1 + \frac{2w}{c^2} - \frac{2w^2}{c^4},$$

$$g_{ii} = 1 + \frac{2}{c^2}w,$$

$$g_{0i} = -\frac{4}{c^3}w_i, \tag{7.1}$$

with

$$w(t, \mathbf{x}) = G\sum d^3x' \frac{\sigma(t, \mathbf{x}')}{|\mathbf{x} - \mathbf{x}'|} + \frac{G}{2c^2}\frac{\partial^2}{\partial t^2}\sum d^3x'\sigma(t, \mathbf{x}')|\mathbf{x} - \mathbf{x}'|$$

$$w_i(t, \mathbf{x}) = G\sum d^3x' \frac{\sigma^i(t, \mathbf{x}')}{|\mathbf{x} - \mathbf{x}'|}, \tag{7.2}$$

where σ and σ^i are the gravitational mass and current densities, respectively, G is the gravitational constant, *t* is the TCB (see Section 5.5.2), and d^3x' is the element of volume.

A similar expression holds for the definition of the GCRS, the time being the TCG. For more details, see Soffel (2000).

7.2 International Celestial Reference Frame (ICRF)

The ICRF is the observed realization of the ICRS for practical applications. It can similarly be geocentric and barycentric. The coordinates of the radio sources are the same in both frames. Coordinates in this frame should be designated *right ascension* and *declination* without any further qualification, except for designating the epoch of observations for applying proper motions.

The ICRF is based on 212 defining sources distributed over the entire sky, but only 22% are in the Southern Hemisphere (Fig. 7.1). The positions, accurate to better than 1 mas, were based on a single solution of 1.6 million VLBI pairs of group delays and phase delay rates data obtained between August 1979 and July 1995. A Working Group selected a list of well-observed and well-behaved sources as the defining sources for this frame and a list of secondary sources (Ma *et al.*, 1998; Johnston *et al.*, 1995). The sources are from three lists:

(i) the most-compact and well-observed 212 defining sources, with a median uncertainty of individual positions of 0.4 mas;

(ii) compact sources (294) whose positions are likely to improve when more observations are accumulated; and

(iii) sources (102) less suited for astrometric purposes, but which provide ties for reference frames at other wavelengths.

These sources became the definition and representation of the International Celestial Reference Frame (ICRF). These sources have the advantages that they are very distant, so they have no apparent proper motion, and they have been observed over a long period of time by Very Long Baseline Interferometry (VLBI, see Section 2.4.2), so they have individual accuracies of better than 1 mas. A realization of the frame axes accuracy is estimated at 0.02 mas (Ma and Feissel, 1997). They have the disadvantage that they are optically faint and cannot be easily observed; the optical magnitudes of the defining sources range between 14th and 23rd magnitude, with the majority at 18–19 visual magnitude. So, other catalogs are necessary as realizations of the ICRS for the optical and other wavelengths.

Another disadvantage is that a number of the extragalactic radio sources display structures on spatial scales in the mas range. The spatial structure is usually more compact at higher frequencies. The sources are variable on time scales of weeks to years. An example is given in Fig. 7.2. In this particular instance, the source

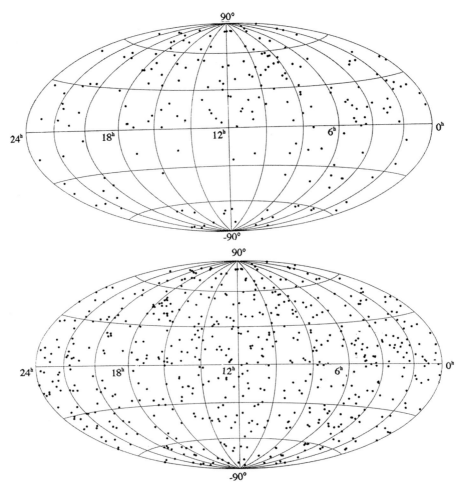

Fig. 7.1. Distribution (above) of the 212 defining sources of the ICRF and (below) distribution of all the 608 ICRF sources of the three lists (Ma *et al.*, 1998).

presents strong modifications of the structure that are not acceptable for astrometric observations. Actually, smaller changes are present in many sources, but generally they do not introduce errors of the order of the uncertainties of the observations. However, this will no longer be the case if observations reach 10 or 20 μas precision. So, owing to this variability of the structure of the sources, the positions must be measured over long periods of time, both to determine the ICRF and to maintain it in the future. Geodetic observational programs will be a source of observations in the future, as they have been in the past, but they concern only a limited number of sources. So, in addition, astrometric observations will be needed from VLBI antennas worldwide. Observations for source structure by the VLA or the VLBA will also be needed. Observations for source structure of sources south of $-20°$

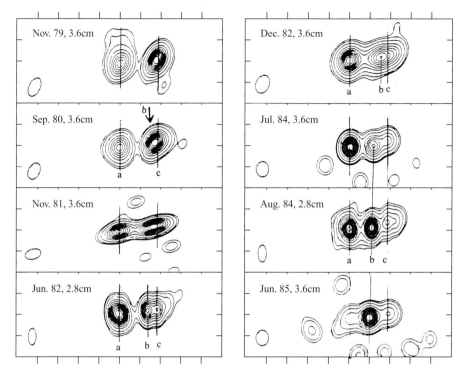

Fig. 7.2. Variation in the observed structure of the radio source 4C39.25. A new component (b) seems to have been ejected in 1980 (arrow) from the component (c) and to move towards the component (a) with a velocity of 0.16 ± 0.02 mas/year. The scale is 0.5 mas per interval of graduation (Shaffer and Marscher, 1987).

will be difficult due to the lack of radio telescopes. There are several causes of measurement uncertainties:

(i) the propagation of signals in the troposphere causes an error that varies as a function of elevation and azimuth (Section 6.1);

(ii) the systematic and random portions of the elevation error that can be reduced by mapping functions;

(iii) the gradients in azimuth cause a North–South asymmetry owing to the greater tropospheric thickness near the equator. This can cause errors of about 0.5 mas in declination.

There are also systematic errors introduced by the software used in the data reduction models for the large set of observational data. Comparisons of solutions in the past have shown discrepancies of about 5 mas. In addition, errors in station locations introduce errors in the source positions of 0.25 mas. The resulting formal uncertainties for ICRF sources are of the order of 0.4–1 mas. This is illustrated by Fig. 7.3, which shows the long history of consistent observations of two ICRF

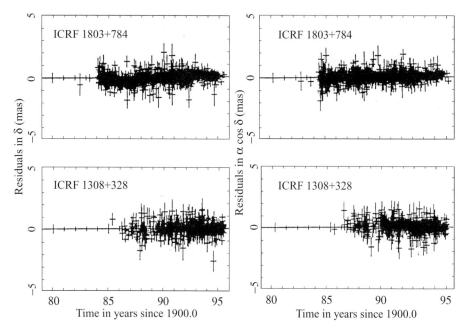

Fig. 7.3. Residuals of observations with respect to the IERS solution of one defining source (1308 + 326) and one candidate source (1803 + 784) (from Ma and Feissel, 1997).

sources. Future observations, better modeling, and improvements in source structure knowledge should reduce these uncertainties. However, source structure and structural motion may set a limit on the precision attainable for the ICRF.

7.3 Optical realizations of the ICRF

Previously the fundamental reference frame was based on optical observations of bright stars, primarily made with transit circle type instruments (Section 1.1), and compiled into fundamental catalogs, the most recent being the FK5 (Section 11.5.2). The accuracies of the observations were generally about 0″.3 and of the catalogs were about 0″.1. Owing to variations in observational histories and accuracies in the Northern and Southern Hemispheres, there were systematic errors in the catalogs. Since these catalogs had been used for a long period, the ICRF was aligned with the FK5 to within the accuracy of the FK5.

The Hipparcos mission (1989–1993) established a global astrometric catalog independent of atmospheric effects and Earth orientation parameters. This catalog is described in Section 11.6.1. It contains 118 218 stars with typical precisions, for stars of magnitude brighter than 9, of 0.8–1 mas in position, proper motion and parallax. This catalog, in addition to being much more accurate than previous catalogs at optical wavelengths, is also not subject to systematic errors based on magnitude and declination zones. However, in contrast to the extragalactic sources, the *Hipparcos Catalogue* stars display proper motions. Thus, the accuracies of the Hipparcos positions are time dependent and degrade with time from the mean observational epoch of 1991.25. In resolution B1.2, the XXIV-th General Assembly of the IAU established the *Hipparcos Catalogue* as the optical realization of the ICRS after removing stars that are double, that have been insufficiently observed, or do not have linear proper motions. These are the stars marked as C, G, V, X, and O. This is now the optical realization of the reference frame (Urban *et al.*, 2000b) and is now called International Hipparcos Reference Frame (IHRF).

Owing to the magnitude limit of the Hipparcos observations, the optical counterparts of the defining sources of the ICRF could not be observed directly. Link procedures were adopted to achieve two small rotations fixing the orientation of the *Hipparcos Catalogue* and removing any global rotation from the proper motions (Lindegren and Kovalevsky, 1995). This link was established using a variety of methods, including VLBI, MERLIN, and VLA observations of radio stars that are also optically bright Hipparcos stars. The most precise contribution was based on VLBI observations of 12 radio stars, which were tied directly to the ICRF. The five astrometric parameters of these stars were determined to better than mas precision (Lestrade *et al.*, 1995). The weighted least-squares solution using all of the methods was able to adjust the *Hipparcos Catalogue* to the ICRF with uncertainties of 0.6 mas in position and 0.25 mas/year in proper motion at epoch 1991.25 (Kovalevsky *et al.*, 1997).

The *Hipparcos Catalogue* is limited in the number of stars and the magnitude of stars. The Hipparcos satellite had a separate observational mode called Tycho, which observed many more stars in the same magnitude range as described in Section 11.6.2. Many further catalogs, based on the IHRF have extended the reference to many fainter stars. They are presented in Section 11.7.

7.4 Dynamical reference frame tie to the ICRF

Before the ICRS, reference frames were dynamical reference frames constructed assuming that the motions of bodies in the Solar System do not present accelerations that would be reflected in some rotation of the reference system. They were, therefore, based on observations of the planets (essentially inner planets) and the Moon. Lately a mixed stellar–dynamical reference frame was used for the FK5 catalog (Section 11.5.2), which was the official celestial reference frame until 1998. The dynamical reference frame and the FK5 reference frame were aligned as well as possible within their errors for the epoch J2000.0. This was done using the bright star reference frame, the adopted precession constant, and the dynamically determined equinox. In practice, since the ephemerides of the inner planets were based on radar observations and the ephemerides of the outer planets were based on optical observations, there was the possibility of discrepancies between the two systems. Actually, there is a difference between the dynamical equinox, determined from the ephemerides, and the origin, 0 hour right-ascension value, of the FK5 catalog. This difference is also time dependent and was a source of errors in the old system.

The ICRF is a fixed reference frame, and the numerical integration of the ephemerides in the ICRF is performed on a fixed reference frame. Thus, radar and laser ranging observations can be reduced on that frame. The advent of VLBI observations of spacecraft at planets and planetary satellites provides positions directly on the radio reference frame at accuracies of 1–3 mas. CCD observations of planets using the *Hipparcos Catalogue* give planet and satellite observations on the ICRF at accuracies of about 30 mas (Stone and Dahn, 1995). A joint analysis of VLBI and lunar laser ranging observations provides a tie between the JPL ephemerides and the IERS radio catalogs (Folkner *et al.*, 1994). Thus, the present JPL ephemerides of the planets and the Moon use the ICRF reference frame (Standish *et al.*, 1995).

7.5 Transformation between GCRS and BCRS

Since the origin of the geocentric system moves nonlinearly along a geodesic in the barycentric system, but has fixed directions with respect to extragalactic sources, there is a Coriolis-like effect from the relativistic theory of the transformation if referred to the BCRS. Thus, the theory of the transformations between reference frames (Section 5.3.4) must be applied (Brumberg and Groten, 2001). The principal effect is a secular effect, called *geodesic precession*, which is a rotation of 1.92 mas/year of the geocentric reference system with respect to the barycentric reference system.

The full theory can be found in Brumberg (1991a,b) who gives the following result in the general case of a weak gravitational field for the rotation of a system B in another system A:

$$\Omega_G = \frac{3}{2} \frac{\mathbf{v} \times U}{c^3},$$ (7.3)

where U is the gravitational potential at A and $\mathbf{v}$ is the velocity of the origin of A. If one assumes the framework of a Schwarzschild field, in which U is a point mass potential of the Sun (Section 5.3.3, Equation (5.26)), the formulation (Ohanian, 1980) reduces to

$$\Omega_G = \frac{3GM}{2r^3} \frac{\mathbf{r} \times \mathbf{v}}{c^2}. \tag{7.4}$$

Applied to the actual case of an elliptic orbit of the Earth with an eccentricity e, a semi-major axis a, and a mean motion n, the mean value of this expression is (Murray, 1983),

$$\Omega_G = \frac{3n^3 a^2}{2r^3 c^2 \sqrt{(1 - e^2)}}. \tag{7.5}$$

The numerical value that is derived for the geodesic precession is

$$\Omega_G = 0\overset{''}{.}019\,194 \text{ per year.}$$

It is to be noted, that in the equation (7.4), $\mathbf{r}$ and $\mathbf{v}$ present periodic variations. They produce periodic changes in the angle of precession and constitute the *geodesic nutation*. Fukushima (1991) gives the following expression for the precession angle:

$$\Delta\psi_G = -0\overset{''}{.}000\,153 \sin l' - 0\overset{''}{.}000\,002 \sin 2l',$$

where l' is the mean anomaly of the Sun.

The transformation also includes a relation between the geocentric and the barycentric coordinate times (TCB and TCG). It is given by equation (5.39) in Section 5.5.2.

Let us finally mention that there is an equivalent to the geodesic precession in comparing the barycentric and the galactic reference systems. In this case the rate of rotation is of the order of 8.5 nanoarcseconds (nas) per year and is, therefore, negligible (Brumberg, 1991a,b).

7.6 ICRF stars at other epochs

The simplistic formulae for transforming a celestial position (α, δ from epoch T_0 in the BCRF) to an arbitrary epoch T are:

$$\alpha = \alpha_0 + (T - T_0)\mu_\alpha$$
$$\delta = \delta_0 + (T - T_0)\mu_\delta, \tag{7.6}$$

where μ_α and μ_δ are the unreduced proper motion in α and δ as given in catalogs. This model describes a curved spiraling motion towards one of the poles, while real stars are expected to move along great circles. The difference with respect to a rigorous model is usually very small, but it may become significant near the celestial pole or over long time periods. Thus, while Equation (7.6) should not be used in general applications, it does provide a first-order approximation and an estimate of position uncertainties.

Note that in (7.6), the parallactic correction does not appear. We have assumed there that the transformation is performed in the BCRF. In the geocentric frame (GCRF), one must also correct for the parallax. This is what is done in the following sections. However, it must be kept in mind that, if the epoch transformation is done in the barycentric reference frame, one should omit from the equation the parallactic correction.

The rigorous transformation of parameters can be formulated based on a standard model of stellar motion. The standard model assumes a uniform space velocity for the object. Its path on the celestial sphere (as seen from the Solar System barycenter, without the displacement due to parallax) is an arc of a great arc circle. The angular velocity (proper motion) along this arc is variable, reaching a maximum when the object is closest to the Sun along its rectilinear path; the distance (parallax) and distance rate (radial velocity, V_r) change also with time. In a rigorous treatment, the variation of these six parameters must be considered. This is presented in Section 11.3, together with the corresponding uncertainty propagation.

7.7 Consequences of adoption of the ICRS and ICRF

What are the practical consequences of the adoption of the new ICRS and ICRF for the general practice of astronomy? For uses with accuracy requirements worse than 50 mas, the adoption of the ICRS has no significant effect. But for better precisions, there are a number of positive impacts on astronomy:

(i) the ICRS and ICRF are fixed for all epochs; however, the Earth's equator still moves kinematically for observations from the surface of the Earth. The new procedures, which include a new precession–nutation model, the geodesic precession and nutation, and a new definition of intermediate axes are given in Chapters 8 and 10;

(ii) there is no epoch attached to the ICRS, therefore, future improvements of the ICRF will not change the ICRS fiducial point or the direction of its axes;

(iii) changes of stellar positions between two epochs are determined from proper motions referred to the ICRF;

(iv) the determination of the directions of celestial objects in the ICRS must be consistent with the terrestrial coordinates in the ITRS (Chapter 9) by use of the IERS orientation parameters of universal time, polar motion, and precession–nutation expressions as given in Chapter 10.

7.8 Transformations to ecliptic and galactic coordinates

Let us establish the notations for these transformations. A direction is given by a unit vector, **a**, written as a column matrix (see Section 3.4). Three systems of coordinates will be considered:

(i) ICRS equatorial system $\mathbf{a}_0$: (x, y, z) with the coordinates α and δ, proper motion components μ_α and μ_δ, and the parallax ϖ,
(ii) ecliptic system $\mathbf{a}_k$: (x_k, y_k, z_k) with the corresponding paramaters λ, β μ_λ, μ_β, ϖ,
(iii) galactic system $\mathbf{a}_g$: (x_g, y_g, z_g) with the corresponding parameters l and b, μ_l, μ_b, ϖ.

Thus, the arbitrary direction **u** may be written in terms of equatorial, ecliptic, and galactic coordinates as

$$
\mathbf{u} = \begin{pmatrix} x \\ y \\ z \end{pmatrix} = \begin{pmatrix} \cos\delta \cos\alpha \\ \cos\delta \sin\alpha \\ \sin\delta \end{pmatrix} = \begin{pmatrix} x_k \\ y_k \\ z_k \end{pmatrix} = \begin{pmatrix} \cos\beta \cos\lambda \\ \cos\beta \sin\lambda \\ \sin\beta \end{pmatrix}
$$

$$
= \begin{pmatrix} x_g \\ y_g \\ z_g \end{pmatrix} = \begin{pmatrix} \cos b \cos l \\ \cos b \sin l \\ \sin b \end{pmatrix}. \tag{7.7}
$$

The transformation between equatorial and ecliptic systems is given by:

$$
\begin{pmatrix} x_k \\ y_k \\ z_k \end{pmatrix} = \mathcal{A}_k \times \begin{pmatrix} x \\ y \\ z \end{pmatrix},
$$

where, using ε as the obliquity of the ecliptic,

$$
\mathcal{A}_k = \begin{pmatrix} 1 & 0 & 0 \\ 0 & \cos\varepsilon & \sin\varepsilon \\ 0 & -\sin\varepsilon & \cos\varepsilon \end{pmatrix} \times \begin{pmatrix} x \\ y \\ z \end{pmatrix}
$$

$$
= \begin{pmatrix} 1 & 0 & 0 \\ 0 & +0.917\,482\,1315 & 0.397\,776\,9958 \\ 0 & -0.397\,776\,9958 & 0.917\,482\,1315 \end{pmatrix}, \tag{7.8}
$$

where ε is the obliquity of the fixed ecliptic with its origin identical as the origin of the ICRS. The values in (7.8) correspond to $\varepsilon = 23°\,26'\,21''.4059$, which is a value given by Fukushima (2001).

The transformation between equatorial and galactic systems is given by:

$$
\begin{pmatrix} x_g \\ y_g \\ z_g \end{pmatrix} = \mathcal{A}_g \times \begin{pmatrix} x \\ y \\ z \end{pmatrix},
$$

where the matrix $\mathbf{A}_g$ relates the galactic pole and the center of the ICRS system. This transformation is not standardized by an IAU definition. If we use the following values for the galactic pole in the ICRS system: $\alpha_g = 192°859\,48$ and $\delta_g = +27°128\,25$, and define the origin of the galactic longitude by the galactic longitude of the ascending node of the galactic plane on the equator of the ICRS, given by $l_\Omega = 32°931\,92$, then the transformation matrix may be computed to any desired accuracy. To 10 decimal places, it is:

$$\mathbf{A}_g = \begin{pmatrix} -0.054\,875\,5604 & -0.873\,437\,0902 & -0.483\,835\,0155 \\ +0.494\,109\,4279 & -0.444\,829\,6300 & +0.746\,982\,2445 \\ -0.867\,666\,1490 & -0.198\,076\,3734 & +0.455\,983\,7762 \end{pmatrix}. \tag{7.9}$$

The galactic coordinates given above are based on adopting coordinates of directions of the north galactic pole and galactic center based on physical features in the Galaxy with respect to the B1950 coordinate frame (Blaauw *et al.*, 1960). The transformation to ICRS values was given by Murray (1989). The resulting change in the directions of the principal axes, compared to the definition by Murray, is at most 18 mas.

The ecliptic longitude and latitude are computed from:

$$\begin{pmatrix} \cos\beta\cos\lambda \\ \cos\beta\sin\lambda \\ \sin\beta \end{pmatrix} = \mathcal{A}_k \times \begin{pmatrix} \cos\delta\cos\alpha \\ \cos\delta\sin\alpha \\ \sin\delta \end{pmatrix}, \tag{7.10}$$

and the galactic longitude and latitude from:

$$\begin{pmatrix} \cos b\cos l \\ \cos b\sin l \\ \sin b \end{pmatrix} = \mathcal{A}_g \times \begin{pmatrix} \cos\delta\cos\alpha \\ \cos\delta\sin\alpha \\ \sin\delta \end{pmatrix}. \tag{7.11}$$

Now, let $\mathbf{p}$ and $\mathbf{q}$ be unit vectors in the ICRF perpendicular to $\mathbf{u}$ in the direction of the increasing α and δ, and similar definitions for $\mathbf{p}_k$ and $\mathbf{q}_k$ in the ecliptic axes, and $\mathbf{p}_g$ and $\mathbf{q}_g$, in the galactic axes. These vectors are parts of the normal $\mathbf{p}, \mathbf{q}, \mathbf{r}$ triads, where $\mathbf{p} = \mathbf{q} \times \mathbf{u}$ and $\mathbf{q} = \mathbf{u} \times \mathbf{p}$. The variation in $\mathbf{u}$ due to small changes in the spherical coordinates can be written as:

$$\Delta\mathbf{u} = \Delta\alpha^*\mathbf{p} + \Delta\delta\mathbf{q} = \Delta\lambda^*\mathbf{p}_k + \Delta\beta\mathbf{q}_k = \Delta l^*\mathbf{p}_g + \Delta b\mathbf{q}_g, \tag{7.12}$$

where

$$\Delta\alpha^* = \Delta\alpha\cos\delta,$$
$$\Delta\lambda^* = \Delta\lambda\cos\beta,$$
$$\Delta l^* = \Delta l\cos b.$$

Using the orthonormality of the tangent vectors, one may solve these equations and get:

$$\Delta\lambda = (\Delta\alpha^* \mathbf{p} + \Delta\delta \mathbf{q}) \cdot \mathbf{p}_k,$$
$$\Delta\beta = (\Delta\alpha^* \mathbf{p} + \Delta\delta \mathbf{q}) \cdot \mathbf{q}_k,$$

and

$$\Delta l = (\Delta\alpha^* \mathbf{p} + \Delta\delta \mathbf{q}) \cdot \mathbf{p}_g,$$
$$\Delta b = (\Delta\alpha^* \mathbf{p} + \Delta\delta \mathbf{q}) \cdot \mathbf{q}_g.$$

The above equations contain the partial derivatives of the ecliptic and galactic coordinates with respect to the equatorial coordinates. They are obtained as scalar products of the relevant tangential vectors expressed in any convenient coordinate system. This can be written in matrix notation as follows:

$$\begin{pmatrix} \Delta\lambda^* \\ \Delta\beta \end{pmatrix} = \begin{pmatrix} \mathbf{p}'_k \cdot \mathbf{p} & \mathbf{p}'_k \cdot \mathbf{q} \\ \mathbf{q}'_k \cdot \mathbf{p} & \mathbf{q}'_k \cdot \mathbf{q} \end{pmatrix} \times \begin{pmatrix} \Delta\alpha^* \\ \Delta\delta \end{pmatrix}, \tag{7.13}$$

and

$$\begin{pmatrix} \Delta l^* \\ \Delta b \end{pmatrix} = \begin{pmatrix} \mathbf{p}'_g \cdot \mathbf{p} & \mathbf{p}'_g \cdot \mathbf{q} \\ \mathbf{q}'_g \cdot \mathbf{p} & \mathbf{q}'_g \cdot \mathbf{q} \end{pmatrix} \times \begin{pmatrix} \Delta\alpha^* \\ \Delta\delta \end{pmatrix}. \tag{7.14}$$

Considering the proper motion components as the time derivatives of coordinates, their transformations are given by:

$$\begin{pmatrix} \mu_{\lambda^*} \\ \mu_\beta \end{pmatrix} = \begin{pmatrix} \mathbf{p}'_k \cdot \mathbf{p} & \mathbf{p}'_k \cdot \mathbf{q} \\ \mathbf{q}'_k \cdot \mathbf{p} & \mathbf{q}'_k \cdot \mathbf{q} \end{pmatrix} \times \begin{pmatrix} \mu_{\alpha^*} \\ \mu_\delta \end{pmatrix}, \tag{7.15}$$

and

$$\begin{pmatrix} \mu_{l^*} \\ \mu_b \end{pmatrix} = \begin{pmatrix} \mathbf{p}'_g \cdot \mathbf{p} & \mathbf{p}'_g \cdot \mathbf{q} \\ \mathbf{q}'_g \cdot \mathbf{p} & \mathbf{q}'_g \cdot \mathbf{q} \end{pmatrix} \times \begin{pmatrix} \mu_{\alpha^*} \\ \mu_\delta \end{pmatrix}. \tag{7.16}$$

The complete transformation of the equatorial position and proper motion into the ecliptic system is given by Equations (7.10), (7.13) and (7.15), the parallax is independent of the coordinate system. Taking the five astrometric parameters in the order α, δ, ϖ, μ_α, and μ_δ, the Jacobian matrix for the transformation is:

$$J = \begin{pmatrix} c & s & 0 & 0 & 0 \\ -s & c & 0 & 0 & 0 \\ 0 & 0 & 1 & 0 & 0 \\ 0 & 0 & 0 & c & s \\ 0 & 0 & 0 & -s & c \end{pmatrix}, \tag{7.17}$$

where $c = \mathbf{p}'_k \cdot \mathbf{p} = \mathbf{q}'_k \cdot \mathbf{q}$, and $s = \mathbf{p}'_k \cdot \mathbf{q} = -\mathbf{q}'_k \cdot \mathbf{p}$, so that the full equation is given by

$$
\begin{pmatrix} \Delta\lambda^* \\ \Delta\beta^* \\ \varpi \\ \mu_{\lambda^*} \\ \mu_\beta \end{pmatrix} = \begin{pmatrix} c & s & 0 & 0 & 0 \\ -s & c & 0 & 0 & 0 \\ 0 & 0 & 1 & 0 & 0 \\ 0 & 0 & 0 & c & s \\ 0 & 0 & 0 & -s & c \end{pmatrix} \times \begin{pmatrix} \Delta\alpha^* \\ \Delta\delta^* \\ \varpi \\ \mu_{\alpha^*} \\ \mu_\delta \end{pmatrix}.
$$
(7.18)

The transformation of the equatorial parameters into the galactic system is given by Equations (7.11), (7.14) and (7.16); the Jacobian matrix is given by (7.17), with $c = \mathbf{p}'_g \cdot \mathbf{p} = \mathbf{q}'_g \cdot \mathbf{q}$, and $s = \mathbf{p}'_g \cdot \mathbf{q} = -\mathbf{q}'_g \cdot \mathbf{p}$.

8

Dynamical reference frame

Until 1998 the reference frame was defined in terms of the Solar System and the dynamics induced by the motions of Solar System bodies and their shapes. The two reference planes were the equator of the Earth and the ecliptic, the mean plane of the Earth's orbit. The intersection of these two planes, the equinox, was the fiducial point. Since both the equator and the ecliptic move owing to solar and lunar gravitational forces on the shape of the Earth and perturbations by the planets on the Earth's motion, the equinox moves with time. Therefore, each dynamical reference frame had to be defined for a specific epoch and observations and predictions were transformed to and from standard epochs to the chosen times of observations.

We have presented, in Chapter 7, the new adopted International Celestial Reference System (ICRS) based on the extragalactic radio sources. Now, the fundamental reference frame is fixed for all times. The IAU introduced in 2000 two *space fixed systems*. The Barycentric Celestial Reference System (BCRS) and the Geocentric Celestial Reference System (GCRS) have been defined in terms of metric tensors and the generalized Lorentz transformation between them, which contains the acceleration of the geocenter and the gravitational potential. The ICRS is to be understood as defining the orientation of the axes of both these systems for each of the origins. The International Celestial Reference Frame, (ICRF), determined from VLBI observations is the realization of the ICRS and can similarly be geocentric and barycentric.

The fiducial point of the ICRS does not have to revolve as the equinox did. It was adopted to be aligned with the equinox of the FK5 and the dynamical equinox at J2000.0 within their errors. For the reference frame of date on the moving equator, there is the option for a fiducial point (Section 8.4). One option is to retain the equinox, which is moving along the equator. The other option, which has been recommended by the IAU, is the use of the Celestial Ephemeris Origin (CEO), which does not have any motion along the instantaneous equator.

However, the motions of the Solar System and its bodies have not changed. Thus, as they are applicable, these motions must be taken into account. Specifically, the kinematics of the Earth must be included for Earth based observations. Similarly, the dynamics and kinematics of other observational platforms must be included as appropriate. Information concerning the celestial mechanics explaining the Solar System dynamics can be found in Brouwer and Clemence (1961), Brumberg (1991a,b), Hagihara (1970–76), Kovalevsky (1967) and Murray and Dermott (1999).

8.1 Our Solar System

Our Solar System, as it is currently known, consists of the Sun; four terrestrial planets, Mercury, Venus, Earth, and Mars; four large gaseous planets, Jupiter, Saturn, Uranus, and Neptune; a distant small planet, Pluto; planetary satellites, one around the Earth, two around Mars, one around Pluto, and many around the gaseous planets; many minor planets, primarily in orbits between the terrestrial and gaseous planets, but with some crossing the orbits of the terrestrial planets and others among the gaseous planets; some minor planets have satellites in orbit around them; Kuiper belt objects beyond the orbit of Neptune; and comets, short-period comets that have been observed at more than one apparition, long-period comets that have not been observed before, and Oort cloud comets, which are beyond the Kuiper belt. There is also a large amount of material consisting of smaller particles in the Solar System, some in rings about the gaseous planets, and some which enters the Earth's atmosphere daily, some observed as meteors, and much that is unobserved.

Each of the objects has gravitational effects on the motions of the other objects, proportional to their masses and inversely proportional to the squares of the distances between them. Some of these effects are negligible, depending upon the accuracy of the computations. Additionally, none of the objects is a homogeneous sphere, so the shapes of the bodies affect the motions of nearby objects and the kinematics of the bodies are affected by the gravitational forces due to other bodies on the irregular shapes. Some quantitative information concerning the orbits and the physical characteristics of the planets is given in the Appendix B.

8.2 Ephemerides

The motions of the Solar System objects can be represented as two-body motion with six Keplerian orbital elements, either by osculating or mean elements. More accurate ephemerides may be achieved by means of algebraic expressions that represent the perturbations due to the many additional bodies. These algebraic expressions are usually in terms of Fourier or Chebyshev series. The most accurate ephemerides are currently achieved by means of numerical integration of the equations of motion, which include all bodies that can have a significant effect on the motions. For the planets, the motions of all the planets are integrated simultaneously. The gravitational forces are computed considering the bodies to be point masses. Included in the equations of motion are the relativistic terms.

For each body i, the point mass acceleration is written in the framework of general relativity in the book by Brumberg (1991a). However, it is generally not necessary to use the full relativistic procedures. It is simpler to use the limited PPN formulation as presented in section 5.3.6. The acceleration, including gravitational perturbations by minor planets, is given by Will (1974) as follows:

$$
\begin{aligned}
\frac{d^2\mathbf{r}_i}{dt^2} = &\sum_{j\neq i} \frac{\mu_j(\mathbf{r}_j - \mathbf{r}_i)}{r_{ij}^3} \Bigg\{ 1 - 2(\beta + \gamma)/c^2 \sum_{k\neq i} \frac{\mu_k}{r_{ik}} - (2\beta - 1)/c^2 \sum_{k\neq j} \frac{\mu_k}{r_{jk}} \\
&+ \gamma(v_i/c)^2 + (1+\gamma)(v_j/c)^2 - 2(1+\gamma)(v_j/c)^2 - 2(1+\gamma)/c^2 \frac{d\mathbf{r}_i}{dt} \cdot \frac{d\mathbf{r}_j}{dt} \\
&- (3/2c^2)(\mathbf{r}_i - \mathbf{r}_j)^2(d\mathbf{r}_j/dt)^2 + 1/2c^2(\mathbf{r}_j - \mathbf{r}_i)\frac{d\mathbf{r}_j}{dt^2} \Bigg\} \\
&+ c^{-2} \sum_{j\neq i} \frac{\mu_j}{r_{ij}^3} \Bigg\{ (\mathbf{r}_i - \mathbf{r}_j) \left[(2+2\gamma)\frac{d\mathbf{r}_i}{dt} - (1+2\gamma)\frac{d\mathbf{r}_j}{dt} \right] \Bigg\} \left(\frac{d\mathbf{r}_i}{dt} - \frac{d\mathbf{r}_j}{dt} \right) \\
&+ (3+4\gamma)/2c^2 \sum_{j\neq i} \frac{\mu_j}{r_{ij}} \cdot \frac{d^2\mathbf{r}_j}{dt^2} + \sum_m \frac{\mu_m(\mathbf{r}_j - \mathbf{r}_i)}{r_{im}^3}.
\end{aligned}
\tag{8.1}
$$

Here, $\mathbf{r}_i$, $d\mathbf{r}_i/dt$ and $d^2\mathbf{r}_i/dt^2$ are the Solar System barycentric position, velocity, and acceleration vectors of body i, $\mu_j = Gm_j$, where G is the gravitational constant and m_j is the mass of body j; β is the PPN parameter measuring the nonlinearity in superposition of gravity; γ is the PPN parameter measuring space curvature produced by a unit rest mass (in general relativity, $\beta = \gamma = 1$); $v_j = |d\mathbf{r}_j/dt|$; c is the velocity of light; and index m is for minor planets included in the integration.

In order to achieve full accuracy in the ephemerides, the effects of the body figures (i.e. the lack of symmetry to be represented by point masses), the tidal effects between the Earth and the Moon, and lunar libration effects must be included.

The numerically computed ephemerides must be fit to observational data for all the bodies on a single consistent reference system. The computation and fitting is an iterative process providing improved values of initial conditions, orbital elements, and masses, as appropriate. As a result, each ephemeris defines its own motion of the Earth and dynamical reference frame. The current ephemeris is Jet Propulsion Laboratory's DE 405, which is described by Standish (1998) and is available from http://ssd.jpl.nasa.gov/eph_info.html.

This will be replaced by a new ephemeris as necessary to maintain the accuracy required for the Solar System ephemerides for the navigation of space-probes.

8.3 Reference planes

There are several natural reference planes, which can be used to define a reference frame. There are reference planes that must be used for observational purposes, there are reference frames that are theoretically desirable, but not observational, and unfortunately most of the planes are in motion, and their accuracies are restricted by observational uncertainties.

The Earth, being our habitat and base for observations, becomes a natural basis for most reference frames. The equatorial plane is the most obvious and necessary reference plane, since the motions of all observatories are with respect to that plane. It can be determined from observational data either by determining the celestial pole or from diurnal motions of celestial sources. Unfortunately, the equatorial plane moves owing to gravitational forces by the Sun, Moon and planets on the bulge of the Earth, and owing to the dynamics of the many components of the interior, surface and exterior of the Earth. These motions are called precession, nutation and polar motion.

Another natural plane of reference is the orbital plane of the Earth. This is subject to perturbations caused by the planets and is changing continuously. The mean orbital plane of the Earth is called the ecliptic. As long as a general theory of Newcomb was used to determine the ephemeris of the Earth, the ecliptic was well defined by the mean elements. With a numerical integration for the ephemeris of the Earth, the time period used to define the mean orbit is arbitrary.

The equinox is defined as the intersection of the ecliptic and the mean equator of the Earth. In determining this intersection from numerically integrated ephemerides, a distinction must be made. The equator can be taken as the instantaneous plane or as a uniformly moving plane. The historical determination was made based on the uniformly moving plane, and that is the definition that has continued to be used. The difference is approximately a tenth of an arcsecond. For calendars, seasons, and holiday definitions the equinox is specified as the time when the Sun passes through either of the intersection points, i.e. when the apparent latitude of the Sun is $0°$.

Another reference plane defined by the Solar System is the invariable, or Laplacian, plane. This is the plane normal to the axis of angular momentum and passing through the barycenter of the Solar System. This plane can be theoretically defined and specified, but it is not observable. It is somewhat dependent upon the completeness of the knowledge of the bodies of the system and their masses.

8.4 New definitions of a reference frame and fiducial point

To transform from the fixed International Celestial Reference Frame (ICRF) to that of date, the position of the moving equator and the location of an origin on that equator must be defined. The IAU has recommended a new system such that the origin of right ascensions on the true, or intermediate, celestial equator is no longer the equinox, but a point called the *Celestial Ephemeris Origin* (CEO), which does not depend on the position of the ecliptic. The true equatorial plane is determined from the precession–nutation model and its pole is called the *Celestial Intermediate Pole* (CIP). The resulting reference system is called the *Celestial Intermediate Frame* (Seidelmann and Kovalevsky, 2002). Note that these names are not consistent. It would have been more appropriate to call the origin on the equator *Celestial Intermediate Origin* (CIO), so that the word *intermediate* would be present in the names of all quantities defining this system. However, in this book, we shall stick to the official designation – CEO.

New precession–nutation models were adopted, designated by IAU 2000A and IAU 2000B, depending on the precision at the milliarcsecond levels. Based on the latest observational data, the IERS provides the most precise precession–nutation values, which are put in the same form as values derived from theoretical formulae. Precession and nutation models are applied together, thus eliminating the distinction between the *mean* and *true* equators. Owing to increased accuracies and frequency of observations of the Earth orientation, sub-daily polar motion and nutation periods can be determined. Thus, the separation of nutation and polar motion terms becomes arbitrary, and the choice was made to designate all terms with periods less than two days as polar motion.

In addition, the pole of the ICRF does not coincide with the mean pole of J2000.0, and there is a difference between the origin of the x-axis in the ITRS and the equinox of J2000.0. These shifts have been included in the precession–nutation models. The numerical developments of the precession–nutation models, including the pole offsets, are changing with the availability of additional observations.

The Celestial Ephemeris Origin (CEO) is defined such that its motion on a fixed celestial sphere has no component along the equator. The position of the CEO on the equator is defined by an integral that involves only the path followed by the precession–nutation pole (CIP) since the reference epoch (Capitaine *et al.*, 1986). For the details concerning the CIP and the CEO, see Sections 8.7 and 10.3.

The ICRS provides a fixed, epoch independent frame. For the moving reference frame of date, at least for an extended transitional period, there will be choices available. These may be described as the classical system, the improved classical system, and the new, or intermediate, system.

(i) The *classical system* is the system (since 1984) that uses the Celestial Ephemeris Pole (CEP) to define the equatorial plane, the Lieske *et al.* (1977) precession for the mean equator and equinox, the IAU 1980 Theory of Nutation (Seidelmann, 1982; Wahr, 1981) for the true positions, and the Aoki *et al.* (1982) expression to relate Greenwich Mean Sidereal Time (GMST) to UT1.

(ii) The *improved classical system* uses the CIP as determined by the IAU 2000 precession and nutation model to define the equatorial plane, the improved precession values to determine the mean equinox, the new nutation model for the true positions, and an improved expression to relate GMST to UT1.

(iii) For the *new, or intermediate, system*, the true equatorial plane and its pole (CIP) are determined from the IAU2000 precession–nutation model as above, the CEO is determined by a formula based uniquely on observations of the CIP, and the rotation of the Earth is determined from the CEO as the stellar angle, or Earth rotation angle, which is linear with time. The stellar angle does not depend on precession or nutation. The equinox is no longer used.

The differences between the systems are that the equinox moves along the equator at about 50 arcseconds per year, while the CEO moves only about 20 milliarcseconds per century. The improved classical system with precession and nutation separated and the new system with the precession–nutation model should lead to the same results within the applied accuracy. The classical system will have a precession error of about $0''.3$ per century, nutation errors of a little more than 1 mas, and uncorrected offsets of the pole. In the classical systems precession terms appear in the formula between GMST and UT1, while nutation terms appear in the formula for Greenwich Apparent Sidereal Time, as the equation of the equinoxes. So there is crosstalk and resulting errors in those expressions. See Section 10.3 for the details.

In the following sections, precession and nutation will be described as necessary for both the classical, old and improved values, and the new systems. Chapter 10 gives Earth orientation details, including the stellar angle, CIP and CEO.

8.5 Generalities on precession and nutation

The motion of the equatorial plane is caused by the torque of the Sun, Moon and planets on the dynamical figure of the Earth. It has been traditionally divided into two parts. The first is the smooth long-period motion of the mean pole of the equator about the pole of the ecliptic with a period of about 25 800 years, discovered by Hipparchus in the second century BC. This is a motion of the rotation axis of the Earth. This is the luni-solar precession. The other part, which comprises a large number of periodic oscillations of the rotation axis with small amplitudes, is the nutation. The most important nutation term, associated with the regression of the lunar orbit's line of nodes, has an amplitude of about $9''.0$ and a period of 18.6 years. It was discovered by Bradley in 1748. The methods of observation of precession and nutation are described in Section 10.2.

The motion of the ecliptic is the result of the gravitational effects of the planets on the Earth's orbit and this makes a contribution to precession known as planetary precession. If the equator were fixed, this motion would cause a precession of the equinox of about 12″ per century and a decrease in the obliquity of the ecliptic of about 47″ per century. The combination of luni-solar and planetary precession is called general precession. The causes of precession and nutation are the same, only the periods are different. The new (IAU 2000) precession–nutation model removes the distinction. The new precession–nutation model has been adopted by the IAU and IAU 2000A is accurate to 0.2 mas, while IAU 2000B is accurate to 1 mas. This model replaces the old precession model based on a constant of precession and the IAU 1980 Theory of Nutation. The precession–nutation model is subject to corrections based on observational data. The real, non-rigid Earth dynamics are not currently fully understood. Thus, for the best accuracies it is recommended that numerical values determined and distributed by the IERS be used for precession–nutation.

The new precession–nutation model is documented in three papers: the new nutation series for non-rigid Earth and insights into the Earth's interior is in Mathews *et al.* (2002); the Very Long Baseline Interferometry results are in Herring *et al.* (2002); and the effects of electro-magnetic coupling are in Buffett *et al.* (2002).

Until the year 2003, the officially adopted constant of general precession in longitude per Julian century at epoch J2000.0 was $p = 5029\rlap{.}{''}0906$. It was known to be incorrect by about $-0\rlap{.}{''}30$ per century. The IAU 1980 Theory of Nutation also needed improvement. For improved accuracy with the classical system and the equinox, the precession constant should be corrected to the value $p = 5028\rlap{.}{''}797$ per Julian century, and a nutation model without the precession terms should be used.

There are some additional changes of which astronomers should be aware. The pole of the ICRS does not coincide with the mean pole of J2000.0 and there is a difference between the origin of the x-axis in the ICRS and the equinox of J2000.0. These shifts have been included in the precession–nutation models. The geodesic precession and nutation have been included in the precession–nutation model, although that is a very different effect that arises from the transformation between the geocentric and the barycentric coordinates, each with fixed directions (see Section 7.5). Free core nutation is not included in the precession–nutation model. Since new observational capabilities provide the possibility of determining motions of the pole with periods less than a day, the arbitrary choice has been made to include all such motions with periods less than two days in polar motion values.

8.6 Precession

The motion of the ecliptic is described using Fig. 8.1 with angles π_A and Π_A, where π_A is the angle between the mean ecliptic at a fixed epoch t_0 and the mean ecliptic of date t, and Π_A is the angular distance from the equinox γ_0 at the fixed epoch to the ascending node where the ecliptic of date meets the fixed ecliptic of epoch. The

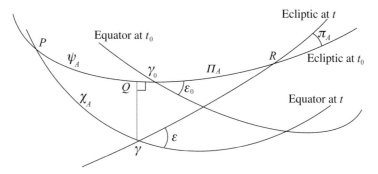

Fig. 8.1. The ecliptic and equator at epoch and date.

expressions for π_A and Π_A are of the form

$$\sin \pi_A \sin \Pi_A = st + s_1 t^2 + s_2 t^3$$
$$\sin \pi_A \cos \Pi_A = ct + c_1 t^2 + c_2 t^3, \tag{8.2}$$

where for continuity $\pi_A > 0$ to $t > 0$ and $\pi_A < 0$ to $t < 0$. The precessional effect due to the motion of the ecliptic is the arc $P\gamma$ in Fig. 8.1 and is the accumulated planetary precession χ_A. Its rate of change at epoch t_0 is

$$\chi = s \operatorname{cosec} \epsilon_0, \tag{8.3}$$

where ϵ_0 is the obliquity of the ecliptic at the epoch t_0.

The precessional constant, P_0, occurs in the dynamical equations of motion for the equator because of the torques produced by the Sun and Moon. The value of the constant is determined from observations, since it can not be accurately calculated from a theoretical dependency on geophysical and gravitational parameters. Luni-solar precession at epoch t_0

$$\psi = P_0 \cos \epsilon_0 - P_g \tag{8.4}$$

is the rate of change in longitude ψ_A along the ecliptic due to the motion of the equator at epoch t_0, where P_g is the geodesic precession, which is a relativistic effect amounting to $1''.92$ per Julian century described in Section 7.5. The combined effect of planetary precession and luni-solar precession is called general precession in longitude. The accumulated general precession p_A in Fig. 8.1 is

$$p_A = \gamma R - \gamma_0 R. \tag{8.5}$$

The rate of precession at epoch t_0 is

$$p_A = \psi - \chi \cos \epsilon_0 = P_0 \cos \epsilon_0 - P_g - s \cos \epsilon_0. \tag{8.6}$$

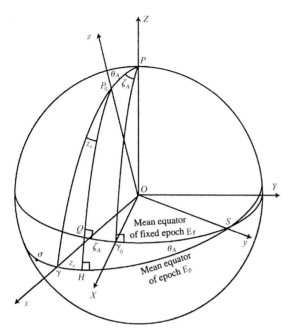

Fig. 8.2. The precession angles ζ_A, z_A, and θ_A. The ecliptic of fixed epoch E_f meets the mean equator of the same epoch in γ_0. Similarly, γ is the intersection of the mean equator and the ecliptic of epoch E_p.

and p may be resolved into general precession in right ascension, m, and in declination n, where

$$m = \psi \cos \epsilon_0 - \chi = (P_0 \cos \epsilon_0 - P_g) \cos \epsilon_0 - s \operatorname{cosec} \epsilon_0$$

$$n = \psi \sin \epsilon_0 = (P_0 \cos \epsilon_0 - P_g) \sin \epsilon_0. \qquad (8.7)$$

So,
$$p = m \cos \epsilon_0 + n \sin \epsilon_0.$$

Precession may be formulated in several ways. The vector directions of the polar axis of the Earth and polar axis of the ecliptic may be expressed in terms of polynomials involving time arguments (Murray, 1983), which are then used to calculate the effects of precession. The following method developed from Newcomb, and adopted by the IAU (Lieske *et al.*, 1977), is useful for practical applications.

The accumulated precession angles ζ_A, z_A and θ_A, which are used to calculate the effect of precession on equatorial coordinates, are referred to a base epoch E_0 and have time arguments that describe precession from an arbitrary fixed epoch E_f to an epoch of date E_p. The angles are shown on the surface of a sphere in Fig. 8.2. In the figure the pole of the equator at E_f is point P_0. At the epoch E_p, P_0 has moved to P. Imagine a right-handed set of three-dimensional, Cartesian coordinate axes with origin at the center of the sphere, X-axis pointing to γ_0, where the equator and ecliptic meet (the equinox) at the fixed epoch E_f. The Y-axis is 90° away in an easterly direction along the equator, and the Z-axis points toward the pole P_0. Initially, precession will move P_0 toward γ_0, but the movement of the ecliptic will cause P_0 to move in a slightly different direction.

A rotation of $-\zeta_A$ about the Z-axis makes $P_0\gamma_0$ pass through P. This great circle meets the mean equator of epoch at right angles at point Q and the mean equator of date at right angles at R. The x-axis is now in the direction Q; the y-axis points toward the node S, where the two equators cross.

A rotation of $+\theta_A$ equal to the angular separation of P from P_0 about the new y-axis brings the mean equator of epoch to the mean equator of date. The z-axis points to the pole of date P, the y-axis to the node S, and the x-axis now points toward R in the plane of the equator of date.

Finally, a rotation of $-z_A$ equal to the angle $\gamma P R$ about the z-axis brings R to γ, so that the x-axis points toward γ, the equinox of date, and still lies in the plane of the mean equator of date.

The precession matrix $\mathcal{P}$, made up of these rotations, precesses equatorial rectangular coordinates from an arbitrary fixed equinox and equator of epoch E_f to one of date E_p and is given by

$$\mathcal{P}(E_f, E_p) = \mathcal{R}_3(-z_A)\mathcal{R}_2(-\theta_A)\mathcal{R}_3(-\zeta_A). \tag{8.8}$$

Rewriting Equation (8.8) in terms of spherical coordinates gives

$$P = \begin{pmatrix} \cos z_A \cos \theta_A \cos \zeta_A & -\cos z_A \cos \theta_A \sin \zeta_A & -\cos z_A \sin \theta_A \\ -\sin z_A \sin \zeta_A & -\sin z_A \cos \zeta_A & \\ \sin z_A \cos \theta_A \cos \zeta_A & -\sin z_A \cos \theta_A \sin \zeta_A & -\sin z_A \sin \theta_A \\ \cos z_A \sin \zeta_A & \cos z_A \cos \zeta_A & \\ \sin \theta_A \cos \zeta_A & \sin \theta_A \sin \zeta_A & \cos \theta_A \end{pmatrix}. \tag{8.9}$$

Having calculated the precession angles for the matrix P, one can calculate the inverse matrix $\mathcal{P}^{-1}$ in various ways; for example,

$$\begin{aligned}
\mathcal{P}^{-1} &= \mathcal{R}_3^{-1}(-\zeta_A) \times \mathcal{R}_2^{-1}(+\theta_A) \times \mathcal{R}_3^{-1}(-z_A), \\
\mathcal{P}^{-1} &= \mathcal{R}_3^{\top}(-\zeta_A) \times \mathcal{R}_2^{\top}(+\theta_A) \times \mathcal{R}_3^{\top}(-z_A), \\
\mathcal{P}^{-1} &= \mathcal{R}_3(\zeta_A) \times \mathcal{R}_2(-\theta_A) \times \mathcal{R}_3(z_A), \\
\mathcal{P}^{-1} &= \mathcal{P}(E_p, E_f),
\end{aligned} \tag{8.10}$$

where use has been made of the property that the inverse of a rotation matrix is its transpose (i.e. it is orthogonal).

8.6.1 *Precession angles and rates adopted by IAU (1976)*

The basis for precession is taken directly from the discussion of Lieske *et al.* (1977) and modified by the small amount discussed in Lieske (1979). The following values are adopted constants at epoch J2000.0:

(i) general precession in longitude: $p = 5029\rlap{.}''0966$ per Julian century,
(ii) geodesic precession: $P_{\mathrm{g}} = 1\rlap{.}''92$ per Julian century,
(iii) obliquity of the equinox: $\epsilon_0 = 23°26'21\rlap{.}''448$.

In addition numerical expressions are provided for a number of quantities used in computing precession in various cases. Let us give the more important:

$$
\begin{aligned}
\sin \pi_A \sin \Pi_A &= (4\rlap{.}''1976 - 0\rlap{.}''75\,250T + 0\rlap{.}''000\,431T^2)t \\
&\quad + (0\rlap{.}''194\,47 + 0\rlap{.}''000\,697T)t^2 - 0\rlap{.}''000\,179t^3, \\
\sin \pi_A \cos \Pi_A &= (-46\rlap{.}''8150 - 0\rlap{.}''001\,17T + 0\rlap{.}''005\,439T^2)t \\
&\quad + (0\rlap{.}''05059 - 0\rlap{.}''003\,712T)t^2 + 0\rlap{.}''000\,344t^3, \\
\pi_A &= (47\rlap{.}''0029 - 0\rlap{.}''066\,03T + 0\rlap{.}''005\,98T^2)t \\
&\quad + (-0\rlap{.}''03\,302 + 0\rlap{.}''000\,598T)t^2 + 0\rlap{.}''000\,060t^3, \\
\Pi_A &= 174°52'34\rlap{.}''982 + 3289\rlap{.}''4789T + 0\rlap{.}''60\,622T^2, \\
\zeta_A &= (2306\rlap{.}''2181 + 1\rlap{.}''396\,56T - 0\rlap{.}''000\,139T^2)t \\
&\quad + (0\rlap{.}''301\,88 - 0\rlap{.}''000344T)t^2 - 0\rlap{.}''017998t^3, \\
z_A &= (2306\rlap{.}''2182 + 1\rlap{.}''39\,656T - 0\rlap{.}''000\,139T^2)t \\
&\quad + (1\rlap{.}''094\,68 + 0\rlap{.}''000\,066T)t^2 + 0\rlap{.}''018\,203t^3, \\
\theta_A &= (2004\rlap{.}''3109 - 0\rlap{.}''853\,30T - 0\rlap{.}''000\,217T^2)t \\
&\quad + (-0\rlap{.}''426\,65 - 0\rlap{.}''000\,217T)t^2 - 0\rlap{.}''041\,833t^3, \\
\epsilon_A &= \epsilon_0 - 46\rlap{.}''8150T - 0\rlap{.}''000\,59T^2 + 0\rlap{.}''001\,813T^3 \\
&\quad + (-46\rlap{.}''8150 - 0\rlap{.}''001\,17T + 0\rlap{.}''005\,439T^2)t \\
&\quad + (-0\rlap{.}''000\,59 + 0\rlap{.}''005\,439T)t^2 + 0\rlap{.}''001\,813t^3.
\end{aligned}
$$

$$(8.11)$$

These equations give expressions for the accumulated precession angles as functions of time, where the base epoch of the equations is $E_0 = J2000.0$ or Julian day $JD(E_0) = 2\,451\,545.0$, and the time arguments are in units of Julian centuries, i.e.

$$
\begin{aligned}
T &= (JD(E_{\mathrm{f}}) - JD(E_0))/36\,525 \\
t &= (JD(E_{\mathrm{D}}) - JD(E_{\mathrm{f}}))/36\,525,
\end{aligned}
$$

$$(8.12)$$

where E_{f} is a fixed epoch and E_{D} is the epoch of date expressed in days. The expressions at epoch E_f for the rates per Julian century of general precession, in longitude (p), right ascension (m) and declination (n), and the rate of rotation of the ecliptic, are

$$p = \frac{d}{dt}(p_A)|_{t=0} = 5029''.0966 + 2''.222\,26T - 0''.000\,042T^2,$$

$$m = \frac{d}{dt}(\zeta_A + z_A)|_{t=0} = 4612''.4362 + 2''.793\,12T - 0''.000\,278T^2,$$

$$n = \frac{d}{dt}(\theta_A)|_{t=0} = 2004''.3109 - 0''.853\,30T - 0''.0002\,17T^2,$$

$$\pi = \frac{d}{dt}(\pi_A)|_{t=0} = 47''.0029 - 0''.066\,03T + 0''.000\,598T^2. \tag{8.13}$$

8.6.2 Improved values of precession angles and rates

Improved values of precession constants based on recent observational data and the methods of Lieske are derived by Hilton (2002) from the best estimate of the obliquity of the ecliptic (Fukushima, 2001), the IAU 2000A value of the precession and value of the change in the obliquity of the ecliptic with respect to the mean ecliptic of date, and the value of the motion of the mean ecliptic (Simon *et al.*, 1994). The values are given for J2000.0.

$$p = 5028''.797 \text{ per Julian century}$$
$$\epsilon_0 = 23° 26' 21''.4059 \text{ with respect to the ICRF,}$$

and

$$\zeta_A = (2306''.0806 + 1''.397\,93T - 0''.000\,137T^2)t$$
$$+ (0''.302\,34 - 0''.000\,344T)t^2 + 0''.018\,005t^3,$$

$$z_A = (2306''.0806 + 1''.397\,93T - 0''.000\,137T^2)t$$
$$+ (1''.095\,59 - 0''.000\,069T)t^2 + 0''.018\,211t^3,$$

$$\theta_A = (2004''.1916 - 0''.852\,20T - 0''.000\,217T^2)t$$
$$+ (-0''.426\,65 - 0''.000\,217T)t^2 + 0''.041\,817t^3,$$

$$\varepsilon_A = 23°26'21''.4412 + (-46''.8402T - 0''.001\,16T^2)t$$
$$+ (0''.000\,59 + 0''.005\,439T)t^2 + 0''.001\,813t^3. \tag{8.14}$$

8.6.3 Rigorous reduction for precession

The most convenient method of rigorously precessing mean equatorial rectangular coordinates is to use the precession matrix $\mathcal{P}$ as follows:

$$\mathbf{r} = \mathcal{P}\mathbf{r}_0 \tag{8.15}$$

and

$$\mathbf{r}_0 = \mathcal{P}^{-1}\mathbf{r},$$

which transforms the position vector $\mathbf{r}_0$ referred to the fixed epoch $E_f = t_0$ to the position vector $\mathbf{r}$ referred to the epoch of date $E_D = t$, and vice versa. The matrix $\mathcal{P}$ is evaluated using the precession angles (ζ_A, z_A, and θ_A) in Equation (8.8) or (8.9) and the appropriate time arguments.

Equations (8.14) can be rewritten in terms of right ascension and declination (α, δ) as

$$\sin(\alpha - z_A)\cos\delta = \sin(\alpha_0 + \zeta_A)\cos\delta_0,$$
$$\cos(\alpha - z_A)\cos\delta = \cos(\alpha_0 + \zeta_A)\cos\theta_A\cos\delta_0 - \sin\theta_A\sin\delta_0$$
$$\sin\delta = \cos(\alpha_0 + \zeta_A)\sin\theta_A\cos\delta_0 + \cos\theta_A\sin\delta_0, \qquad (8.16)$$

and the inverse is

$$\sin(\alpha_0 + \zeta_A)\cos\delta_0 = \sin(\alpha - z_A)\cos\delta,$$
$$\cos(\alpha_0 + \zeta_A)\cos\delta_0 = \cos(\alpha - z_A)\cos\theta_A\cos\delta + \sin\theta_A\sin\delta$$
$$\sin\delta_0 = -\cos(\alpha - z_A)\sin\theta_A\cos\delta + \cos\theta_A\sin\delta. \qquad (8.17)$$

8.7 Nutation

We present two theories of nutation. The IAU 1980 theory is associated with the classical reference system (Section 8.4). The IAU 2000 theory is to be used in the improved classical and the new systems.

8.7.1 IAU 1980 Theory of Nutation

The IAU 1980 Theory of Nutation was the officially adopted theory from 1984 until 2003. It is adequate to accuracies of 1 mas. The IAU 1980 Theory of Nutation (Seidelmann, 1982; Wahr, 1981) was based on a modification of a rigid Earth theory published by Kinoshita (1977) and on the geophysical model 1066A of Gilbert and Dziewonski (1975). It included the effects of a solid inner core and a liquid outer core and a distribution of elastic parameters inferred from a large set of seismological data values. The corresponding pole was called the Celestial Ephemeris Pole (CEP) and is the axis of figure of the mean surface of the Earth. The nutation matrix $\mathcal{N}$ is a function of the nutation in longitude, $\Delta\psi$, and the nutation in obliquity, $\Delta\epsilon$, as follows:

$$\mathcal{N} = \mathcal{R}_1(-\epsilon_A) \times \mathcal{R}_3(\Delta\psi) \times \mathcal{R}_1(\epsilon_A + \Delta\epsilon). \qquad (8.18)$$

The currently observed differences to the IAU 1976 precession and the IAU 1980 Theory of Nutation $\delta\Delta\psi$ and $\delta\Delta\epsilon$ are given as *dpsi* and *deps* in the IERS Bulletins.

These are with respect to the conventional celestial pole positions defined by the models and reported as *celestial pole offsets* by the IERS. Using these offsets, the corrected nutation is given by

$$\Delta\psi = \Delta\psi(\text{IAU}1980) + \delta\Delta\psi,$$
$$\Delta\epsilon = \Delta\epsilon(\text{IAU}1980) + \delta\Delta\epsilon. \tag{8.19}$$

This is practically equivalent to replacing $\mathcal{N}$ by

$$\mathcal{N}' = \begin{pmatrix} 1 & -\delta\Delta\psi\cos\epsilon_t & -\delta\Delta\psi\sin\epsilon_t \\ \delta\Delta\psi\cos\epsilon_t & 1 & \delta\Delta\epsilon \\ \delta\Delta\psi\sin\epsilon_t & \delta\Delta\epsilon & 1 \end{pmatrix} \times \mathcal{N}, \tag{8.20}$$

where $\epsilon_t = \epsilon_A + \Delta\epsilon$ and $\mathcal{N}$ represents the IAU 1980 Theory of Nutation.

8.7.2 The precession–nutation model IAU 2000A and IAU 2000B

The precession–nutation model IAU 2000A has been adopted by the IAU 2000 (Resolution B1.6) to replace the IAU 1976 Precession and the IAU 1980 Theory of Nutation described above. This model, developed by Mathews *et al.* (2002), is based on the solution of linearized dynamical equations of the wobble nutation problem and makes use of estimated values of seven of the parameters appearing in the theory, obtained from a least-squares fit of the theory to an up-to-date precession–nutation VLBI data set. The series of nutation relies on the Souchay *et al.* (1999) rigid Earth nutation series, rescaled by 1.000 012 249 to account for the change in the dynamical ellipticity of the Earth implied by the observed correction to the luni-solar precession of the equator. The non-rigid Earth transformation is the MHB2000 model of Mathews *et al.* (2002), which improves the IAU 1980 Theory of Nutation by taking into account the effect of mantle anelasticity, ocean tides, electro-magnetic couplings produced between the fluid outer core and the mantle as well as between the solid inner core and fluid outer core, and the consideration of the linear terms which have hitherto been ignored in this type of formulation.

The resulting nutation series includes 678 luni-solar terms and 687 planetary terms and provides the direction of the celestial pole in the GCRS with an accuracy of 0.2 mas. The series includes the geodesic nutation (Section 5.3.4). This inclusion means that when using the precession–nutation model, the resulting position of stars are in the BCRS. If one insisted in having them in GCRS, one should subtract it from the precession–nutation series.

On the other hand, the free core nutation (FCN) being a free motion that cannot be predicted rigorously, is not considered as being part of the IAU 2000A model.

The IAU 2000 series of nutation is associated with improved numerical values for the precession rate of the equator in longitude and obliquity:

$$\delta\psi_A = (-0\overset{''}{.}299\,65 \pm 0\overset{''}{.}0004)/\text{century},$$
$$\delta\omega_A = (-0\overset{''}{.}025\,24 \pm 0\overset{''}{.}0001)/\text{century}, \tag{8.21}$$

as well as with the following offset of the direction of the CIP at J2000.0 from the direction of the pole of the GCRS:

$$\psi_0 = (-0\overset{''}{.}016\,617 \pm 0\overset{''}{.}000\,01),$$
$$\eta_0 = (-0\overset{''}{.}006\,819 \pm 0\overset{''}{.}000\,01). \tag{8.22}$$

As recommended by IAU 2000 (Resolution B1.6), an abridged model, designated IAU 2000B, is available for those who need a model only at the 1 mas level. Such a model has been developed by McCarthy and Luzum (2003); it includes 80 luni-solar terms plus a planetary bias to account for the effect of the planetary terms in the time period under consideration. It provides the celestial pole motion with an accuracy that does not result in a difference greater than 1 mas with respect to that of the IAU 2000A model during the period 1995–2050.

The precession quantities ψ_A, θ_A, ϵ_A, and χ_A are necessary for computing the coordinates X and Y of the CIP in the GCRS using the current nutation series. On the other hand, the precession quantities ζ_A, θ_A, and z_A are necessary for computing the precession matrix using the procedure described in Section 8.6.1. The expressions are those given in (8.11), together with:

$$\omega_A = \epsilon_0 + 0\overset{''}{.}051\,27t^2 - 0\overset{''}{.}007\,726t^3,$$
$$\epsilon_A = \epsilon_0 - 46\overset{''}{.}8150t - 0\overset{''}{.}000\,59t^2 + 0\overset{''}{.}001\,813t^3. \tag{8.23}$$

with $\epsilon_0 = 84\,381\overset{''}{.}448$.

To derive the position (X, Y, Z) of the CIP, the following procedure can be applied.

The inclination, ω, of the true equator of date on the ecliptic of J2000.0 and the longitude, ψ, on the ecliptic of J2000.0, of the node of the true equator of date, are computed from the precession quantities by:

$$\omega = \omega_A + \delta\psi \sin\epsilon_A \sin\chi_A + \delta\epsilon \cos\chi_A,$$
$$\psi = \psi_A + \frac{\delta\psi \sin\epsilon_A \cos\chi_A - \delta\epsilon \sin\chi_A}{\sin\omega_A}. \tag{8.24}$$

From these, one computes the intermediate coordinates of the pole (Capitaine, 1990),

$$\bar{X} = \sin\omega \sin\psi,$$
$$\bar{Y} = -\sin\epsilon_0 \cos\omega + \cos\epsilon_0 \sin\omega \cos\psi, \tag{8.25}$$

and, finally,

$$X = \bar{X} + \xi_0 - \bar{Y} d\alpha_0,$$
$$Y = \bar{Y} + \eta_0 + \bar{X} d\alpha_0, \tag{8.26}$$

where $\xi_0 = -0{.}''016\,617$ and $\eta_0 = -0{.}''006\,819$ are the celestial pole offsets at the epoch J2000.0 and $d\alpha_0$ is the right ascension of the mean equinox. It is a very small quantity, of the order of 0.2 mas (Chapront *et al.*, 1999) and can therefore be neglected.

The IAU 2000 nutation model is given by series for nutation in longitude $\delta\psi$ and obliquity $\Delta\epsilon$, referred to the mean equator and equinox of date, with t measured in Julian centuries from epoch J2000.0:

$$\Delta\psi = \sum_{i=1}^{N}((A_i + A_i't)\sin(\text{argument}) + (A_i'' + A_i'''t)\cos(\text{argument})),$$

$$\Delta\epsilon = \sum_{i=1}^{N}((B_i + B_i't)\sin(\text{argument}) + (B_i'' + B_i'''t)\cos(\text{argument})).$$

The IAU 2000A subroutine, provided by T. Herring, is available electronically on the USNO website at: ftp://maia.usno.navy.mil/conv2000/chapter5/IAU2000A.f.

It produces celestial pole offsets based on the MHB2000 model with the exception of the free core nutation (FCN). The software can also be used to model the expected FCN based on the most recent astronomical observations. The IAU 2000B subroutine is available at:

ftp://maia.usno.navy.mil/conv2000/chapter5/IAU2000B.f.

The complete matrix algorithms for transformation between the terrestrial reference frame (ITRF) and ICRF, including precession, nutation, polar motion, and Earth's rotation, are given for the classical, the improved classical, and the new CEO-based systems, in Section 10.6.

9

Terrestrial coordinate systems

In contrast with celestial reference frames, which have existed since the time when catalogs of stars over the whole sky became available, a global reference system for the positions on the Earth did not exist until direct geodesic links could be performed between regions separated by oceans and, more generally, by geographically or politically impassable barriers. So there were a number of local, geodetic coordinate systems, called *datums*, to which the positions of terrestrial sites were referred. They were given under the form of parameters defining the shape and the size of a reference ellipsoid, as well as its orientation with respect to some conventional features such as a mean pole and a zero meridian. The ellipsoid parameters were determined to best fit the local geoid (equipotential surface corresponding to the mean ocean level), and attached to the Earth by conventional coordinates of an *initial point*.

9.1 Introduction

At the beginning of the space age, positional and, later, laser or Doppler observations of satellites were used to link the individual datums and to place them in a unique terrestrial coordinate system. However, locally, countries continued (and many still do) to use their own datums for surveying and legal objectives. But for scientific purposes, this was a much too complex system, and a global terrestrial reference system had to be developed. As a first step, it was agreed to orient the reference frame with respect to a conventional terrestrial pole and zero meridian as defined by the Bureau International de l'Heure, and to adopt a *Geodetic Reference System* (GRS), (IAG, 1980), based upon an Earth equatorial radius of a = 6 378 137 m and a polar radius of b = 6 356 752.3 m. These values were used to build an absolute (that is centered at the center of mass of the Earth) datum called the *World Geodetic System* (WGS 84) (see DMA, 1987). This datum is in wide use, particularly in the USA. In the most recent datum, called *World Geodetic Datum*

2000 (*WGD 2000*), the parameters of the reference ellipsoid including the permanent deformation due to the Earth's tides (Section 9.7.2), are: $a = 6\,378\,136.602 \pm 0.053$ m and $b = 6\,356\,752.860 \pm 0.052$ m (Grafarend and Ardalan, 1999).

On the other hand, for actual use in astrometry, one needs very accurate positions of observing instruments with respect to the center of mass of the Earth. For this reason, as well as for many applications in geodesy and geophysics, global terrestrial reference systems and frames in rectangular coordinates are built in an analogous way to its celestial counterpart (Section 7.2). These goals are realized by a certain number of fiducial points on the surface of the Earth, which are difficult to interconnect. However, it is necessary to have a continuous reference frame all over our planet. The role of the geodetic systems is to provide a frame that is readily accessible (Section 9.3).

9.2 Terrestrial reference systems

As in the case of celestial reference systems, it is first necessary to specify how such a system must be defined. Evidently it should be based on positions of a certain number of points on the Earth's crust, but, in contrast with the extragalactic sources, their positions in any terrestrial coordinate system are not fixed, because the various parts of the crust move with respect to each other as the result of global tectonics, and because the crust as a whole undergoes periodic (Earth tides, atmospheric or ocean loading, etc.) and secular (post-glacial rebound, orogenic upheavals) deformations. As a consequence, there are no *a priori* markers that could be considered as fiducial points driven only by the global motions of the Earth. Another condition was that the definition of the *Conventional Terrestrial Reference System* (CTRS) should be compatible with the IAU resolutions on reference systems (Section 7.1).

So, in 1991, the International Union for Geodesy and Geophysics (IUGG) approved these conditions (IUGG, 1992) and endorsed the resolutions on reference systems of the XXI-st General Assembly of the IAU. It declared that the reference system so defined will be considered as the CTRS, also known as the *International Terrestrial Reference System* (ITRS). In this resolution, the IUGG also stated that the five items defining the ITRS should be:

    (i) CTRS to be defined from a geocentric non-rotating system by a spatial rotation leading to a quasi-Cartesian system,

    (ii) the geocentric non-rotating system to be identical to the Geocentric Reference System (GRS) as defined in the IAU resolutions,

    (iii) the coordinate time of the CTRS as well as the GRS to be the Geocentric Coordinate Time (TCG),

    (iv) the origin of the system to be the geocenter of the Earth's masses including oceans and atmosphere, and

(v) the system to have no global residual rotation with respect to horizontal motions at the Earth's surface.

These considerations implied the following characteristic for the International Terrestrial Reference System (ITRS) adopted by the IERS (International Earth Rotation Service), which is in charge of realizing the corresponding reference frames. They are the following (Boucher, 1990):

(i) the terrestrial reference system is to be geocentric, the center of mass being defined for the whole Earth, including oceans and atmosphere,
(ii) its scale is that of a local Earth frame as meant by a relativistic theory of gravitation,
(iii) its orientation is the one initially given by the Bureau International de l'Heure (BIH) in 1984.0, and
(iv) its time evolution in orientation will follow no global net rotation or translation condition.

The materialization of the ITRS is by a terrestrial reference frame through a network of station reference points, or ground marks, specified by Cartesian equatorial coordinates. The prime meridian OXZ is the international meridian, also called *Greenwich meridian* and the OY axis is on the equator towards the east. It was decided that the permanent solid Earth tidal deformations be included in the coordinates, so that the adopted coordinates will differ from the instantaneous coordinates by only periodic terms. In addition, as a consequence of item (iii) above, in defining the OZ direction of the frame, care was taken that it coincides with the reference axis that defines the polar motion (Section 10.4.1).

This materialization was realized by successive approximations. Actually, several such frames were constructed and published by the IERS under the names *International Terrestrial Reference Frame XX* (ITRF-XX) where XX is the year of the last data used. This work was started in 1984 by the BIH, which had constructed a terrestrial reference frame under the conditions specified above. When the IERS was created, it took over the job and published successive solutions of ITRFs. This process led, in 1999, to the publication of ITRF-97 (Boucher *et al.*, 1999). It was later superceded by the ITRF-2000 (Altamimi *et al.*, 2002). This frame is to be considered as the terrestrial counterpart of the ICRF described in Section 7.2.

9.3 The ITRF series

Similar to the construction of the ICRF, the ITRF is the result of a compilation of geocentric station positions, **OP**, and velocities, **V**, provided for a given epoch, *t,* by individual terrestrial reference frames constructed using different observing techniques. These are VLBI (Section 2.4.2), GPS (Section 2.4.1), laser ranging on satellites (Section 2.4.3), and DORIS (see Cazenave *et al.*, 1992). The connection to the center of mass is ensured by a dynamical analysis of observations of satellites by lasers and GPS. Whenever a station is equipped with different instruments, an accurate local geodetic survey provides their connection to a well defined fiducial point on the site.

9.3.1 Construction of the ITRF-97

The bases of the computation of the ITRF-97 were 19 independent catalogs of station positions and velocities derived from observations spanning up to 23 years by one, or the other, of the techniques mentioned above. Then, they are reduced to a common epoch (1997.0) and a shift of the origin, a rotation, and scale factor are applied to put them in the ITRS. This is actually a method that is very similar to what was done in constructing the ICRS (Section 7.1), with the additional compli- cation that the positions of the stations are drifting. If there are several instruments (stations) at the same site, the station velocities are constrained to be the same, while their relative positions are constrained by the results of the local survey. The description and position of the fiducial point adopted in each site are given in the publication (Boucher *et al.*, 1999).

Finally, the ITRF-97 contained the rectangular components in the ITRS of the positions at epoch 1997.0 and the velocities expressed in meters and meters per year together with their standard uncertainties of 550 stations located on 325 sites. The velocities are constrained by a model of tectonic plate motions (DeMets *et al.*, 1994), described in Table 9.1. These motions are defined by the components Ω_X, Ω_Y, and Ω_Z of the rotation vector Ω. The positions of the plates considered in this table are shown in Fig. 9.1.

9.3.2 The ITRF-2000

For the first time in the history of the ITRFs, this new realization of the International Terrestrial Reference System combines unconstrained Space Geodesy solutions that are free from any tectonic plate motion model. The only condition that is retained is to ensure consistency with the series of Earth rotation parameters that link terrestrial and celestial frames as determined by IERS (see Chapter 10). On the whole, 30 individual solutions for the terrestrial reference frame were used in the construc- tion of the ITRF-2000. They include solutions derived from individual techniques computed by different teams, multitechnique derivations, and GPS densifications, also determined by different methods and groups.

The ITRF-2000 is geocentric. The effect of the geocenter motion (Section 9.9.1) is deduced from the determination of the geocenter as sensed by satellite ranging

Table 9.1. *Components of the Cartesian rotation vectors for the main plates in milliarcseconds per year*

Plate name	Ω_X	Ω_Y	Ω_Z
Africa	+0.184	−0.639	+0.809
Antartica	−0.169	−0.351	+0.764
Arabia	+1.379	−0.107	+1.394
Australia	+1.628	+1.057	+1.296
Caribbean	−0.037	−0.698	+0.326
Cocos	−2.150	−4.456	+2.253
Eurasia	−0.202	−0.494	+0.650
India	+1.376	+0.008	+1.401
Juan de Fuca	+1.072	+1.776	−1.200
Nazca	−0.316	−1.769	+1.982
North America	+0.053	−0.742	0.032
Pacific	−0.311	+0.998	−2.056
Philippine	+2.081	−1.477	−1.994
Rivera	−1.937	−6.386	+2.485
Scotia	−0.085	−0.549	−0.262
South America	−0.214	−0.312	−0.179

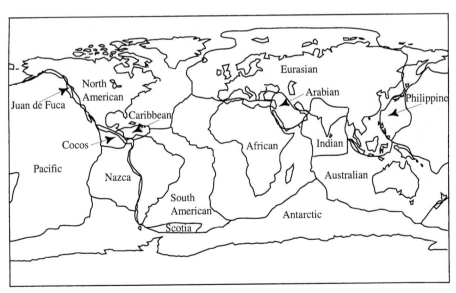

Fig. 9.1. The 16 plates with their boundaries whose motions are given in the NNR-Novelia model (DeMets *et al.*, 1994). Several minor plates are not shown.

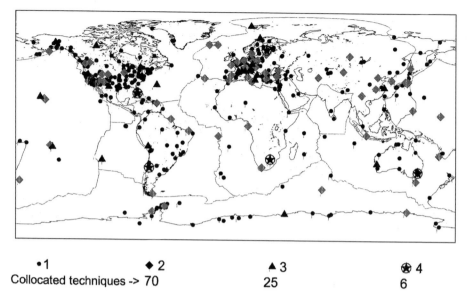

•1   ◆ 2   ▲ 3   ✪ 4
Collocated techniques -> 70   25   6

Fig. 9.2. ITRF-2000 sites. The shape of the spots indicates the number of stations on the site (Altamimi *et al.*, 2002).

and GPS solutions, but was afterwards averaged over the period of the observations (1991–2000). The scale is a weighted mean of VLBI, satellite ranging and GPS solutions. However, it was corrected by 0.7 parts per billion to meet the requirements to be in TCG time frame instead of TT used in the analyses. The orientation is aligned to ITRF-97 using only 50 continuously observed sites located on rigid parts of tectonic plates. The station positions are reduced to epoch 1997.0. The definition of the system is estimated to be at the level of 1 mm/year (Altamimi *et al.*, 2002).

On the whole, positions and velocities are given for about 500 sites (see Fig. 9.2). The positions of about half of them are determined to better than 1 cm, and the uncertainties of the velocities of about 100 sites are estimated to be equal to, or smaller than 1 mm/year.

9.3.3 *Position of a point in the ITRF*

Let **OP** and **V** be the position and velocity of a station in the ITRF-2000, its coordinates at a time *t* are given by:

$$\mathbf{OP}(t) = \mathbf{OP}(1997.0) + \mathbf{V}(t - 1997.0).$$

All the secular motions are included in **V**, while all periodic terms are subtracted from the observations using models and additional local measurements (effects of underground water, variations of the atmospheric loading, local discrepancies from the model, etc.). In the absence of better local measurements, models described in Section 9.7 have to be applied to raw observations before obtaining the coordinates in ITRF.

Note that it is not possible to obtain the coordinates of the position of a point on the Earth, which is not part of the ITRF references, by some kind of interpolation. If a rather limited precision is needed – and this is the case for all astronomical observations of objects outside the Solar System – it is sufficient to use geodetic coordinates as shown in the next section. But this is not the case for observations in the vicinity of the Earth, for instance lunar laser observations. Here, centimeter or even millimeter accuracies are required. Then, one must proceed in establishing geodetic links with ITRF stations using, for instance, simultaneous GPS observations. If this cannot be repeated and no velocity can be determined, as a default, a kinematic tectonic plate motion model must be used under the form of a global rotation Ω. The IERS Conventions (2001) advise the use of the NNR-Novelia model (DeMets *et al.*, 1994). Table 9.1 gives the rotation angles around the *OX*, *OY*, and *OZ* axes in milliarcseconds per year, the same as for ITRF-97.

9.4 Geodetic coordinates

As indicated in Section 9.2, it is preferable to use geocentric Cartesian coordinates *X, Y, Z* to define the position of a point *P* on the Earth. However, when one needs to represent some parameter over the Earth, it is necessary to have a continuous reference. The solution is the geodetic coordinates defined with respect to a reference ellipsoid whose center is at the geocenter.

Let us consider the meridian plane of the point (Fig. 9.3) and draw the intersection ellipse defined by its equatorial radius a and its flattening f, or its polar radius $b = a(1 - f)$. Let x and z be the geocentric coordinates of *P* in the plane

$$x = \sqrt{X^2 + Y^2} = r \cos \Phi,$$
$$z = Z = r \sin \Phi,$$

where r is the geocentric distance and Φ is the geocentric latitude

$$\Phi = \tan^{-1}(z/x).$$

Let us now compute x and z by using the reference ellipsoid. We assume here that the equatorial plane of the ellipsoid is the *XOY* plane of the ITRF. If this is not the case, one should first rotate the ITRF in such a way that these planes coincide.

The geodetic latitude ϕ_g is the angle between the *Ox* axis and the normal *PQ* to the ellipse as drawn from *P*. The geodetic height is $h = PH$ along the normal, $H(\xi, \zeta)$, being its intersection with the ellipse.

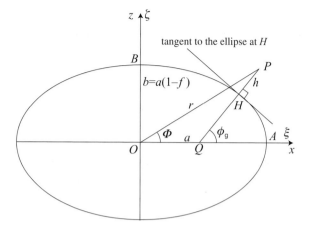

Fig. 9.3. Geodetic latitude ϕ_g and geodetic height h of a point P.

Noting that the equation of the ellipse is

$$\frac{\xi^2}{a^2} + \frac{\zeta^2}{a^2(1-f)^2} = 1,$$

(9.1)

one gets by differentiation the slope ϕ_g of the normal

$$\tan \phi_g = -\frac{d\xi}{d\zeta} = \frac{\zeta}{\xi(1-f)^2}.$$

(9.2)

The rectangular coordinates of P are

$$x = \xi + h \cos \phi_g,$$
$$z = \zeta + h \sin \phi_g.$$

(9.3)

Equations (9.2) and (9.3) can be used to get ξ and ζ as functions of a and ϕ_g, so that one finally gets:

$$\xi = \frac{a}{\sqrt{1 + (1-f)^2 \tan^2 \phi_g}},$$

$$\zeta = \frac{a \tan \phi_g (1-f)^2}{\sqrt{1 + (1-f)^2 \tan^2 \phi_g}},$$

which, with the expression (9.3), gives the final expressions which are usually written using an intermediate function C:

$$C = [\cos^2 \phi_g + (1-f)^2 \sin^2 \phi_g]^{-1/2},$$

(9.4)

$$x = r \cos \Phi = a(C + h/a) \cos \phi_g,$$
$$z = r \sin \Phi = a(C(1-f)^2 + h/a) \sin \phi_g.$$

(9.5)

These equations link geodetic and geocentric coordinates, assuming that they have the same coordinate axes. To determine ϕ and h, several methods exist and can be found, for instance, in Borkowski (1989) or in Fukushima (1999). Let us note, that, neglecting the deflection of the vertical that will be introduced in Section 9.7.2, we have assumed that the longitudes are the same in both systems.

9.5 Physical geodesy

Up to now, we have presented a purely geometrical approach to the description of the Earth. This approach particularly suits space geodesy. However, neither the geocenter, the axes of ITRS, nor the reference ellipsoid are readily available on an observing site, particularly on an astronomical observatory. The ground-based instruments, like transit instruments, or astrolabes (see Section 1.1) refer to the local vertical. The first available coordinate system (Section 3.8) is the horizontal system defined by the azimuth and zenith distance. The vertical direction is defined by a physical property of the Earth, its gravity field. In many respects, there is a tight relationship between the geometrical and the dynamic description of the Earth which is, by nature, in the area of classical geodesy.

9.5.1 Earth's gravitational field

On the surface of the Earth, gravity is actually the result of the gravitational acceleration produced by the planet Earth as a whole, the centrifugal acceleration due to the rotation of the Earth, and several additional components due to the attraction by the Moon and the Sun (tidal accelerations), as well as to the particular mass distribution in the environment, giving rise to gravity anomalies.

The theory of the gravitational field of the Earth is based upon the classical potential theory that provides the tools to compute the gravitational potential outside the Earth. Let G be the constant of gravitation, and $\rho(\xi, \eta, \zeta)$ the density of a point Q with coordinates ξ, η, ζ. According to Newton's law, each element Q produces on an external point P, with coordinates x, y, z, an attractive force

$$\mathbf{dF} = -G\rho(\xi, \eta, \zeta)\mathrm{d}\xi\,\mathrm{d}\eta\,\mathrm{d}\zeta\,\frac{\mathbf{QP}}{QP^3},$$

whose three components are the partial derivatives of the potential

$$\mathrm{d}U = G\frac{\rho(\xi, \eta, \zeta)\mathrm{d}\xi\,\mathrm{d}\eta\,\mathrm{d}\zeta}{\sqrt{(x-\xi)^2 + (y-\eta)^2 + (z-\zeta)^2}}. \tag{9.6}$$

Integrating it over the whole volume of the Earth, one obtains the Earth's full gravitational potential U. There is no finite expression for U, and the integration is possible only for points outside a sphere, which includes all elements of the Earth, so that the denominator of (9.6) does not vanish. Under this assumption, at a point P defined by its spherical coordinates, r, ϕ, λ, in a terrestrial reference system centered at the geocenter, the potential U is, as a rule, written as a development in spherical harmonics with the following notations:

$$
U = \frac{GM}{r} \left[1 - \sum_{k=2}^{\infty} \left(\frac{a}{r}\right)^k J_k P_k(\sin\phi) + \sum_{n=1}^{\infty} \sum_{m=1}^{\infty} \left(\frac{a}{r}\right)^n (C_{nm} \cos m\lambda \right.
$$
$$
\left. + S_{nm} \sin m\lambda) P_{nm} \cos\phi \right],
\tag{9.7}
$$

where the following conditions hold.

(i) GM is the geocentric gravitational constant whose conventional best value is

$$GM = (3.986\,004\,418 \pm 8) \times 10^5 \mathrm{m}^3/\mathrm{s}^2.$$

(ii) a is a scaling constant, conventionally taken as equal to the equatorial radius of the Geodetic Reference System: $a = 6\,378\,136.6$ m (see Section 9.1).

(iii) P_k are Legendre polynomials and P_{nm} are Legendre functions of the first kind that are of degree n and order m. They may or may not be normalized; the normalization coefficient being, or not, put into the coefficients J_n, C_{nm} and S_{nm}.

(iv) J_k are called the zonal harmonics, and C_{nm}, S_{nm} are the tesseral harmonics.

Remarks.

- By construction, since the center of mass is at the origin of coordinates, $J_1 = C_{11} = S_{11} = 0$.
- The values of C_{21} and S_{21} depend on the shift between the polar axis of ITRF and the axis of rotation of the Earth (see Section 10.4.1).
- The coefficient J_2 is called the dynamical form factor of the Earth and is connected to its mean flattening. A conventional value has been adopted by both the IUGG and the GRS:

$$J_2 = 0.001\,082\,6395.$$

This actually varies with time, essentially because of the post-glacial rebound and the melting of Antarctic ice (Section 9.9.2).

The values of the zonal and tesseral harmonics are determined by space geodesy methods, that essentially model gravitational and other forces acting on the motions of artificial satellites, which are observed by various techniques (Doppler, laser ranging, altimetry, etc.). The coefficients of the models – such as Equation (9.7) for the gravitational effects – are determined to minimize the observing residuals (see, for instance, Zarrouati, 1997). A review of various integration methods is to be found in Kinoshita and Kozai (1989). The results are given in the form of lists of values of the spherical harmonics, and constitute *Earth potential models*. Three series of such models exist and they are periodically improved by adding new data. These are:

- the GEM series (Goddard Earth Models) constructed at NASA, Goddard Space Flight Center,
- the JGM models (University of Texas, Austin), and
- the GRIM series (GRGS – Institute of Munich) developed jointly by CNES (Toulouse) and DGFI (Munich).

The models differ in their details, but are roughly of the same quality. The maximum order and degree now reach about 70, representing some 10 000 coefficients in (9.7). The accuracy with which they represent the motion of a satellite depends upon the satellite. For most of the distant satellites, the uncertainty in the representation is in the range of one to a few centimeters.

9.5.2 Gravity on the surface of the Earth

It is not legitimate to extrapolate directly formula (9.7) toward the Earth and derive the potential at the level of the Earth's surface. It has to be done stepwise, by removing the effects of the atmosphere and surrounding ground effects. This is currently being done, and one obtains the potential at any point in the vicinity of the Earth. The differences with the coefficients of the gravitational Earth models are small, and the form of Equation (9.7) can be kept, provided that the centrifugal potential

$$V = \frac{1}{2}\omega_0^2(X^2 + Y^2),\qquad(9.8)$$

where ω_0 is the angular velocity of the Earth, is added.

There is a particular equipotential surface that is of great importance for geodesy. It is the *geoid*, a surface from which the heights are evaluated. In the past, it was defined as the shape of the equipotential surface that best fits the mean surface of the oceans. It represented the shape the Earth would have, if it were entirely covered with oceans at rest. This definition is too imprecise for the needs of very accurate geodesy and terrestrial time definition (Section 5.5.1). It is now defined as the surface on which the geopotential is

$$U_G = 62\,636\,856\,\mathrm{m}^2/\mathrm{s}^2.$$

As an illustration, we present in Fig. 9.4 the heights of the geoid with respect to the GRS ellipsoid. At the global level of this figure, all the models are alike, and

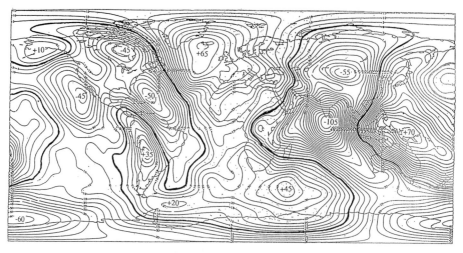

Fig. 9.4. Map of the largest features of the geoid as derived from one of the versions of the GRIM4 model and referred to the GRS ellipsoid. The approximate heights, in meters, of the main features are indicated. The heavy line corresponds to zero height.

differ only in detailed local features. The large-scale deformations that can reach ± 100 m are not correlated with surface relief, and actually result from large mass anomalies deep in the mantle. Over the oceans, very detailed maps of the geoid are now obtained with satellite radar altimetry (Topex-Poseidon, Jason, or ERS missions). Individual height measurements with respect to the ITRF are obtained with uncertainties of the order of a few centimeters. As a result, the ocean geoid is now mapped with a resolution of 50–100 km, and a height uncertainty of 2–5 cm. The small structures observed are highly correlated with the sea-floor topography. These observations also give a refined view of the oceanographic contributions to the heights of the oceans, owing to tides, currents, and associated eddies, winds, atmospheric pressure, and water density.

At this point, it is necessary to stress the practical importance of the geoid. It is the surface to which all geographic heights are referred. The extension of the system of heights is made by geodetic leveling calibrated on tidal gauges. Lately, gravimetric methods replaced leveling. On the ground, the main origin of gravity data is gravimetry. Various campaigns have covered the major part of the continental crust. The results are assembled and discussed by the Bureau Gravimétrique International (BGI) at Toulouse and transformed into geoid height maps as well as surface gravity data files. These corrections are to be applied to get the gravity at the reference ellipsoid: the free air correction for topography to take into account the crust in the environment, and the isostasic correction to compensate for the effect of the weight of the relief on the reaction of the mantle on the crust. These corrections,

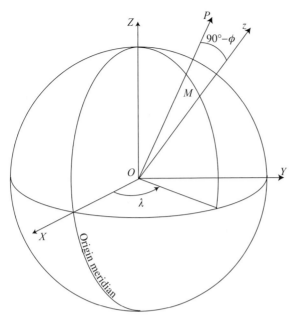

Fig. 9.5. General definition of the longitude (λ) and the latitude (ϕ) of a point M.

and the way to obtain the direction of the vertical on the ellipsoid, are part of physical geodesy, a classical science described in text books on ground-based geodesy (see, for instance, Moritz, 1980).

9.6 Coordinates of a point on the Earth

In Section 9.4 we introduced two types of terrestrial coordinates. However, the principal parameter that is required for classical ground-based astrometry is the direction, in the terrestrial reference frame, of the vertical at the observatory site. This involves a third type of coordinates. So, let us take up again the definition of angular coordinates on the Earth.

The position of a point M is defined by two spherical coordinates: its latitude and its longitude (Fig. 9.5). The general definition of latitude is

$$90° - (\mathbf{MP}, \mathbf{Mz}),$$

where $\mathbf{MP}$ is the direction of the pole and $\mathbf{Mz}$ the direction of a particular vertical. The angle $(\mathbf{MP}, \mathbf{Mz})$ is the colatitude. The other coordinate is the longitude, defined as the angle, reckoned eastward between the plane of the origin meridian and the plane PMz. Although in the past, the opposite sign convention was sometimes adopted, we shall use this convention: the East is positive, so the longitudes in Asia are positive; in America they are negative.

Because of the different approaches to the vertical as described in the preceding sections, there are three different types of coordinate, each of them having a specific use in describing the positions of points on the Earth's surface. Two of them have been already defined in Section 9.4.

- *The geocentric coordinates.* Let XYZ be the rectangular coordinates of M in the ITRF. The latitude and the longitude are respectively defined by

$$\Phi = \tan^{-1}(Z/\sqrt{X^2 + Y^2}),$$
$$\Lambda = \tan^{-1}(Y/X) \quad \text{for} \quad X > 0, \text{ and}$$
$$\Lambda = 180° + \tan^{-1}(Y/X) \quad \text{for} \quad X < 0. \tag{9.9}$$

- *The geodetic coordinates.* They are defined with respect to the reference ellipsoid. The geodetic latitude is the angle ϕ_g between the equatorial plane and the normal to the reference ellipsoid, defined in Section 9.4 and illustrated by Fig. 9.3. We assume, as it is generally the case, that the equator of this ellipsoid is the XOY plane of the ITRF. Then, the geodetic longitude, λ_g, is equal to Λ.

Let us now introduce a third type of coordinate. The vertical is defined as the direction perpendicular to the local potential surface or, in other words, the direction of the gravity acceleration. The corresponding coordinates are called *astronomical coordinates,* or *geographic coordinates.* The vertical, which defines them does not coincide with the direction of the normal to the reference ellipsoid, whose direction is given by Equation (9.2), nor is it the direction **OP** in Fig. 9.3. The astronomical latitude, ϕ, is the angle between the physical vertical and the equatorial plane, which is perpendicular to the direction of the celestial pole (Fig. 9.6). The astronomical longitude, λ, is the angle between the origin meridian and the (**PZ**, **z**) plane.

The angle between the direction of the vertical and the normal to the reference ellipsoid is the *deflection of the vertical,* which has a North–South component, ξ, and an East–West component, η. If we assume – and this point will be discussed in Chapter 10 – that the celestial pole is the celestial intermediate pole, defined as the extension on the celestial sphere of the OZ axis of the ITRF, then by definition of the North–South component of the deflection of the vertical, one has

$$\phi = \phi_g + \xi. \tag{9.10}$$

Here, ϕ is the latitude actually measured by astronomical instruments such as meridian circles or astrolabes, after correction for polar motion with respect to the celestial intermediate pole (Section 10.4.2). Similarly, the East–West deflection of the vertical modifies the longitude, λ_g, which is shifted by an angle corresponding to the East–West vertical deflection, η,

$$\lambda = \lambda_g - \eta/\cos\phi. \tag{9.11}$$

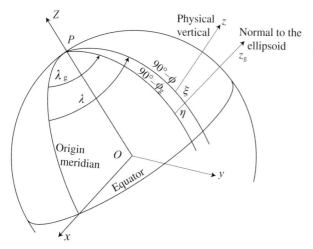

Fig. 9.6. Astronomical and geodetic coordinates and the deflection of the vertical (ξ,η). z is the physical vertical and z_g is the normal to the ellipsoid.

This formula defines the sign of η. Applying these corrections to the computable normal to the reference ellipsoid, one gets the components of the on-site vertical direction in the terrestrial reference frame (see Section 9.7.2).

We have mentioned that the geodetic system of terrestrial coordinates is very much used for various purposes. The geodetic coordinates are those that appear on maps. But the astronomical coordinates are necessarily present in local astronomical observations, which are referred to the physical vertical and the local meridian, defined as the vertical plane containing the direction of the celestial pole.

9.7 Topocentric astronomical coordinates

In what follows, we shall assume that all corrections to observations, in particular for astronomical refraction, are made, and assume that celestial directions are those from which the light arrives when reaching the vicinity of the Earth. Two topocentric coordinate systems, already presented in Section 3.10.3, are used.

The horizontal system of celestial coordinates, also called the alt-azimuth system, is based on the direction of the zenith, Z, and the horizontal plane, perpendicular to the local vertical. The origin on this plane is its northern intersection with the meridian plane. In Fig. 9.7, the position of a point, M, on the celestial sphere is defined by two angles. One is the zenith distance

$$z = (OZ, OM)$$

or the altitude $h = 90° - z$.

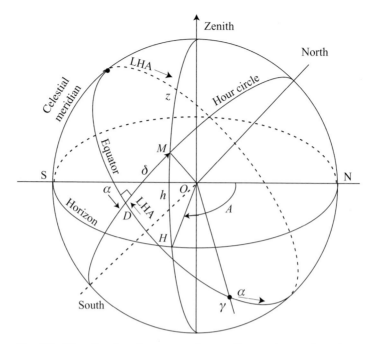

Fig. 9.7. The alt-azimuth topocentric coordinates A, z, or h, and associated angles in the celestial sphere.

The other angle is the azimuth, A, of the half vertical plane ZOM with the northern half meridian reckoned clockwise (towards East). Actually there are several other conventions used, such as counterclockwise (direct) reckoning, and/or with an origin at the southern point of the meridian. In this book, we use the most generally accepted definition given above (Fig. 9.7).

The other commonly used topocentric coordinate system is the local equatorial system (see Fig. 9.8) which, like the celestial reference system, is defined by the direction of the pole and the intermediate equator as defined by its pole, the Celestial Intermediate Pole (CIP, see Section 10.4.2). The two coordinates are the intermediate declination δ_i, defined as in the celestial system, and the local hour angle H, reckoned westward from the meridian plane. It is related to the right ascension by the equation

$$H = T_1 - \alpha_i, \tag{9.12}$$

where T_1 used to be the local sidereal time, which was defined as the hour angle of the equinox, or γ point. Now, with the definition of the Celestial Ephemeris Origin (CEO, see Section 10.3.1) as the origin of the right ascensions in the intermediate celestial reference system (α_i), T_1 is the local stellar angle (see Section 10.4.4), and the corresponding declination is the intermediate declination δ_i. One

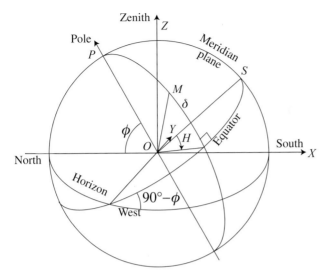

Fig. 9.8. The equatorial H, δ, topocentric coordinates; ϕ is the astronomical latitude of the observer.

has also

$$H = T + \lambda - \alpha_i,$$

where λ is the longitude, and T was the Greenwich sidereal time of the prime meridian of the ITRF and now is the stellar angle of the same meridian.

With the exception of the declination δ, the three other coordinates, z, A and H vary with time because of the rotation of the Earth. In addition, together with δ, they may be biased by the possible deflection ξ, η of the vertical and its time variations due, for instance, to underground hydrology.

As the first step to comparisons between observations made from different locations, it is necessary to transform them into coordinates as seen from a common reference point, chosen to be the geocenter. This is done in two or three steps depending upon the system in which the observations are performed. They are described in the following sub-sections.

9.7.1 Transformation from horizontal to equatorial coordinates

This transformation corresponds, as seen on Fig. 9.8, to a rotation around the West–East axis by the angle (OZ, OP) equal to $90° - \phi$, where ϕ is the astronomical latitude (Section 9.6). In the right-handed, local horizontal rectangular coordinates $O(xyz)$, the components of a unit vector $\mathbf{V}$ are

$$\mathbf{V} \begin{cases} -\sin z \cos A, \\ \sin z \sin A, \\ \cos z. \end{cases}$$

In the equatorial local system, one has

$$\mathbf{V'} \begin{cases} \cos \delta \cos H, \\ -\cos \delta \sin H, \\ \sin \delta. \end{cases}$$

One obtains $\mathbf{V'}$ by applying to $\mathbf{V}$ the rotation $\mathcal{R}_y(\phi - 90°)$, or the following direct relations

$$\left. \begin{aligned} \cos \delta \cos H &= \cos \phi \cos z - \sin \phi \sin z \cos A, \\ \cos \delta \sin H &= -\sin z \sin A, \\ \sin \delta &= \sin \phi \cos z + \cos \phi \sin z \cos A. \end{aligned} \right\} \tag{9.13}$$

From this, one gets unambiguously δ and H. The inverse relation can readily be obtained by a similar method:

$$\left. \begin{aligned} \sin z \cos A &= \cos \phi \sin \delta - \sin \phi \cos \delta \cos H, \\ \sin z \sin A &= -\cos \delta \sin H, \\ \cos z &= \sin \phi \sin \delta + \cos \phi \cos \delta \cos H. \end{aligned} \right\} \tag{9.14}$$

9.7.2 Correction for the deflection of the vertical

Because of the local deflection of the vertical, the direct transformation from the local equatorial system to the geocentric system is not possible. It is necessary first to align the local coordinate axes with the geodetic coordinates. From Equations (9.10) and (9.11), the correction is equivalent to a change in latitude such that

$$\phi_g = \phi - \xi.$$

The rotation of $\eta / \cos \phi$ in longitude is equivalent to a change in the hour angle, so that the new value is

$$H_g = H + \eta / \cos \phi.$$

The actual values of the deflection of the vertical can be computed from the regional gravity anomalies by formulae derived by Vening Meinesz (see Heiskanen and Moritz, 1967). It can also be determined by comparing the geodetic latitude and longitude, as deduced from observations of the site position by GPS, which gives the geocentric coordinates, and the astronomical coordinates as deduced from observations of stars with known positions in the celestial reference frame, for instance using transit instruments. This method is also advantageous because it can be repeated, and time-dependent deflections of the vertical, owing for instance to modifications of the level of underground water, can be monitored.

9.7.3 Computation of the geocentric coordinates

The transformation from the geodetic to geocentric coordinates is straightforward as far as directions are concerned. The longitude is unchanged ($\Lambda = \lambda_g$) and the geocentric latitude, Φ, can be obtained from formulae (9.5). The problem is more complicated for the observations of bodies in the Solar System, for which the shift from topocentric to geocentric coordinates changes significantly with the directions in which they appear. The correction for this is the *diurnal parallax*. For the Moon, and even more so for artificial satellites, it is a major change in coordinates, which has to be computed with a very high accuracy.

We suppose that the geocentric coordinates of the observer, **OP**, are known, for instance from GPS observations. If the topocentric coordinates of the object **PS** were also known, the solution would be straightforward through applying the vectorial addition

$$\textbf{OS} = \textbf{OP} + \textbf{PS}. \tag{9.15}$$

This is not generally the case. The distance from the observing site to the observed body is not known, and one is led to use the geocentric distance as found in ephemerides. Note also that this concerns the position of the object at the time at which the light was emitted, and not the actual position the body may have reached at the time of the observation. The additional correction for this effect is the correction for light-time (Section 6.3.7).

Let ρ', H', δ' be the spherical topocentric coordinates of the object, and ρ, H, δ', its geocentric coordinates. Since we have already corrected for the deflection of the vertical, H' and H are reckoned from the same meridian plane. The geocentric coordinates of P with the meridian plane as the *ZOX* plane are $r, 0, \Phi$. In these geocentric rectangular coordinates, the components of the three vectors of Equation (9.15) are

$$\textbf{PS} \begin{cases} \rho' \cos \delta' \cos H', \\ -\rho' \cos \delta' \sin H', \\ \rho' \sin \delta', \end{cases} \quad \textbf{OS} \begin{cases} \rho \cos \delta \cos H, \\ -\rho \cos \delta \sin H, \\ \rho \sin \delta, \end{cases} \quad \textbf{OP} \begin{cases} r \cos \Phi, \\ 0, \\ r \sin \Phi, \end{cases}$$

where δ' and H' are observed and ρ is taken from the ephemerides. Equation (9.14) becomes

$$\begin{cases} \rho \cos \delta \cos H = \rho' \cos \delta' \cos H' + r \cos \Phi, \\ \rho \cos \delta \sin H = \rho' \cos \delta' \sin H', \\ \rho \sin \delta = \rho' \sin \delta' + r \sin \Phi. \end{cases} \tag{9.16}$$

The solution is straightforward. In a first step, one computes ρ' which is given, adding the squares of the three equations, by

$$\rho'^2 + 2\rho'(r \cos \Phi \cos \delta' \cos H' + r \sin \Phi \sin \delta') + r^2 - \rho^2 = 0$$

taking the positive solution. Then, replacing ρ' by the value just computed, one can solve (9.16) for δ and H, and obtain the geocentric alt-azimuth coordinates as well as the topocentric distance, assuming, of course, that the ephemerides for ρ are correct. One may remark that the Z component does not depend on H. It is the fixed part of the diurnal parallax, while the X and Y components are functions of time with the rotation of the Earth. This latter part is the diurnal parallax proper.

9.8 Time variations

The position of an apparently *fixed point* on the surface of the Earth is actually not stable. Several crustal motions occur that modify, at a level of a few millimeters to a few decimeters, the geocentric coordinates of the site, and it is necessary, depending upon the accuracy expected, to take them into account. The most important are the Earth tides, but many other smaller effects due to ocean and atmospheric loading, and polar tides are to be considered. Secular effects due to tectonic motions and post glacial rebound are generally taken into account in the ITRF.

9.8.1 The tidal potential

Since the Earth has finite dimensions and is not spherical, the gravitational forces exerted by an external body, such as the Moon or the Sun, are different at the center of mass of the Earth than at some other point on the planet, in particular on its surface. Let us consider, in a first approximation, a spherical Earth, centered at O, and the Moon, the center of mass, M, of which produces an attractive acceleration on a point P on the surface of the Earth equal to

$$\mathbf{f} = Gm' \frac{\mathbf{PM}}{PM^3},$$

where m' is the mass of the Moon (see Fig. 9.9) and G is the gravitational constant. In the system of reference centered at the center of mass of the Earth, O, the acceleration A is equal to the effect of the gravitational interaction between the Earth and the Moon, assuming that their masses are concentrated at O and M

$$\mathbf{A} = Gm' \frac{\mathbf{OM}}{OM^3}.$$

Hence, the apparent force, produced by the perturbing action of M on a unit mass

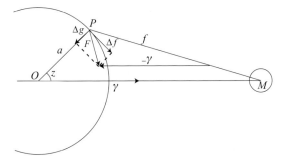

Fig. 9.9. Attraction of the Moon on P and its components Δg and Δf.

(which is, evidently, equal to the acceleration) is, in this reference system,

$$\mathbf{F} = \mathbf{f} - \mathbf{A} = Gm' \left[\frac{\mathbf{PM}}{PM^3} - \frac{\mathbf{OM}}{OM^3} \right].$$

It is called the tidal force. Now, both forces are conservative and their field can be written in terms of the gradient of a potential, the *tidal potential*,

$$W = Gm' \left(\frac{1}{PM} - \frac{\mathbf{OP} \cdot \mathbf{OM}}{OM^3} \right). \tag{9.17}$$

Let us express W by introducing the geocentric zenith distance z of M (we assume as a first approximation that $\mathbf{OP}$ is the vertical of P). Let us call $OP = a$, $PM = r'$ and $OM = r$. Then one has

$$\frac{1}{r'} = \frac{1}{r} \left(1 - \frac{2a}{r} + \frac{a^2}{r^2} \right)^{1/2},$$

or

$$\frac{1}{r'} = \frac{1}{r} + \frac{a}{r^2} \cos z + \frac{1}{r} \sum_{n=2}^{\infty} \left(\frac{a^2}{r^2} \right) P_n(\cos z),$$

where the functions P_n are the Legendre polynomials already introduced in Section 9.5.1. The additional term in (9.17) is precisely $-a \cos z / r^2$, so that the expression of W is:

$$W = \frac{Gm'}{r} \sum_{n=2}^{\infty} \left(\frac{a}{r} \right)^2 P_n(\cos z) = \sum_{n=2}^{\infty} W_n. \tag{9.18}$$

The tidal acceleration can be split into its components along the vertical and along the horizontal plane of P.

Along the vertical, this acceleration corresponds to a variation of the local gravity

$$\Delta \mathbf{g} = -Gm' \frac{\partial W}{\partial a}. \tag{9.19}$$

In the vertical plane containing the Moon, the horizontal force is

$$\Delta \mathbf{f} = -\frac{\partial W}{a \partial z}. \tag{9.20}$$

The horizontal acceleration modifies the direction of the vertical; the tides produce an angular deflection $\Delta \mathbf{v}$ of the vertical, in the direction of the vertical plane containing the Moon, and whose value is

$$\Delta \mathbf{v} = \frac{-1}{g} \frac{\partial W}{a \partial z}.$$

In practice, in computing W, one will add the effects due to the Moon and the Sun. Let us note that the variation of the local gravity also modifies the shape of the geoid. The deformation of the geoid due to W – which is time dependent because of the rotation of the Earth – is called the static oceanic, or geoidal, tide. The radial shift at the point P is

$$\Delta h = \frac{W}{g} = \frac{Gm'}{gr} \sum_{n=2}^{\infty} \left(\frac{a}{2}\right)^2 P_n(\cos z).$$

It is generally accepted that for the computation of tides, the second and third harmonics of W represent the actual displacements to a few millimeters. The amplitude of Δh may be as large as 80 cm, while the angular deviation of the vertical around its mean value can reach 25 mas.

Let us now take into account the fundamental feature in the theory of tides, the rotation of the Earth which causes the plane OPM to move around OP. For this, let us express W in the equatorial geocentric coordinates α, δ of the Moon, M. The components of $\mathbf{OP}$ and $\mathbf{OM}$ are, respectively,

$$\mathbf{OP} \begin{cases} a \cos \Phi \\ 0 \\ a \sin \Phi \end{cases} \qquad \mathbf{OM} \begin{cases} r \cos \delta \cos H \\ r \cos \delta \sin H \\ r \sin \delta \end{cases}$$

from which, one gets,

$$\cos z = \sin \Phi \sin \delta + \cos \Phi \cos \delta \cos(T + \lambda - \alpha),$$

where T is the Greenwich sidereal time, or the stellar angle, and λ is the longitude of P. The second-order term of W includes two terms, $i = 1$ for the Moon and $i = 2$ for the Sun:

$$W_2^i = \frac{Gm'a^2}{r^3} \left(\frac{3 \cos^2 z - 1}{2}\right),$$

which become, after some calculations,

$$
\begin{aligned}
W_2^i = \frac{Gm'a^2}{r^3} \Bigg[& \left(\frac{3}{2} \sin^2 \Phi \sin^2 \delta + \frac{3}{4} \cos^2 \Phi \cos^2 \delta - \frac{1}{2} \right) \\
& + 3 \sin \Phi \cos \Phi \sin \delta \cos \delta \cos(T + \lambda - \alpha) \\
& + \frac{3}{4} \cos^2 \Phi \cos^2 \delta \cos 2(T + \lambda - \alpha) \Bigg].
\end{aligned}
\tag{9.21}
$$

Similarly, W_3^i can be developed in $\sin k(T + \lambda - \alpha)$ with $1 \leq k \leq 3$.

The quantities δ and λ must be obtained from the theories of the motions of the Moon and of the Earth. For the Moon the various expressions of δ and α are expressed as functions of four trigonometric time arguments (Brouwer and Clemence, 1961).

(i) l: mean anomaly of the Moon,
(ii) g: longitude of the lunar perigee,
(iii) h: longitude of the node of the Moon's orbit,
(iv) l': longitude of the Sun.

For the solar tides, l' is the only significant argument. After the substitution of these expressions into (9.21), one obtains general, multi-argument, trigonometric series in T, l, g, h, and l' from which significant terms are identified and kept for computing the tides. These terms can be arranged into several classes.

- Terms including $2T$ correspond to semi-diurnal tidal waves.
- Terms including T correspond to diurnal tidal waves. There are also small third diurnal terms in $3T$.
- Monthly and semi-monthly effects are produced by terms which include no T term, but respectively l and $2l$ terms.
- Long periodic terms are essentially semi-annual and annual, arising from terms in l' and $2l'$.
- Very long period terms are functions only of g and h, while secular terms give a constant tide, which is included in the shape of the geoid and in the coordinates provided by the ITRF. They are not to be considered in the theory.

9.8.2 Solid Earth tides

If the Earth were a perfectly rigid body, the above description would be sufficient. This is not the case and, under the influence of the tidal force field, the Earth is globally deformed. It reacts essentially like an elastic body, so that the deformation is proportional to the amplitude of the forces. The actual response is governed by Love and Shida numbers.

- The Love number h represents the ratio of the height of the Earth tide to the height of the corresponding static geoidal tide at the surface.
- The Love number k represents the ratio of the additional potential produced by this deformation to the deforming potential.
- The Shida number l represents the ratio between the horizontal displacement of the crust and that of the corresponding static geoidal tide at the surface.

Before the advent of very accurate astrometry and satellite geodesy, these numbers were considered as unique, and the deflection of the vertical was given by its two components for a point with coordinates ϕ and λ (Melchior, 1966):

- along the meridian, towards North

$$v_N = -\frac{l}{g}\frac{\partial W_2}{\partial \phi},$$

- along the East–West direction towards West

$$v_W = \frac{l}{g \cos \phi}\frac{\partial W_2}{\partial \lambda}.$$

Similarly, the vertical component of the tide was limited to

$$\Delta h = (1 + k)\frac{W_2}{g}.$$

With respect to the instantaneous position of the geoid affected by the geoidal tide, the vertical component was

$$\Delta h' = (1 + k - h)\frac{W_2}{g},$$

where only the second-order part W_2 of the development of W was kept. The values adopted for these numbers were (Derr, 1969):

$$h = 0.615, \quad k = 0.085, \quad l = 0.303.$$

Now that an accuracy of 1 mm is desired, a much more sophisticated representation is necessary, and there are as many numbers as there are periods in the time series representing the tidal potential. In addition, they depend on station latitude Φ (Wahr, 1981). This dependence is a consequence of the ellipticity, the rotation of the Earth, and the resonance with the nearly diurnal free wobble in the Earth's rotation. For long periods, the mantle anelasticity is to be taken into account and leads to corrections to the elastic Love, or Shida, numbers.

It is outside the scope of this book to present the details of the computation of the effects of the Earth tides on the displacement of a station. They are described in IERS (2003).

9.8.3 Polar tides

In writing the expression (9.8) for the Earth centrifugal potential, it was assumed that the angular velocity $\vec{\Omega}$ was a vector supported by the OZ axis. Because of polar motion (Section 10.4.1), this is not true and the actual potential is

$$V = \frac{1}{2}(r^2|\vec{\Omega}|^2 - (r \cdot \vec{\Omega})^2),$$

where the components of $\vec{\Omega}$ are set to be

$$\vec{\Omega} \begin{cases} \omega_0 m_1, \\ \omega_0 m_2, \\ \omega_0 (1 + m_3), \end{cases}$$

where the parameters m_1 and m_2 describe polar motion with respect to the mean pole adopted for the ITRF, and m_3 – a negligible quantity – is a variation of the rotation ratio ω_0. The formulation for the displacement of a site is given in Wahr (1985) and, calling S_r the upward radial, S_ϕ the meridian southward, and S_λ the eastward displacements, and using the mean values (quite sufficient here) of the Love and Shida numbers, one gets in millimeters as functions of ΔV, the effect on V of m_1 and m_2:

$$S_r = \frac{h}{g}\Delta V = -6.6 \times 10^6 (m_1 \cos \lambda + m_2 \sin \lambda) \sin 2\Phi$$

$$S_\phi = -\frac{l}{g}\frac{\partial \Delta V}{\partial \Phi} = -1.9 \times 10^6 (m_1 \cos \lambda + m_2 \sin \lambda) \cos 2\Phi$$

$$S_\lambda = \frac{1}{g \sin \Phi}\frac{\partial \Delta V}{\partial \lambda} = 1.9 \times 10^6 (m_1 \sin \lambda - m_2 \cos \lambda) \cos 2\Phi. \quad (9.22)$$

Despite the apparently large coefficients m_1 and m_2 being expressed in radians, the values are smaller than $1''$, the maximum radial displacement is of the order of 25 mm and the maximum horizontal displacement is around 7 mm. There is also a variation of the gravity with a maximum of 15 μGal,

$$\delta g = -3.9 \times 10^6 (m_1 \cos \lambda + m_2 \sin \lambda) \sin 2\Phi.$$

9.8.4 Ocean loading

We have not yet described the ocean tides. They are produced by the same tidal potential (Equation (9.19)) as the Earth tides. The effect on a given point of the ocean is, however, much more complicated to compute, because the variety of local limiting conditions, such as the depth and floor topography of the seas, and the shape of the coastlines, induce a very complex response of the oceans to the

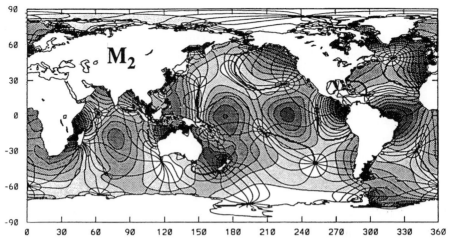

Fig. 9.10. Amplitudes of the M_2 tidal wave and the phases with an interval of 30° referred to the passage of the Moon through the local meridian. The amplitude is proportional to the levels of grey, and varies from 10 to 180 cm. (From Le Provost *et al.*, 1994, Spectroscopy of the world ocean tides from a fluid element hydro-dynamic model, *J. Geophys. Res.*, **99**, 24 777–24 797. Copyright 1994 American Geophysical Union. Reproduced by permission of American Geophysical Union.)

tidal potential. Models based upon tidal gauges and satellite altimetry observations compute the phase with respect to the lunar, or solar hour angle, and the amplitude of various components of the tidal potential. Since they have different periods, the actual tidal heights are time dependent. Figure 9.10 gives, as an example of the complexity, a map of the amplitudes produced by the M_2 semi-diurnal wave, the main wave produced by the lunar attraction and having a period of 12 h 25 min.

The variations of the height of the water surface with respect to the mean ocean surface produce a direct effect, owing to the gravitational attraction of the raised water, or a negative effect, owing to the lowered surface of the oceans. There is a second effect, a loading effect, because the changing surface exerts a time-variable load on the solid Earth below the ocean bottom, and deforms the elastic Earth all around. In addition, this deformation produces a variation of the gravitational attraction, called the indirect effect of tidal water.

In order to compute the sum of these effects, it is necessary to introduce the sea tide potential, W_{w}, produced by the additional or missing water layers with respect to the mean surface of the oceans (that is, in practice, the geoid). In order to compute W_{w}, one can integrate the tidal water effect through the whole Earth, or within some limited area, S, of the ocean close to the point, A, of coordinates ϕ_A, λ_A, for which the deformation is being computed:

$$W_{\mathrm{w}}(\phi_{\mathrm{A}}, \lambda_{\mathrm{A}}) = -Ga^2 \int \int_S \frac{Z(\phi, \lambda)}{\rho(\phi_{\mathrm{A}}, \lambda_{\mathrm{A}}, \phi, \lambda)} \frac{\mathrm{d}\phi\mathrm{d}\lambda}{\cos \phi}, \tag{9.23}$$

where a is the radius of the Earth, Z is the height (positive or negative) of the tide at the point $P(\phi, \lambda)$, ρ is the distance PA, and where it is assumed that the density of water is unity. This potential may be used, as already done in Section 9.8.2, to compute the effect on the position of A. More details on these computations will be found in Vaniček and Krakiwsky (1982). To facilitate computations, it is possible to develop W_w in a spherical harmonic series. However, when determining the actual effects on A, one has to know load numbers, similar to Love and Shida numbers, which express the proportionality between the potential and the deformation. They are functions of the local rheological structure of the crust and should be determined for each site from the observed ocean loading effect. Theoretical values were computed by Goad (1980). The amplitudes and phases of the vertical motions are tabulated for each tidal wave in the *Explanatory Supplement to the Astronomical Almanac* (Seidelmann, 1992). The amplitudes are generally smaller than 10 mm, but the effects of all tidal waves must be summed up, taking into account at each instant, the values of phases.

9.8.5 Atmospheric loading

The atmosphere exerts a variable load on the Earth's surface, depending linearly on the pressure. But, because of the Earth's elasticity, one has to consider the total load over a large surface, typically 2000 km in diameter. The effect can be modeled by different empirical formulae, the simplest being a linear function of the pressure, p_A, on the site, A, and a mean pressure, $\bar{p}$, in the surrounding region. Second-order polynomial representations have also been proposed. However, one has to rely only on the pressure actually measured at A. So, in practice, a linear function of p_A is used. The slope depends upon the relationship that may exist between the local pressure and the mean pressure $\bar{p}$ as a mean throughout a year or a season as well as on the load numbers. So the atmospheric load effect must be investigated at each site, and this was done in particular for some VLBI and GPS sites, in which observations are continuously performed.

The most accurate results were obtained for 11 VLBI sites (van Dam and Herring, 1994). The slope, expressed in millimeters per hectopascal, is generally of the order of -0.3 to -0.5 for height variations. Following theoretical calculations by van Dam and Wahr (1987), the peak-to-peak displacement may frequently be of the order of 20–25 mm, with accompanying gravity perturbations of 4.6 µGal. There are also horizontal motions that are of the order of one third of the vertical displacements.

In conclusion, as in the case of ocean loading, the atmospheric loading effects must be calibrated at each site, whenever observing precisions call for a knowledge of the position of the instrument to better than 2 cm.

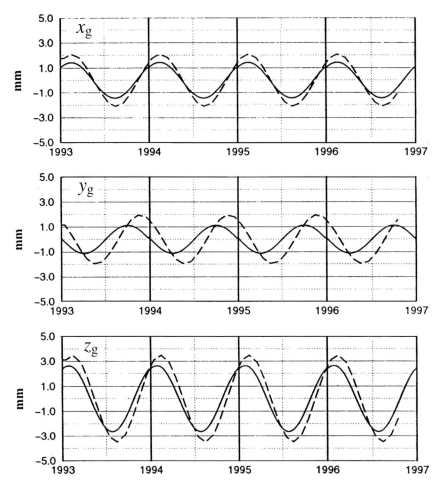

Fig. 9.11. Annual geocenter motion from space geodesy results (dashed curve) and a geophysical model including the main loading effects. (Reproduced with permission from Bouillé *et al.*, 2000, *Geophys. J. Int.*, **143**, 71–80.)

9.9 Dynamical consequences

The different local effects described in the previous section, when integrated over the Earth, modify the global distribution of matter in our planet and consequently its gravitational field. This surface mass redistribution is secular, owing to plate tectonics, post-glacial rebound, or melting of ice sheets. It also presents shorter, but not necessarily periodic, cycles produced by atmospheric or oceanic mass transfers. These changes in mass distribution produce changes in the Earth's gravitational potential (9.7). The result is perceptible by space geodesy techniques, insofar as the lowest-degree harmonics are concerned.

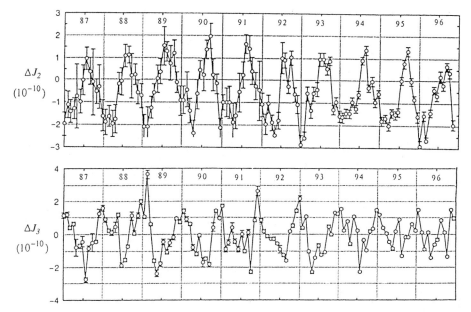

Fig. 9.12. Variations of J_2 and J_3 during 10 years from the observations of the satellites Starlette and LAGEOS1. (Adapted from Cheng and Tapley, 1999, Seasonal variations of low degree zonal harmonics of the Earth's gravity field from satellite laser ranging observations, *J. Geophys. Res.*, **104**, 2661–2681. Copyright 1999 American Geophysical Union. Reproduced by permission of American Geophysical Union.)

9.9.1 Motion of the geocenter

The appearance of first-order harmonics in the residuals of the determination of an Earth potential model is an indication that the reference system is not centered at the geocenter. The actual coordinates of the center of mass of the Earth are (Heiskanen and Moritz, 1967):

$$x_g = aC_{1,1}, \quad y_g = aS_{1,1}, \quad z_g = aJ_1. \tag{9.24}$$

The most important periodic effects are annual variations due to atmospheric and ocean loading and the variations of soil and underground water. The amplitude is of the order of a couple of millimeters and, as Fig. 9.11 shows, the geophysical model represents the measured effect within the uncertainties of the space geodesy determinations.

Actually, the spectrum of time variations of the first-order harmonics is quite complex, sensitive to shorter, and sometimes random, modifications of the atmospheric and hydrologic environments. It is, therefore, not possible to give a clear picture of these variations; it is to be compared with similar features in the motion of the polar motion (Section 10.4.1). It seems however, that annual, and smaller semi-annual motions of the geocenter are now well assessed (Chen *et al.*, 2000).

9.9.2 Variations of the Earth potential

Similarly, and for the same reasons, other lower terms of the geopotential undergo periodic variations. Such changes are now detected in the zonal harmonics J_2, J_3, J_4, J_5, J_6, and J_8 (Cheng and Tapley, 1999). They are of the order of a few units in 10^{-10}. The best-determined variations concern J_2 and J_3 as shown in Fig. 9.12.

Secular variations of some zonal harmonics have been detected. They are due to the consequences of the deglaciation of the late Pleistocene glacial cycle, and the resulting post-glacial rebound (Mitrovica and Peltier, 1993), together with the present ice melting (James and Ivins, 1997). The values found for the time derivatives are (Cheng *et al.*, 1997):

$$\dot{J}_2 = (-2.7 \pm 0.4) \times 10^{-11}/\text{year}, \quad \dot{J}_3 = (1.3 \pm 0.5) \times 10^{-11}/\text{year},$$
$$\dot{J}_4 = (-1.4 \pm 1.0) \times 10^{-11}/\text{year}, \quad \dot{J}_5 = (2.1 \pm 0.6) \times 10^{-11}/\text{year}.$$

10

Earth orientation

Without any internal or external forces acting on it, the Earth would rotate at a uniform rate around its center of mass and around an axis inclined by $23°26'$ with respect to the ecliptic pole. The mean rotation period is $23^h56^m4\overset{s}{.}10$. In reality, the Earth is composed of non-rigid parts in an oblate spheroidal shape and it is not isolated in space. These conditions produce variations in the rotation speed and in the direction of its axis. They will be described in the next section.

10.1 Earth orientation presentation

The orientation of the Earth in space is a complex combination of several types of motions. The axis of rotation moves with respect to the celestial reference frame. This is the precession–nutation already studied in Chapter 8. But this axis also moves with respect to the Earth's crust as represented by the ITRF. This is the polar motion. Finally, the Earth rotates at a variable rate about this axis. In this chapter, we study the last two effects, but in association with precession–nutation for a global description of the Earth orientation.

10.1.1 General description and history

The diurnal rotation axis is perpendicular to the equatorial plane, but the equatorial plane moves as described in Sections 8.5–8.7. It is not regular, and the variations in the rotation of the Earth were detected in the 1800s from the observed motion of the Moon, but they were not confirmed until correlations in the variations of the motions of the Moon and Mercury were established. De Sitter (1927), Stoyko (1937), and Spencer Jones (1939) discovered the secular deceleration and seasonal variations of the Earth's rotation rate. The variations in the Earth's rotation rate are now known to be of three types: secular, irregular, and periodic. The secular variation of the rotational speed is the approximately linear increase in the length-of-day

(LOD), primarily because of tidal friction. This effect causes a slowing of the Earth's rotational speed with a lengthening of the day by about 0.0005–0.0035 s per century. The irregular changes appear to be random, but may be correlated with physical processes within the Earth. These changes cause variations in the LOD by as much as 0.01 s over the past 200 years and consist of the so-called decade fluctuations with periods of 5–10 years and short interval variations. The higher-frequency variations are related to the changes in the total angular momentum of the atmosphere, oceans, ice caps, and interior of the Earth. Periodic variations are the result of repeatable physical processes, such as tides in the solid Earth due to the Sun and Moon. These produce variations in the LOD as large as 0.0005 s with periods of 1 year, 1/2 year, 27.55 days, 13.66 days, etc. Also, season changes in the global weather patterns will cause similar amplitude changes with annual and semi-annual periods. Finally, very-short-period (in particular diurnal and sub-daily) variations in the speed of rotation of the solid Earth are produced by an exchange of angular momentum between the atmosphere and the crust.

In addition to the variable Earth rotation, there is also polar motion. This is the motion of the Earth's axis of rotation with respect to the surface of the Earth. In 1765, Euler predicted that, if the axis of rotation is not coincident with the principal axis of inertia, the rotation axis would have a motion with respect to the Earth-fixed reference frame. In 1891, Chandler discovered the motion of geographical latitudes of astronomical observatories from observational data. However, the period is 433 days, while Euler predicted 305 days owing to the Earth's ellipticity. The difference is the result of the non-rigidity of the Earth. A second motion follows an approximately elliptical path with an annual period and is caused by the redistribution of the atmospheric and ice masses during the year. There also exist short period components sensitive to the global meteorological situation, and, in particular, to jet streams and high-altitude winds.

Polar motion is distinct from precession and nutation, which are motions of the axis of rotation with respect to the stars, or the International Celestial Reference Frame (ICRF), and can almost entirely be developed theoretically (see Section 8.7). This is not the case for the rotation of the Earth and polar motion, which must be determined from observations. From the 1890s until 1988, a regular program of optical observations, using zenith tubes and astrolabes, was conducted by the International Polar Motion Service (IPMS) and the Bureau International de l'Heure (BIH) to determine polar motion and the Earth's rotation. In the 1980s an IAU Working Group investigated methods of achieving greater accuracy by means of Very Long Baseline Interferometry (VLBI), lunar laser ranging, and other techniques. This led to the regular use of VLBI, and the progressive suppression of optical observations for Earth orientation parameter determination. Now, the use of Global Positioning System (GPS) measurements has made available almost continuous determinations of Earth Orientation Parameters (EOP). Thus, short-term (sub-daily) variations of both Earth rotation and polar motion can now be investigated. Since the GPS is a system around the Earth with orbits determined from Earth observations, it is not tied to an independent reference frame, and VLBI observations are still necessary.

This extension of techniques, and the necessity to merge the different results into a single solution, led to the establishment in 1988 of the International Earth Rotation Service (IERS) with broad responsibility for Earth orientation parameters. The IERS is organized into a central bureau, now in Germany, and various functional centers. Technical Notes are published as needed, including periodic editions of the IERS Conventions, which specify constants, reference frames, and methods to be used for Earth orientation. More information is available at http://www.iers.org.

10.1.2 Dynamical causes

The dynamical causes of Earth orientation changes can be classified as astronomical and geophysical. Most of them are common with those causing the precession and nutation (Section 8.5), but some are specific to the Earth's rotation and polar motion. The astronomical causes are tidal torques due to gravitational attractions of celestial bodies on the Earth. The geophysical causes are mass displacements in the Earth's layers (solid core, fluid core, mantle, oceans, and atmosphere), mechanical (gravitation, pressure, and viscosity), and electro-magnetic interactions between the Earth's layers.

It has been established that conservation of angular momentum is preserved, but it is the angular momentum of the complete Earth–Moon system. In the Earth's case the angular momentum is divided between the Earth's crust, the atmosphere, the oceans, the mantle, and the fluid and solid cores. So the observed variations of the Earth's surface are caused by redistributions of the angular momentum with the other components (see Section 9.8). The correlations have been well established, but to date the variations are not predictable to meaningful levels of accuracy. There are also observed seasonal short- and long-term variations.

The Moon raises tides (Section 9.8.1) in the ocean, and tidal friction carries the maximum tide ahead of the line joining the center of the Earth and Moon. The resulting couple diminishes the speed of rotation of the Earth and this reacts on the Moon to increase its angular momentum. The sum of the angular momentum on the Earth and the orbital momentum of the Moon remains constant to the extent that one neglects the very small contribution due to solar torques. This produces an increase in the size of the orbit of the Moon, and a reduction of the angular speed of the Earth. Let us note at this point that tidal friction is distinct from changes in the moments of inertia caused by tides.

10.1.3 Geophysical causes

Geophysical effects introduce complications into the astronomical forcing motions of Earth orientation. Gravitational and centrifugal forcing cause deformations and moment of inertia changes. These are responsible for modifications of nutation amplitudes and variations of the Earth's rotation, at periods the same as nutations and with amplitudes up to 1 ms. These effects are scaled by the elasticity factors of the mantle, the Love and Shida numbers (Section 9.8.2). The elastic deformations of centrifugal origin cause the modification of the period from Euler's theoretical to Chandler's observed value.

The fluid core has a clockwise, nearly diurnal wobble with respect to the crust, because the inertia axes do not coincide. The effect on the mantle is a motion independent of rotation, called *free core nutation*, with a period dependent on the flattening of the core–mantle boundary. In a celestial frame, the free nutation would appear as a clockwise oscillation. The present model of precession–nutation (IAU 2000 precession–nutation model, see Section 8.6), takes into account the rigid solution and non-rigid effects, but not the free core nutation which has to be derived from the observations. The underlying theories agree with observations at the sub-milliarcsecond level.

The displacement of masses inside the Earth's superficial layer, triggered by thermal, tidal, and electro-magnetic influences, have long-period effects on the Earth's rotation. Variations in the centrifugal forces due to wobble of the rotation axis cause a change in water height in the oceans, called "pole tide". This changes the period of the Chandler wobble.

Decade-long fluctuations of the length-of-day (LOD) might be caused by the superficial layers, such as the El Niño cycle, in the atmosphere and ocean, and coupling between the core and mantle. The secular change in the LOD is attributed to tidal friction, which transfers the Earth's angular momentum to the lunar orbit.

An observed secular motion of the pole with respect to the crust of about $0.''3$ per century toward $80°$ West is attributed to the slow viscous mass redistribution (post-glacial rebound) associated with the ice cap melting after the termination of the Pleistocene glaciation 20 000 to 15 000 years ago.

While qualitative, if not quantitative, explanations exist for the variations in the Earth's orientation due to geophysical effects, there remains an unexplained aspect to the Chandler motion of the pole. The theory is that, in the absence of any external torque, the rotation axis turns around the axis of figure in the terrestrial frame. However, the Earth is not a perfect elastic body, so this motion should be dissipated in thermal energy. No damping of the Chandler motion has been observed. Thus, there must be one or several excitation mechanisms, which have not been identified.

Earth's rotation plays an important role in physical theories. While an extra-galactic reference frame is very close to being inertial (see Chapter 7), the rotating crust is non-inertial. Non-inertial frames, undergoing acceleration with respect to the inertial frame, are subject to centrifugal and Coriolis effects. The consequence is that the Earth has an equatorial bulge and lower gravity at the equator than at the poles. Historically, this has been confirmed by geodetic measurements and latitude dependent pendulum recordings in the eighteenth century. Also, the Foucault pendulum, designed in 1851, exhibits the rotation of the oscillation plane from the Coriolis effect due to the Earth's rotation.

There are many reciprocal influences between the Earth's rotation and the behavior of various components of the Earth, in particular the atmosphere and the oceans. For a complete description of geophysical causes and consequences of the Earth's rotation, see Lambeck (1980).

10.2 Observation of the Earth's rotation

The methods and accuracy of observations of Earth rotation, polar motion, precession, and nutation have changed. With the availability of VLBI observations, the Earth orientation parameters and precession–nutation can now be determined much more accurately. The optical observations were limited to accuracies of about 200 mas and reference frame accuracies of about 30 mas. VLBI observations are close to milliarcsecond accuracies and the frame can be determined to 0.020 mas. Thus, the accuracies of precession–nutation and Earth orientation parameters have improved accordingly. The present daily EOP determinations are of the order of 0.2 mas, or 1 cm, accuracy. In addition, the observations can be related directly to a pseudo-inertial reference frame. The observational techniques are summarized in the following sections.

10.2.1 Classical techniques

The variations in the rotation of the Earth and the position of the pole have been monitored since the latter part of the nineteenth century. Observations by photographic and visual zenith tubes, astrolabes, lunar occultations, eclipses, and transits of stars and Solar System observations have been used. Many studies over the years have been complicated by their dependence on the position and motion of the Moon and whether there has been a change in the acceleration of the Moon's mean motion and any change in the rate of Ephemeris Time based on the Moon's motion. These astronomical observations indicate that the Moon's mean motion is constant. Results of these studies are given by Stephenson and Morrison (1984), McCarthy and Babcock (1986), and Stephenson and Lieske (1988). A review of optical observations for solutions of Earth orientation parameters was done by Vondrak and Ron (2000). But recently, using lunar laser ranging data, Chapront *et al.* (2002) have detected the tidal acceleration of the lunar longitude and found it equal to

$\Gamma = -25''\!.858 \pm 0''\!.003$ per century squared. Historical summaries of length of day and polar motion determinations are given in the *Explanatory Supplement* (Seidelmann, 1992).

10.2.2 Very Long Baseline Interferometry (VLBI)

The availability of VLBI observations (see Section 2.4.2) provides much more accurate observations that are tied to an extragalactic reference frame, which has no apparent motion. Thus, this source of observations became the preferred method for determination of Earth orientation parameters. VLBI has disadvantages in that there is a time lag due to the transportation of the observational tapes from the observation site to a correlator and the correlation time. Thus, there is a delay in the availability and currency of the observations. In addition, VLBI is an expensive observational technique.

10.2.3 Laser ranging

The installation of corner cube reflectors on the Moon and on artificial satellites provided targets for laser ranging from sites on Earth (see Section 2.4.3 and Kovalevsky, 2002). Although lunar laser ranging measurements are few in number, they have the advantage of being available immediately without the delay required to correlate the VLBI observations. Satellite ranging observations can be used to determine the motions of satellites, motions of the observing sites, and Earth orientation parameters. The observations are complicated by the multitude of parameters that are required in the solution, but can provide very accurate distance measurements between the observatory and the reflector. The quantity of such observations has never been sufficient to allow the technique to be a primary source for Earth orientation parameters, but the observations can be very timely.

10.2.4 Global Positioning System (GPS)

The availability of the Global Positioning System (GPS) for navigation and timing provided a powerful geological observational tool (see Section 2.4.1). Observations can be made continuously from many sites for very little cost. The observations from well-determined sites are combined for determinations of geodetic datums and plate motions. They have the disadvantage that they are not independent of the Earth, since, as with satellite ranging, the observations link the surface of the Earth to satellites in motion around the Earth. For Earth rotation they have the further complication that a part of the variable Earth rotation that is being sought is coplanar with a variable solar radiation pressure that is affecting the motions of the satellites. However, a method has been developed using models for the motion of the nodes of the orbits (Kammeyer, 2000). Therefore, from GPS observations variations in rotation and polar motion over short time periods of minutes can be measured and are precise to $\pm 20\,\mu s$/day. This has provided a powerful new tool in the investigations of very-short-period components of polar motion, nutation, and Earth rotation. However, GPS observations are tied to a reference frame that is not inertial, so VLBI observations are still necessary with respect to the ICRF. Details concerning GPS observations can be found in Kovalevsky (2002) and Hofmann-Wallenhof *et al.* (1992).

10.3 Celestial and terrestrial moving frames

To describe the rotation of the Earth, one needs to specify the rotation axis and the rotation angle of a reference system linked to the solid Earth with respect to some celestial reference. However, a direct link between the terrestrial reference frame (now the ITRF) and the celestial reference frame (now the ICRF) that would be described by a single transformation matrix, $\mathcal{M}$, would involve too many parameters and a variety of different physical causes, only some of which can be modeled. So, an intermediate moving reference frame is introduced and the transformation is made in two steps.

First, the correction for *polar motion* and the Earth's rotation proper, described by a matrix $\mathcal{R}^\star$ transforms ITRS into the *Intermediate Reference System* as defined in Section 8.4. Then, a precession–nutation matrix $\mathcal{P}N$ (Section 8.7) transforms the intermediate system into the celestial reference system. One has,

$$\mathcal{M} = \mathcal{P}N \times \mathcal{R}^\star. \tag{10.1}$$

Traditionally the equinox, the intersection of the ecliptic and moving mean equatorial plane, has been used as the fiducial point for measurements along both the equator and the ecliptic. With the transition from the dynamical system to the kinematic system a new fiducial point was desired. A number of possibilities were considered as origins on the moving equator. These included the intersections of the fixed ecliptic, with the zero-line meridian of the International Celestial Reference System (ICRS), and with the instantaneous prime meridian, the node on the celestial reference great circle, and the so-called non-rotating origin (Kovalevsky and McCarthy, 1998).

It is desirable, in defining the origin for the Earth's rotation on the moving equator, to eliminate the spurious rotation due to the motion of the equator with respect to the ICRS due to precession and nutation. In considering each of the various geometrical definitions, the instantaneous motion of the considered origin becomes added to the Earth's angular rotation. This led to the choice of an origin defined by the kinematical property of having no instantaneous rotation about the Geocentric Celestial Reference System (GCRS) axis of rotation. This means that the instantaneous movement of the CEO is always at right angles to the instantaneous equator.

This is the so-called non-rotating origin introduced by Guinot (1979). It is designated as the Celestial Ephemeris Origin (CEO) with respect to the GCRS. Similarly, it was necessary to define a Terrestrial Ephemeris Origin (TEO), which designates the terrestrial non-rotating origin in the International Terrestrial Reference System (ITRS) (Seidelmann and Kovalevsky, 2002).

In this approach to the description of the rotation of the Earth in the ICRS, one also has to define the Intermediate Reference Frame. Its pole, the Celestial Intermediate Pole (CIP), is no longer defined by a physical property or a model, but it is just a conventional intermediate pole for the transformation between the geocentric ICRS (GCRS) and the ITRS. The CIP separates the motion of the ITRS pole into a *celestial part* including the long-period precession–nutation motion with periods greater than two days in the GCRS, and the *terrestrial part* including the long-period, daily and sub-daily components of the polar motion. This terrestrial part contains also the variations corresponding to the prograde diurnal and semi-diurnal nutations as well as the daily and sub-daily tidal variations in polar motion. The motion of the CIP in the GCRS is determined from the IAU 2000A model for precession and forced nutation for periods greater than two days, plus the celestial pole offsets as estimated from observations. The motion of the CIP in the ITRS is determined from Earth orientation observations and takes into account the modeled high-frequency variations in both the GCRS and ITRS.

10.3.1 Celestial Ephemeris Origin

For the concept of the Celestial Ephemeris Origin (CEO), consider a rigid body represented by a celestial sphere, with fixed ICRS reference Cartesian axes $OXYZ$ and Σ_0, the origin of the equator of the ICRS (intersection of OX with the celestial sphere), as in Fig. 10.1. Let P represent the pole of the intermediate system at time t, defining the Cartesian coordinates $O\xi\eta\zeta$, such that $O\zeta$ is along OP and σ is the point on the equator of P where $O\xi$ pierces the sphere. The condition is imposed that for any infinitesimal displacement of P there is no rotation around $O\zeta$. Then σ would be the non-rotating origin, that is the CEO on the moving equator of P.

Let, similarly, P_0 be the pole of the intermediate system at time t_0 and the corresponding CEO (σ_0). The motion of point P between dates t_0 and t is determined by evaluating the quantity s

$$s = [\sigma N] - [\Sigma_0 N] - ([\sigma_0 N_0] - [\Sigma_0 N_0]), \tag{10.2}$$

where quantities in brackets are arcs measured on oriented great circles shown in Fig. 10.1; N_0 and N are the ascending nodes of the equators at t_0 and t on the equator of the ICRS. Then s is given by

$$s = -\int_{t_0}^{t} \frac{X\dot{Y} - Y\dot{X}}{1 + Z} dt - ([\sigma_0 N_0] - [\Sigma_0 N_0]), \tag{10.3}$$

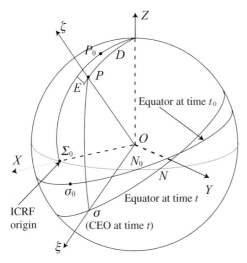

Fig. 10.1. Definition of the non-rotating origin (σ) on ICRF with motion of the pole P from t_0 to t.

where the dot denotes the time derivative. In polar coordinates, if D is the angle (**OZ**, **OP**) and E is the angle of the plane **OZP** with respect to the principal plane **OXZ**, the equation is

$$s = \int_{t_0}^{t} (\cos D - 1)\dot{E}\,dt - ([\sigma_0 N_0] - [\Sigma_0 N_0]).$$

The convention is adopted that t_0 corresponds to the epoch when the true (or intermediate) equator coincided with the equator of the ITRF (J2000.0). Then, $[\sigma_0 N_0] = [\Sigma_0 N_0]$ and is the constant $s(t_0)$ in the integral. Then,

$$s = \int_{t_0}^{t} (\cos D - 1)\dot{E}\,dt. \tag{10.4}$$

For the period 1975–2025, the development of $s(t)$ including all terms exceeding $0.5\,\mu\text{as}$ is, in μas (Capitaine *et al.*, 2002):

$$
\begin{aligned}
s(t) = {} & -\frac{XY}{2} + 2004 + 3812t - 121t^2 - 72\,574t^3 \\
& + \sum_k C_k \sin \alpha_k + 2t \sin \Omega + 4t \sin 2\Omega \\
& + 744t^2 \sin \Omega + 57t^2 \sin 2L \\
& + 10t^2 \sin 2L' - 9t^2 \sin 2\Omega,
\end{aligned}
\tag{10.5}
$$

where t is expressed in centuries since J2000.0 ($t_0 = 0$) and where α_k and C_k are given in Table 10.1. To extend the development for the periods 1900–1975 and

Table 10.1. *Additional terms in $s(t)$ and $\Delta s(t)$ for Equations (10.5) and (10.6) in μas units (Capitaine et al., 2002)*

α_k	C_k	α_i	D_i
Ω	-2641	l'	-6
2Ω	-63	$l' + 2L$	$+2$
$\Omega + 2L$	-12	l	-3
$\Omega - 2L$	$+11$	$2L' + \Omega$	$+2$
$2L$	$+5$	$l+2L'$	$+1$
$\Omega + 2L'$	-2		
$\Omega - 2L'$	$+2$		
3Ω	$+2$		
$l' + \Omega$	$+1$		
$l' - \Omega$	$+1$		

2025–2100 for terms exceeding 0.5 μas, add in μas:

$$\Delta s(t) = +28t^4 + 15t^5 - 22t^3 \cos \Omega - t^3 \cos 2L + \sum_i D_i t^2 \sin \alpha_i, \tag{10.6}$$

where α_i and D_i are given in Table 10.1. In this table and in Equations (10.5) and (10.6), the notations, referred to the ICRF axes and the corresponding fixed ecliptic are:

Ω: mean longitude of the node of the lunar orbit,
L: mean longitude of the Sun,
L': mean longitude of the Moon,
l: mean anomaly of the Sun ($l = L - \varpi$) where ϖ is the mean longitude of the Sun's perigee,
l': mean anomaly of the Moon ($l' = L' - \varpi'$), ϖ' being the mean longitude of the Moon's perigee.

At present, the uncertainty of these expressions is smaller than 1 μas and is insensitive to expected modifications of the nutation–precession theory to about this level. A maximum error of 3 μas could be reached in 2100 (Capitaine *et al.*, 2000).

10.3.2 Terrestrial Ephemeris Origin

For Earth orientation, the concept of the non-rotating origin must be applied to both the CRS and the TRS. The definitions are quite similar.

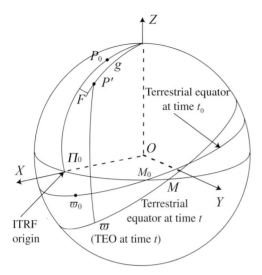

Fig. 10.2. Definition of the non-rotating origin (NRO) on ITRF with motion of the pole P' from t_0 to t.

Let us call M the ascending node of the instantaneous equator corresponding to the terrestrial pole P' on the ITRF equator (Fig. 10.2). The non-rotating origin ϖ on the instantaneous equator is derived from the origin Π_0 of the terrestrial longitudes by the evaluation of s', with similar conventions as for s concerning the subscript 0 corresponding to $t = 0$, one has:

$$s' = [\varpi M] - [\Pi_0 M] - ([\varpi_0 M_0] - [\Pi_0 M_0]). \tag{10.7}$$

If $u(t)$ and $v(t)$ are the geocentric angles of the components x and y of the position of P' in the ITRS, then,

$$s' = -\int_{t_0}^{t} \frac{u\dot{v} - v\dot{u}}{2} dt - ([\varpi_0 M_0] - [\Pi_0 M_0]). \tag{10.8}$$

Here, because of the very small magnitude of the polar displacements, the third component of $\mathbf{OP'}$ may be taken as equal to 1 to better than 1 μas in several centuries (Capitaine *et al.*, 2000). Now, adopting the convention that ϖ_0 is on the ITRF equator at $t = 0$, the second part of the equation is cancelled by the constant part s'_0 of the integral. Then, if F and g are the polar coordinates of P', one obtains the final expression analogous to (10.4):

$$s' = \int_{t_0}^{t} (\cos g - 1)\dot{F} dt. \tag{10.9}$$

This is a very small correction that must however be taken into account when microarcsecond accuracies are sought. The quantities F and g are provided by the IERS.

10.4 Description of the Earth's orientation

The matrix $\mathcal{R}^\star$ entering in (10.1) includes two components. One is the polar motion, which represents the displacement of the axis of rotation with respect to the surface of the Earth realized by the ITRF (Section 10.4.1). Let us call the corresponding rotation matrix, $\mathcal{W}$. The other component describes, using the matrix, $\mathcal{R}$, the Earth's rotation with respect to some celestial meridian (or a fiducial point on the equator). This rotation is described by the *sidereal time* (Section 10.4.3) in the old equinox–equator system and by the *stellar angle* (Section 10.4.4) in the new system. So, one has,

$$\mathcal{R}^\star = \mathcal{R}(t) \times \mathcal{W}(t). \tag{10.10}$$

10.4.1 Polar motion

In the past, the terms nutation and polar motion were defined as the forced and free motions of the reference pole. However, these separations were not completely clear. The forced solution accounts for the external forces, i.e. gravitational forces due to the Sun, Moon, and planets. The free solution is the solution of the differential equations that results from setting the external forces to zero. For a rigid Earth this results in the Eulerian free motion, which is a component of polar motion with annual periods in the terrestrial frame. For a non-rigid Earth the period is lengthened to about 14 months, and is called the Chandler component of polar motion. Thus, the forced, annual, polar motion due to climate and motions resulting from geophysical strains and meteorological effects might logically be included in nutation, but are included in polar motion, because these effects cannot be calculated as accurately as the external forcing effects. In addition, with the Global Positioning System (GPS) observations being made on an almost continuous basis, short period effects are detectable. Now, with the definition of the intermediate reference frame, an arbitrary separation has been introduced that anything with a period shorter than two days is considered as polar motion.

For an Earth with a fluid core there exists a second free solution, the nearly diurnal free polar motion, also called the nearly diurnal, free wobble, or free core nutation. This is not in the nutation series and the observed values are provided by the IERS.

The pole and zero meridian of the International Terrestrial Reference Frame (ITRF, see Section 9.3) are defined implicitly by the ITRF coordinates for instruments used to determine the Earth orientation parameters. Polar motion is specified in terms of two small angles, x and y, which correspond to the coordinates of the Celestial Intermediate Pole (CIP) with respect to the ITRF pole, and are measured along the meridians at longitudes $0°$ and $270°$ ($90°$ west). If we take also into account the shift s' from the zero meridian to the Terrestrial Ephemeris Origin (10.9), then the polar motion matrix is

$$W(t) = \mathcal{R}_3(-s') \times \mathcal{R}_1(-y) \times \mathcal{R}_2(x). \tag{10.11}$$

The values for polar motion are given by the IERS and are available at its web site.

10.4.2 Rotation of the Earth around the CIP axis

The Earth rotates about its axis and at the same time moves in its orbit around the Sun. Thus, while the Earth rotates with respect to an inertial frame, or the ICRF, it moves in its orbit so that additional rotation with respect to the Sun must be included. As said in the introduction of this section, there are two methods to express the angle of rotation with respect to some celestial meridian taken as origin. Both provide the parameter for the time-dependent matrix $\mathcal{R}(t)$ as a function of time.

(i) The celestial reference fiducial point is the true equinox, γ, on the instantaneous equator. The parameter is the *sidereal time* (Section 10.4.3).
(ii) The celestial fiducial point is the Celestial Ephemeris Origin (CEO). The corresponding parameter is the *stellar angle* (Section 10.4.4).

10.4.3 Sidereal time

Historically, sidereal time was the hour angle of the catalog equinox, and was subject to the motion of the equinox due to precession and nutation (see Section 8.5).

The local hour angle of the equinox is the local sidereal time (τ_L) and increases by $360°$ or 24 h in a sidereal day. It is to be remarked that, strictly speaking, the sidereal time is an angle, not a time. The local hour angle of an object (H_L) equals the local sidereal time minus the right ascension, α, of the object.

$$H_L = \tau_L - \alpha. \tag{10.12}$$

The local sidereal time is equal to the right ascension of the local meridian, and is determined by the techniques described in Section 10.2. Local sidereal time (τ_L) is calculated from sidereal time on the International Meridian ($\tau = $ GST = Greenwich sidereal time), when the geographic longitude λ (eastward) is known; thus

$$\tau_L = \tau + \lambda. \tag{10.13}$$

One must distinguish the sidereal time proper, which contains periodic as well as secular terms of the precession–nutation, from a *mean sidereal time*, which is an expression in which only secular terms are present. Finally, the matrix $\mathcal{R}(t)$ representing the rotation as measured from the equinox γ at a time t is

$$\mathcal{R}(t) = \mathcal{R}_3(-\tau). \tag{10.14}$$

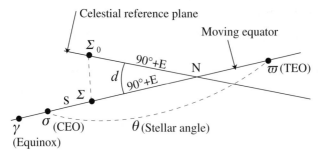

Fig. 10.3. Definition of the stellar angle.

10.4.4 Stellar angle

In the CEO based system the equivalent of sidereal time is the hour angle of the Celestial Ephemeris Origin (CEO), and hence is subject only to the motions of that origin. It is called the *stellar angle* or *Earth rotation angle*, and is noted θ. The stellar angle in Fig. 10.3 is defined by the angle measured along the equator of the Celestial Intermediate Pole (CIP) between the Terrestrial Ephemeris Origin (TEO) and the CEO, positively in the retrograde direction. The main property of the CEO, being a non-rotating origin, ensures that the derivative $\dot{\theta}$ is strictly equal to the instantaneous angular velocity ω of the Earth around the CIP. Thus, θ represents rigorously the sidereal rotation of the Earth around its axis, and the matrix $\mathcal{R}(t)$ at a time t is given by

$$\mathcal{R}(t) = \mathcal{R}_3(-\theta). \tag{10.15}$$

The numerical relation between t and θ is given by Equation (10.20) in the next section.

There is also a relation that links the Greenwich sidereal time τ and the stellar angle θ. Capitaine and Gontier (1993) gave an analytical expression as a function of the nutation referred to the ecliptic of epoch and of precession. Inserting numerical values given by the IAU 2000 precession–nutation theory, one obtains (Capitaine *et al.*, 2002),

$$\tau = \theta + 4612\!''\!.157\,482\,t + 1\!''\!.396\,678\,41\,t^2 - 0\!''\!.000\,093\,44\,t^3$$
$$+ 0\!''\!.000\,0188\,t^4 + \Delta\psi \cos\epsilon_A - \sum C_k \sin\alpha_k - 0\!''\!.002\,012, \tag{10.16}$$

where the quantities C_k and α_k are given by Table 10.1, $\Delta\psi$ is the nutation in longitude referred to the mean equator and equinox of date provided by the IAU 2000 nutation–precession series, and the obliquity of the ecliptic ϵ_A is given by

$$\epsilon_A = 84381\!''\!.448 - 44\!''\!.8150T - 0\!''\!.000\,59T^2 + 0\!''\!.001\,813T^3, \tag{10.17}$$

where T is in Julian centuries since 2000 January 1, 12 UT1.

10.5 Universal Time

The rotation of the Earth with respect to the Sun is called Universal Time. Thus, the natural unit is the day. The definition of the day has naturally been based on the rotation of the Earth with respect to the Sun. Owing to the orbital motion of the Earth around the Sun in about 365 days, the Earth must rotate about four minutes longer to be in the same angular position with respect to the Sun than to be in the same angular position with respect to the stars. Actually, there is a relation between Universal Time and either the sidereal time or the stellar angle.

10.5.1 Solar time

The time defined in terms of the Sun is the solar time. The local apparent solar time (LAT) is related to the local hour angle of the Sun, $H_\odot$ but is, by definition, 12 hours larger. Now, using (10.12), one gets

$$\text{LAT} = \text{LST} - \alpha_\odot - 12 \text{ hours}, \tag{10.18}$$

where LST is the local sidereal time and $\alpha_\odot$ is the right ascension of the Sun.

Owing to the eccentricity of the Earth's orbit about the Sun and its inclination with respect to the equator, the right ascension of the Sun does not vary uniformly with time. To define a more uniform time scale, the concept has been introduced of a point U, called *mean Sun*, that moves along the celestial equator at a uniform rate and whose right ascension is α_U. This is the Universal Time (UT), defined in terms of Greenwich sidereal time (τ) by an expression of the form

$$\text{UT} = \tau - \alpha_U - 12 \text{ hours}, \tag{10.19}$$

where the coefficients in the expression of the right ascension of the mean Sun, RA(U), can be chosen so that UT may be regarded as mean solar time on the International (Greenwich) meridian.

The variation in the apparent motion of the Sun introduces a difference between local mean time (LMT) and the local apparent solar time (LAT) known as the equation of time. This is the difference between the time given by a sundial and the time given by a clock for a given location.

$$\text{LAT} = H_\odot + 12 \text{ hours} = \text{LMT} + \text{equation of time}, \tag{10.20}$$

where $H_\odot$ is the local hour angle of the Sun. The principal contributions to the equation of time arise from the eccentricity of the Earth's motion around the Sun and the inclination of the plane of the ecliptic to the plane of the equator. The equation of time varies through the year in a smoothly periodic manner by up to 16 minutes (see Fig. 10.4). This is also the reason why in June and December, when

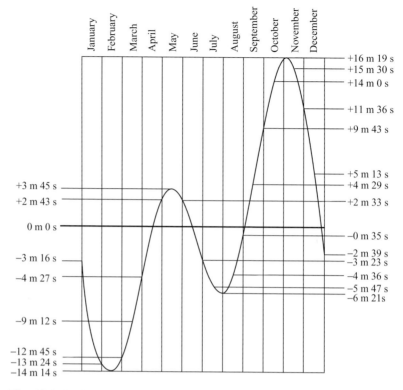

Fig. 10.4. Variation in the equation of time through the year.

the longest and shortest periods of sunlight occur, the times of sunrise and sunset do not reach their extreme values on the shortest and longest days.

10.5.2 Universal Time

As just mentioned, Universal Time conforms closely to the mean diurnal motion of the Sun. The apparent diurnal motion of the Sun involves both the non-uniform diurnal rotation of the Earth and the motion of the Earth in its orbit around the Sun. Universal Time is related to sidereal time or the stellar angle by means of a numerical formula. It can be derived at any instant from observations of the diurnal motion of stars or radio sources. The uncorrected, observed, rotational time scale, which is dependent on the place of observation, is designated UT0. When this is corrected for the shift in longitude of the observing station caused by polar motion, it is designated UT1. This is influenced by the variable rotation of the Earth and is the usual version of Universal Time.

Thus, UT1 is proportional to the angle of rotation of the Earth in space, around the true position of the rotation axis. The rate of UT1 is chosen so that the day of

UT1 is close to the mean duration of the solar day, and the phase of UT1 is chosen so that 12^h UT1 corresponds approximately to the instant when the Sun crosses the international meridian. In practice UT1 is determined from observations of sources defining the extragalactic reference frame (ICRF) and thus varies with the true rotation of the Earth.

Since January 1, 1984, Greenwich mean sidereal time (GMST) has been related to UT1 by the equation (Aoki *et al.*, 1982):

$$\text{GMST for 0h UT1} = 24\,110\overset{s}{.}548\,41 + 8640\,184\overset{s}{.}812\,866\,T_u'$$

$$+ 0\overset{s}{.}093\,104\,T_u'^2 - 6\overset{s}{.}2 \times 10^{-6}T_u'^3 , \qquad (10.21)$$

where T_u' is expressed in Julian centuries, $(T_u' = d_u/36\,525$, d_u being the number of days of Universal Time elapsed since JD 2 451 545.0 UT1 (2000 January 1, 12h UT1) taking on values of $\pm 0.5, \pm 1.5, \pm 2.5, \pm 3.5, \ldots$.

Let us note that the IAU 2000 precession–nutation series uses a different precession rate and an offset in obliquity. This produces a change in GMST, which can be written in microseconds, as:

$$dGMST = (-18\,330.266\,T_u' + 7.87\,T_u'^2 + 1.27\,T_u'^4)\,\mu s. \qquad (10.22)$$

Now, with the introduction of the new intermediate reference system, UT1 is linearly proportional to the stellar angle (or Earth rotation angle). The definition of UT1 by its relationship with mean sidereal time is replaced by a similar definition as a function of the stellar angle. It is proportional to the angle θ defined in Section (10.4.4) as the angle between the TEO and the CEO. Thus, the definition of UT1 is such that the time derivative of UT1 is proportional to ω.

The numerical relationship between the stellar angle and UT1 has been derived (Capitaine *et al.*, 2000) to be consistent with the conventional relationship between GMST and UT1. This provides the definition of UT1 for the ICRF. To an accuracy of 1 μas, the relationship is

$$\theta(T_u) = 2\pi(0.779\,057\,273\,2640 + 1.002\,737\,811\,911\,354\,48\,T_u), \qquad (10.23)$$

where θ is expressed in radians, and T_u in days:

$$T_u = (\text{Julian UT1 date} - 2\,451\,545.0).$$

One can also express θ in *seconds*, that is in units analogous to those used for GMST in Equation (10.19):

$$\theta_s(T_u) = (67\,310\overset{s}{.}548\,410\,01 + 86\,636\overset{s}{.}546\,949\,141\,027\,086\,T_u).$$

This linear expression shows that, as implied by its definition, and in contrast with GMST, the stellar angle for 0h UT1 is a constant.

10.5.3 Universal time scale

The variable rotation of the Earth being the basis for the UT1 time scale, the difference between UT1 and the uniform time scale TAI (Section 5.5.1) increases while the Earth rotation is slowing down, thus leading to the introduction of leap seconds into a uniform time scale also based on TAI, called UTC (Coordinated Universal Time). UTC is the basis for all civil timekeeping. The introduction of a leap second is decided by the IERS, when the difference, UT1−UTC, tends to be larger than 0ˢ9, and the introduction takes place as a sixty-first second in the last minute of June or December.

The length of the SI second, in terms of the cesium atom, was determined in the mid 1950s from observations of the motion of the Moon with respect to the new cesium atomic time standard. The lunar orbit was related to the solar ephemeris of Newcomb that was developed in 1898. That ephemeris was based on observations of the Sun from the nineteenth century. The mean epoch of the observations was about 1850. From this ephemeris the second of Ephemeris Time had been defined, and it was to that ephemeris second that the cesium frequency was related. So, the SI second was really defined in terms of the second of time appropriate to the rotation of the Earth in 1850. Thus, a century and a half of Earth slowing is reflected in the difference between the current rate of rotation of the Earth and the current rate of the uniform time. This difference is expected to continue to increase and require more frequent leap seconds.

There is also the complication that modern electronic navigation and communication systems are based on the Uniform Time system, and they are not capable of accepting leap seconds. A system like the GPS navigation system cannot have a step in its time scale at critical times, such as when an aircraft is in final approach for a landing. This combination of more frequent leap seconds in the future and systems that cannot tolerate leap seconds has led to the initiation of studies of alternative means of relating a variable rotating Earth and a uniform, accurate time scale.

10.6 ITRF to ICRF transformations

The Earth rotation parameters define the coordinate transformation from a terrestrial reference (ITRF) frame (coordinates x, y, z) to a celestial reference (ICRF) frame (coordinates X, Y, Z); taking into account the two components of $\mathcal{R}^{\star}$ defined by (10.10) and the transformation from the intermediate system to ICRS as described in Chapter 8, and which can be represented by the precession–nutation matrix $\mathcal{P}N(t)$. The latter is the product of the precession matrix $\mathcal{P}$ given by Equation (8.8) by the nutation matrix $\mathcal{N}$ or $\mathcal{N}'$ given by Equations (8.18) or (8.20) respectively.

The overall transformation is generally written in the following form:

$$\begin{pmatrix} X \\ Y \\ Z \end{pmatrix} = \mathcal{P}N(t) \times \mathcal{R}(t) \times \mathcal{W}(t) \times \begin{pmatrix} x \\ y \\ z \end{pmatrix}, \tag{10.24}$$

where the fundamental components $\mathcal{P}N(t)$, $\mathcal{R}(t)$, $\mathcal{W}(t)$ are the transformation matrices arising from the motion of the celestial pole in the CRS owing to precession and nutation, from the rotation of the Earth around the same pole, and from polar motion, respectively.

The $\mathcal{P}N(t)$ matrix is the product of the $\mathcal{P} \times \mathcal{N}$ of the precession matrix $\mathcal{P}$ given by (8.8) by the nutation matrix $\mathcal{N}$ given by (8.19).

10.6.1 Classical and improved classical systems

The difference between these two systems lies only in the choice of different precession–nutation theories and poles (CEP, see Section 8.6.1, or CIP, see Section 8.4). The transformation algorithms are the same, so that the two cases are presented together.

In Equation (10.11), the matrix $\mathcal{W}(t)$ is determined by the two components of the polar motion in terms of two small angles, x and y, which correspond to the coordinates of the celestial pole with respect to the terrestrial pole. The first matrix $\mathcal{R}_3(s')$ is not introduced in this case.

$$\mathcal{W}(t) = \mathcal{R}_1(y) \times \mathcal{R}_2(x). \tag{10.25}$$

The matrix $\mathcal{R}(t)$ is, similarly,

$$\mathcal{R}(t) = \mathcal{R}_3(-\tau) \tag{10.26}$$

In these equations, the axis of the rotation is defined by the conventions of Section 3.4.1. The quantity τ is the Greenwich sidereal time at date t and is obtained in three steps.

   (i) Take the relationship between Greenwich mean sidereal time (GMST) and Universal Time UT1 as given by Equation (10.21), possibly corrected by (10.22)
   (ii) Interpolate for any time of the day, and obtain GMST(t). One must use the ratio r of universal to sidereal time with the same notations as in (10.21):

$$r = 1.002\,727\,909\,350\,795 + 5.9006 \times 10^{-11} T'_{\mathrm{u}} - 5.9 \times 10^{-15} T'^{2}_{\mathrm{u}}.$$

   (iii) Add the accumulated precession and nutation in right ascension as given by Aoki and Kinoshita (1983):

$$\tau = \mathrm{GMST} + \Delta\psi \cos \epsilon_A + 0''.002\,64 \sin \Omega + 0''.000\,063 \sin 2\Omega. \tag{10.27}$$

Here, $\Delta\psi$ is the nutation in longitude and ϵ_A is the obliquity of the ecliptic given by (10.17).

Thus, the angle τ refers to the ecliptic of date and includes Earth rotation, accumulated precession and nutation along the equator, crossed terms between precession and nutation, and crossed nutation terms (Capitaine and Gontier, 1993). The secular term in the GMST and UT1 relationship mixes UT1 and Barycentric Coordinate Time (TCB).

The third matrix of (10.24) $\mathcal{PN}(t)$ is described above.

The above procedure conforms to the position and motion of the equinox specified in the IAU 1976 System of Astronomical Constants, the 1980 IAU Theory of Nutation, and the positions and proper motions of stars in the FK5 catalog. Improved values for constants in the above expressions may be available in the future.

10.6.2 CEO-based system

The ICRS eliminates any possible global rotation and any link to the orbital motion of the Earth. So in the new system the Earth Orientation Parameters (EOP) are based on the intermediary system defined by the CIP and the CEO. We have seen, in Chapter 8, that the position of the CIP is presented under the form of series for nutation in longitude $\Delta\psi$ and obliquity $\Delta\epsilon$. The derivation of the position (X, Y, Z) of the CIP, using the IAU 2000 precession–nutation series is described in Section 8.7.2.

The coordinate transformation from the ITRS to the ICRS at date t is given by (10.24). The matrix $\mathcal{W}(t)$ is given by (10.11) and the matrix $\mathcal{R}(t)$ is given by (10.15). The precession–nutation matrix, $\mathcal{PN}(t)$, can be obtained in the following manner. Using the notations of Section 10.3.1, one has

$$X = \sin D \cos E,$$
$$Y = \sin D \sin E. \tag{10.28}$$

Let us call

$$a = \frac{1}{(1 + \cos D)}, \tag{10.29}$$

which can also be written with sufficient accuracy as

$$a = \frac{1}{2} + \frac{1}{8}(X^2 + Y^2).$$

Under these conditions, one has

$$\mathcal{PN}(t) = \mathcal{R}_3(-E) \times \mathcal{R}_2(-D) \times \mathcal{R}_3(E) \times \mathcal{R}_3(s). \tag{10.30}$$

The notation $\mathcal{R}_i$ follows the notations of Section 3.4.1. One may note, that $\mathcal{P}N(t)$ can also be given in an equivalent form directly from X and Y as:

$$\mathcal{P}N(t) = \begin{pmatrix} 1 - aX^2 & -aXY & X \\ -aXY & 1 - aY^2 & Y \\ -X & -Y & 1 - a(X^2 + Y^2) \end{pmatrix} \cdot \mathcal{R}_3(s). \qquad (10.31)$$

10.7 Other effects

The accurate determination of the Earth orientation parameters requires the consideration of a number of forces and effects. In order to achieve consistent results, the IERS has established conventions that should be used for all data reduction procedures. These conventions are revised periodically and published as an IERS Technical Note by the IERS Central Bureau. Items that need to be considered in the different data reduction procedures include the geopotential model, site displacement due to ocean and atmospheric loading and solid Earth tides, tidal variations in the Earth's rotation, the tropospheric model, atmospheric angular momentum, ocean currents, and plate tectonics.

11

Stars

The main objective of stellar astrometry is to determine the positions of stars in space at some epoch and to describe their displacements in time. The instruments used for observations, and how the positions are determined, are given in Kovalevsky (2002). The reference frames have been discussed in Chapters 7–10 of the present book. The observational reduction procedures and corrections for apparent displacements were described in Chapter 6.

Star positions are not an objective *per se*. What is of interest are the motions and distances of celestial bodies. The first are the proper motions that describe the apparent displacements due to the actual motion of stars with respect to the barycenter of the Solar System. As we have already seen in Chapter 6, in order to transfer Earth-based observations to a barycentric position, one has to correct for annual parallax. However, the value of the parallax coefficient has a major importance in astronomy, because it is the basic source of distances in the Universe. This is why the determination of distances is discussed in detail in this chapter. By adding radial velocity to a combination of proper motion and distance, one obtains the space motion of a star. In addition, a section on magnitudes and spectra is given, not only because they provide important information to be used together with astrometric parameters, but also because they enter in the reduction of astrometric observations.

The last sections of this chapter give an overview of the most important astrometric star catalogs and their accuracies, with an emphasis on the *Hipparcos Catalogue*, which brought a dramatic change in the accuracies and triggered the compilation of a large number of star catalogs.

11.1 Star positions

Traditionally, the position of a star at some time t is given by two spherical coordinates. The most generally used system of coordinates is the equatorial system, so that the position on the celestial sphere is given by:

- the right ascension α
- the declination δ

as already defined in Section 3.8. The unit vector of the direction of the star in this system is:

$$S = \begin{pmatrix} \cos\delta\,\cos\alpha \\ \cos\delta\,\sin\alpha \\ \sin\delta \end{pmatrix}. \tag{11.1}$$

However, two other systems of coordinates are sometimes used: ecliptic coordinates and galactic coordinates (Section 7.8).

11.1.1 Proper motion

The proper motion is the time derivative of the coordinates of a star at an epoch t_0, typically the epoch of the catalogue. The proper motion in equatorial coordinates is composed of two quantities:

- the proper motion in right ascension $\mu_\alpha = (d\alpha/dt)_{t=t_0}$,
- the proper motion in declination $\mu_\delta = (d\delta/dt)_{t=t_0}$.

While μ_δ corresponds to an actual angle on the sky, μ_α is reckoned on the equator, so that the actual angular component along the local small circle is $\mu_\alpha\,\cos\delta$. For consistency reasons, we shall assume that μ_δ and μ_α are both expressed in arcseconds. This means that if μ_α is given in seconds of time, it must be multiplied by 15 to give arcseconds.

The proper motion is a vector on the tangent plane to the celestial sphere at the position α_0, δ_0 of the star at time t (see Section 3.9.1). Its modulus is

$$\mu = \sqrt{\mu_\alpha^2\cos^2\delta + \mu_\delta^2}$$

and its position angle ϕ reckoned from the North, Eastward from $0°$ to $360°$ is given by

$$\mu_\alpha\,\cos\delta = \mu\,\sin\phi$$
$$\mu_\delta = \mu\,\cos\phi.$$

Another way of describing this is to introduce the normal orthogonal vector triad at epoch $(\mathbf{p_0}, \mathbf{q_0}, \mathbf{r_0})$ such that

$$\mathbf{p_0} = \begin{pmatrix} -\sin\alpha_0 \\ \cos\alpha_0 \\ 0 \end{pmatrix}; \quad \mathbf{q_0} = \begin{pmatrix} -\sin\delta_0\,\cos\alpha_0 \\ -\sin\delta_0\,\sin\alpha_0 \\ \cos\delta_0 \end{pmatrix},$$

and $\mathbf{r_0}$ is directed outwards from the celestial sphere. Then, the proper motion vector, $\mathbf{V}$, is

$$\mathbf{V} = \mathbf{p_0}\mu_\alpha \cos\delta + \mathbf{q_0}\mu_\delta.$$

We shall make use of this standard space triad (Section 3.9.3), and regard it as a fixed set of coordinate axes, strictly and completely defined by α_0 and δ_0 and, in particular, we shall consider that it is not affected by the observational uncertainty of α_0 and δ_0.

11.2 Distance of stars

Stars are not infinitely distant. This is revealed by the annual parallax already defined and studied in Section 6.2. By definition, the annual parallax ϖ expressed in parsecs is related to the distance ρ by

$$\rho = \frac{1}{\varpi}$$

where ϖ is expressed in arcseconds. Let us assume that the parallax was found to be ϖ_0 with a standard uncertainty σ, and that its probability density function (pdf) is normal (formula (4.12))

$$f(\varpi) = \frac{1}{\sqrt{2\pi}\sigma} \exp\left[\frac{-(\varpi - \varpi_0)^2}{2\sigma^2}\right]. \tag{11.2}$$

Let ρ_0 be the corresponding value of the distance ($\rho_0 = 1/\varpi_0$). The pdf of ρ is no longer Gaussian in $\rho - \rho_0$, and in some cases is so different that the treatment of uncertainties must take this fact into account. Noting that

$$\varpi - \varpi_0 = \frac{1}{\rho} - \frac{1}{\rho_0} = \frac{\rho_0 - \rho}{\rho\rho_0}$$

and

$$d\varpi = -\frac{d\rho}{\rho^2},$$

the transformed pdf for ρ is

$$g(\rho) = \frac{1}{\sqrt{2\pi}\sigma\rho^2} \exp\left[\frac{-(\rho - \rho_0)^2}{2\sigma^2\rho^2\rho_0^2}\right]. \tag{11.3}$$

Another way of considering this distribution is to express it as a function of a relative shift, x, in distance

$$x = \frac{\rho - \rho_0}{\rho_0},$$

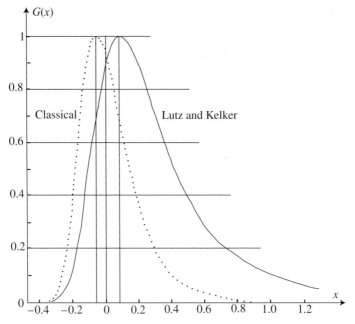

Fig. 11.1. Example of the pdf of the distance corresponding to an observed parallax $\varpi_0 = 1/\rho_0$ with an r.m.s. equal to σ. In this example $\alpha = \rho_0\sigma = 0.2$. The other line presents, for the same value of x, the pdf resulting from the assumptions of Lutz and Kelker (1973).

or $\rho = \rho_0(1 + x)$. Then formula (11.3) becomes, setting $\alpha = \rho_0\sigma$,

$$G(x) = \frac{1}{\sqrt{2\pi}\alpha(1 + x)^2} \exp\left[\frac{-x^2}{2\alpha^2(1 + x)^2}\right]. \tag{11.4}$$

The structure of $G(x)$ depends only on the product of $\alpha = \rho_0\sigma$, or better, the ratio $\alpha = \sigma/\varpi_0$ of the r.m.s. of the observed parallax to its value given by this observation. The function is not Gaussian as shown in Fig. 11.1. In particular, the maximum is shifted towards negative x, meaning that the value $1/\varpi_0$ is somewhat over-estimated. Actually, the curve has one maximum and one minimum, given by the following expression, obtained by setting $dG/dx = 0$,

$$x^2 + \left(2 + \frac{1}{2\alpha^2}\right)x + 1 = 0. \tag{11.5}$$

Figure 11.2 gives the value of the positive root as a function of α. It can be seen that the bias is negligible for values of $\alpha < 0.1$, which, in the case of Hipparcos data ($\sigma = 1$ mas) corresponds to stars closer than 100 parsecs.

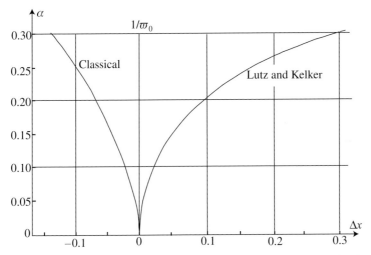

Fig. 11.2. Values of the maximums of the pdf resulting from classical statistical reasoning and those derived with the assumptions of Lutz and Kelker (1973).

Another approach was proposed by Lutz and Kelker (1973), and widely used in the literature. These authors assume that the stars are uniformly distributed in space. Then, the number of stars at distances between ρ and $\rho + d\rho$ is

$$N(\rho)d\rho = 4\pi\rho^2 d\rho,$$

or, in terms of parallaxes,

$$N(\varpi)d\varpi = 4\pi\frac{d\rho}{\varpi^4}.$$

As a consequence, (11.3) becomes

$$F^*(\varpi) = \frac{\sqrt{8\pi}}{\sigma\varpi^4}\exp\left[\frac{-(\varpi - \varpi_0)^2}{2\sigma^2}\right],$$

or

$$G^*(\rho) = \frac{\sqrt{8\pi}\rho^2}{\sigma}\exp\left[\frac{-(\rho - \rho_0)^2}{2\sigma^2\rho^2\rho_0^2}\right]. \tag{11.6}$$

One can again use x as the new variable and it is found that the maximum of the distribution is given by the smallest root of

$$x^2 + \left(2 - \frac{1}{2\alpha^2}\right)x + 1 = 0.$$

An example of the Lutz and Kelker distribution is given in Fig. 11.1, and the value of the maximum as a function of α is given in Fig. 11.2. In this case, the value

$\rho_0 = 1/\varpi_0$ appears to be an under-estimation of the parallax. Again the bias is negligible for small values of α.

It is necessary to point out, that, in this second approach, the parallax is no longer a result of an observation, but rather a new random variable with an *a priori* distribution law. It is to be used, assuming that the distribution is realistic, only in the case where the parallax is used as one of the parameters to search for some common feature, for example the absolute magnitude of a set of stars having the same spectral or photometric features, as was the case in the original paper by Lutz and Kelker (1973).

Another consequence of the special pdf for distances is that truncating a sample using the observed parallaxes necessarily induces a systematic error. The truncation must be performed using an independent parameter, for instance the apparent magnitude, but not the absolute magnitude determined from the value of the distance. Some pernicious effects of doing so are described in Luri and Arenou (1997).

In conclusion, whenever the distances are used as a parameter related to a given star, as in the case of kinematics or dynamics of the Galaxy, the function to be used is the one given by $g(\rho)$, and no Lutz and Kelker correction should be applied. However, this brings up another difficulty.

11.2.1 Uncertainties in distance determination

An essential feature of the functions $g(\rho)$ or $G(x)$ described in the previous section is that their integral from $-\infty$ to $+\infty$ does not converge, so that they cannot be considered as being actual probability distributions around the observed value of the parallax. In particular, the second-order moments cannot be computed, because the integrals also do not converge. Therefore, it is necessary to transform $g(\rho)$ or $G(x)$ in such a way that the new function has all the properties of a pdf, and choose among the infinity of solutions those that best fit these functions in some given intervals of ρ and x, that have the largest probabilities of including the true values. If, in addition, this new pdf is Gaussian, it would have the advantage of giving an estimate of the uncertainty, a well-defined maximum that could be chosen as the best value of the distance, and provide the means of pursuing the calculation of any quantity that is a function of the distance, together with its uncertainty. This would be the case, for instance, of absolute magnitudes, transverse velocities expressed in kilometers per second, etc.

A simplistic approach to the problem of assigning an r.m.s. to the distance is to consider the two values of ρ,

$$\rho_1 = \frac{1}{\varpi_0 + \sigma}; \quad \rho_2 = \frac{1}{\varpi_0 - \sigma},$$

and declare that the probability to have $\rho_1 < \rho < \rho_2$ is the same as to have $\varpi_0 - \sigma < \varpi < \varpi_0 + \sigma$. This leads one to declare that $\sigma_\rho = (\rho_2 - \rho_1)/2$. This is actually an acceptable approximation for the condition that σ is small with respect to ϖ_0. But this is not the case if α is large. For instance, if one has $\varpi_0 = 4 \pm 2$ mas, $\rho_1 = 167$ pc and $\rho_2 = 500$ pc, while $\rho_0 = 250$ pc, the interval $\rho_1 - \rho_2$ is not centered at ρ_0, and adopting σ_ρ as given above, one gets $83 < \rho < 416$ pc, which is nonsense. For smaller ϖ_0, one may be in a position to consider $\rho = \infty$ as probable.

Before going further, let us remark that if α is large, there remains, statistically speaking, a large probability for negative values of the distance. This is of course physically not acceptable, but for statistical consistency, one must take them into account. For this reason, in the *Hipparcos Catalogue*, some parallaxes are given that have been found negative and not replaced by zero. A possible approach is to filter out smoothly the lowest values of parallaxes by assigning them a weight that approaches zero gradually with ϖ. Two techniques to do so are proposed by Smith and Eichhorn (1996). The simpler one realizes this weighting by the following functions:

$$w(\varpi, \sigma) = \exp\left(-\eta \frac{\sigma^2}{\rho^2}\right),$$

where η is a free parameter that determines the width of the excised region around $\varpi = 0$. This renders the function $g(\rho)$ integrable and, provided that an adequate scaling is applied, it is transformed into a *bona fide* pdf. The major disadvantage is the presence of an adjustable parameter, whose value directly influences the value of the second moment. So, these weighting methods should be applied with great care, only for large α and when there is no additional external model, which can constrain the limits of acceptable distances such as, for instance, the Lutz and Kelker assumption on the distribution of stars or a distribution of absolute magnitudes (Smith, 1987).

The approach announced in the beginning of this section consists first of cutting $g(\rho)$ at places beyond which the probability becomes practically negligible. The fit to a Gaussian pdf is realized within an interval of distances defined by

$$A = \frac{1}{\varpi_0 + k\sigma} \leq \rho \leq \frac{1}{\varpi_0 - k\sigma} = B, \tag{11.7}$$

where k is a number between 2 (probability to be outside the interval equal to 0.0455) and 3 (probability 0.0027), and ϖ_0 is the observed value of the parallax. Evidently, this approach is possible only if ϖ_0/σ, which is $1/\alpha$, is not too large, namely that $k \ll \varpi_0/\sigma = 1/\alpha$. But in many studies, one does not consider remote stars for which the distance is too badly determined. This is the case of all local kinematic or astrophysical studies.

Let us consider the function $G(x)$ as given by equation (11.5), and define

$$G'(x) = \frac{G_0}{\sqrt{2\pi}\alpha(1+x)^2} \exp\left[\frac{-x^2}{2\alpha^2(1+x^2)}\right],$$

where $G_0 = \sqrt{2\pi}\alpha$ is chosen in such a way that $G'(0) = 1$ and let us represent it by

$$F(x) = \frac{G_0}{\sqrt{2\pi}s} \exp\left[\frac{-(x-x_m)^2}{2s^2}\right], \tag{11.8}$$

where s is the new r.m.s. and x_m the maximum of the function. The determination of these two parameters is made by minimizing

$$I_A = \int_a^0 (G(x) - F(x)) \, dx,$$

and

$$I_B = \int_0^b (G(x) - F(x)) \, dx,$$

where a and b are the values of x corresponding to A and B:

$$a = \frac{A - \rho_0}{\rho_0}; \quad b = \frac{B - \rho_0}{\rho_0},$$

with the additional condition that $I_A = I_B$. Since $G'(x)$ is a one-parameter function, the resulting values of s and x_m are functions only of α, so that they can be determined in advance. Figure 11.3 illustrates this fit in the case of $\alpha = 0.2$. In Fig. 11.4, the values of m_0 and s are given as a function of α.

11.3 Space motions

The positions of stars are not fixed. After correcting the apparent positions for various parallaxes, what remains is their motion with respect to the barycenter of the Solar System. This motion is divided into two components.

(i) The motion perpendicular to the line of sight is the proper motion (Section 11.1.1) expressed in arcseconds either per century or per year. We shall assume in what follows that they are yearly proper motions expressed in arcseconds per year.
(ii) The motion along the line of sight is the radial velocity. Obtained from measurements of the Doppler shift of spectral lines, it is expressed in kilometers per second. We must, however, ascertain that the observations are corrected for the motions of the Earth and reduced to the value at the barycenter of the Solar System.

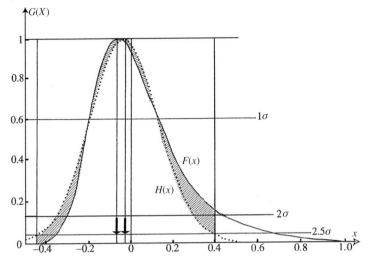

Fig. 11.3. Representation of $G'(x)$ by a Gaussian pdf $F(x)$ for $\alpha = 0.2$. The limits correspond to $k = 2.5$ (probability to be outside the interval is 1.3%). Arrows show the values of x corresponding to the maximums of distributions. In this computation, the shaded areas almost cancel out. The integrated weighted probability of this area is 3.3%.

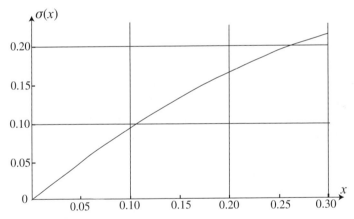

Fig. 11.4. Plot of the r.m.s. σ in x as a function of α for the best-fit Gaussian, computed assuming $k = 2.5$.

11.3.1 Radial velocities

Let us consider a source moving with respect to an observer with a velocity v whose direction has an angle θ with the direction source–observer, and emits a wavelength λ_0. The apparent wavelength as perceived by the observer is

$$\lambda = \lambda_0\left(1 - \frac{v}{c}\cos\theta\right),$$

where c is the speed of light. If the source recedes from the observer, $\cos\theta < 0$ and $\lambda > \lambda_0$ and the light seems redder. If it approaches the observer, $\cos\theta > 0$ and $\lambda < \lambda_0$, the light seems to be bluer. This is the classical Doppler effect.

Remark. The atom is a small clock, which in its proper reference system has a period T_0. Following the general relativity theory, in a reference frame with respect to which it moves, with a velocity v, the period is (Section 5.2.5):

$$T = T_0/\sqrt{1 - v^2/c^2}.$$

This modifies the actual wavelength, so that the modified formula for Doppler effect is

$$\lambda = \lambda_0 \left(1 - \frac{v}{c}\cos\theta\right)/\sqrt{1 - v^2/c^2}.$$

For stars in the Galaxy, $v < 500$ km/s, the correcting relativistic factor is negligible, taking into account that star's radial velocities are observed with a precision at the best of 0.1 km/s. The Doppler shift, which is measured by comparing the stellar spectrum with the spectrum of a fixed emitter, is

$$\Delta\lambda = \lambda - \lambda_0 = \lambda_0\left(\frac{V_r''}{c}\right),$$

where $V_r'' = -v\cos\theta$ is called the *radial velocity*,

$$V_r'' = \frac{c\Delta\lambda}{\lambda_0},$$

chosen to be positive when the star is receding from the observer along the $\mathbf{r_0}$ axis.

The observed radial velocity V_r'' is topocentric. One has to transform it into the barycentric system of coordinates by subtracting the projection on the direction of the object, of the observer's velocity around the axis of rotation of the Earth, $\mathbf{u'}$, and of the velocity of the geocenter with respect to the barycenter of the Solar System, $\mathbf{u}$.

In equatorial coordinates, the components of the radial velocity vector, $\mathbf{V_r''}$, are

$$\mathbf{V_r''} \begin{cases} V_r'' \cos\delta\cos\alpha \\ V_r'' \cos\delta\sin\alpha \\ V_r'' \sin\delta. \end{cases}$$

The velocity $\mathbf{u'}$ of the observer on the Earth is perpendicular to the local meridian. Its equatorial components are

$$\mathbf{u'} \begin{cases} r\omega\sin T_1 \\ -r\omega\cos T_1 \\ 0, \end{cases}$$

where r is the distance, in kilometers, from the observer to the axis of rotation of the Earth, ω the rate of rotation of the Earth expressed in radians per second ($\omega = 0.000\,072\,523\,489$) and T_1 the local sidereal time.

The velocity $\mathbf{u}$ of the geocenter with respect to the barycenter of the Solar System is derived from the ephemerides of the Earth's motion. For instance in the JPL series of Development Ephemerides, the current one in 2002 is DE 405, which gives the rectangular coordinates in the ICRS from which the time derivatives may be computed in kilometers per second.

The actual barycentric radial velocity vector is, therefore,

$$\mathbf{V}_r = (\mathbf{V}_r'' - \mathbf{u}' - \mathbf{u}),$$

and the barycentric radial velocity is

$$V_r = \mathbf{V}_r \cdot \mathbf{S},$$

where $\mathbf{S}$ is the unit vector of the barycentric direction of the star.

11.3.2 Motions in space

Both proper motions and radial velocities are expressed in different units. To obtain the three components of the motion in space, one has to write the proper motion in kilometers per second. To do this, it is necessary to know the distance ρ of the star. However, in what follows, we shall use $\varpi = 1/\rho$, whether or not the corrections for ρ presented in Section 11.2 are made.

The annual proper motion, μ, can be readily expressed in astronomical units per year. From the actual definition of the parallax, it is μ/ϖ. In order to express it in kilometers per second, one has to divide it by the number of seconds in a Julian year:

$$s = 0.315\,576 \times 10^7,$$

and to multiply it by the number of kilometers in an astronomical unit:

$$a = 0.149\,597\,87 \times 10^9.$$

So, the transverse velocity (in km/s) is:

$$V_T = 4.740\,47\frac{\mu}{\varpi} = A\frac{\mu}{\varpi}. \tag{11.9}$$

Often, it is sufficient to take $A = 4.74$. Other values are sometimes given in the literature, because the usage was to consider the tropical rather than the Julian year. The correct value to use is $A = 4.740\,47$.

The standard expression for the space velocity is:

$$V = A \frac{\mu_\alpha}{\varpi} \cos \delta \, \mathbf{p}_0 + A \frac{\mu_\delta}{\varpi} \mathbf{q}_0 + V_r \, \mathbf{r}_0.$$

However, a more accurate formula must be used in some occasions. This is obtained, if one no longer neglects the light-time effect described in Section 6.5, by dividing the preceding expression by $k = 1 - V_r/c$. So, that the actual formula is:

$$V = \frac{1}{k} \left(A \frac{\mu_\alpha}{\varpi} \cos \delta \, \mathbf{p}_0 + A \frac{\mu_\delta}{\varpi} \mathbf{q}_0 + V_r \, \mathbf{r}_0 \right). \tag{11.10}$$

For most of the cases in galactic astronomy, with the exception of close high-velocity stars, one may neglect the factor k, but it is an important correction for large radial velocities, for instance for extragalactic objects.

11.3.3 Variability of proper motions

In their motions in the Galaxy, stars undergo gravitational forces that may modify their velocities, so that their proper motions and radial velocities vary with time. There is no general theory, and the effect is to be studied case by case from a dynamical point of view. However, there is a class of objects where neglecting this effect may lead to serious errors.

To assume that the space motion of a star is linear means that the center of mass of the system represented by the star coincides with the star. This is often not the case, many stars are double or multiple systems and each component moves around a common barycenter, which is the point that has a linear motion in space, while the apparent proper motion of each component is curved. To detect such variations in proper motion is one of the methods used to detect so-called astrometric binaries. This problem will be described in Section 12.4. It is necessary to have this possibility in mind when dealing with proper motions and radial velocities, and be cautious that the apparent motion may be a combination of the space motion of the stellar system and the orbital motion of the observed star.

Another case is the *secular* or *perspective acceleration*, which is a purely geometric effect due to the variation of the proper motion of an approaching or receding object. Let V be the space velocity and θ its angle with the direction of the star, whose distance is ρ_0 (Fig. 11.5). The modulus μ of the proper motion is

$$\mu = \frac{V \sin \theta}{\rho}.$$

Taking its derivative with respect to time, one gets

$$\frac{d\mu}{dt} = -\frac{V}{\rho^2} \sin \theta \frac{d\rho}{dt} + \frac{V}{\rho} \cos \theta \frac{d\theta}{dt}.$$

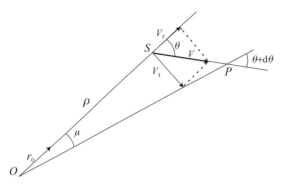

Fig. 11.5. Definition of the angle θ and its variation $d\theta$ when the star moved by μ.

From the triangle OSP in Fig. 11.5, one can see that $d\theta/dt = -\mu$, and $V\cos\theta = V_r = d\rho/dt$. One has therefore

$$\frac{d\mu}{dt} = -\frac{\mu}{\rho}\frac{d\rho}{dt} - \frac{\mu}{\rho}V_r = -\frac{2\mu}{\rho}V_r, \qquad (11.11)$$

or, if μ is expressed in arcseconds per year, ϖ in arcseconds, V_r in kilometers per second,

$$\frac{d\mu}{dt} = -0.000\,002\,05\,\mu\varpi\,V_r.$$

This correction is significant only for nearby stars with high radial velocities.

11.3.4 Epoch transformation

Independent of the physical importance of proper motions and space velocities in the kinematic and dynamic studies of the Galaxy and other star systems, formula (11.11) is the basis of the computation of star positions at some epoch T, different from the basic epoch T_0 at which positions and proper motions are given. If one assumes that the parameters are uncorrelated, one has, putting $t = T - T_0$,

$$\alpha = \alpha_0 + t\mu_{\alpha_0}$$
$$\delta = \delta_0 + t\mu_{\delta_0}. \qquad (11.12)$$

However, this is an approximate solution. Let us make a rigorous treatment of the problem. At the time t_0, the barycentric direction of the star S is $\mathbf{u_0} = \mathbf{r_0}$ and the radius vector is

$$\mathbf{BS} = \mathbf{b}(0) = \mathbf{r_0}\frac{A}{\varpi_0}.$$

The proper motion vector is:

$$\mathbf{M_0} = \mathbf{p_0}\mu_\alpha^* + \mathbf{q_0}\mu_\delta,$$

where we have assumed again, that the proper motion components are in arcseconds per year, and introduced the simplified notations

$$\alpha^{\star} = \alpha \cos \delta,$$
$$\mu_{\alpha}{}^{\star} = \mu_{\alpha} \cos \delta.$$

The space velocity $\mathbf{V}$, which is assumed not to vary with time, is given by (11.11) in which we neglect the coefficient k

$$\mathbf{V} = \mathbf{p_0} \frac{A}{\varpi_0} \mu_{\alpha^{\star}} + \mathbf{q_0} \frac{A}{\varpi_0} \mu_{\delta} + \mathbf{r_0} V_{\mathrm{r}},$$

that we write somewhat differently in order to have A/ϖ as a factor. Putting

$$\zeta_0 = V_{\mathrm{r}} \frac{\varpi_0}{A},$$

one gets

$$\mathbf{V} = \frac{A}{\varpi_0} (\mathbf{p_0} \mu_{\alpha^{\star}} + \mathbf{q_0} \mu_{\delta} + \mathbf{r_0} \zeta_0). \tag{11.13}$$

At time $t = T - T_0$, the radius vector is

$$\mathbf{b}(t) = \mathbf{b}(0) + \mathbf{V} t.$$

The unit vector at time t is

$$\mathbf{u_0}(t) = (\mathbf{r_0}(1 + \zeta_0 t) + \mathbf{M_0} t) f, \tag{11.14}$$

where f is a normalizing factor equal to

$$f = |\mathbf{b}(0)| / |\mathbf{b}(t)| .$$

Recalling that

$$|\mathbf{b}(0)| = \left| \mathbf{r_0} \frac{A}{\varpi_0} \right| = \frac{A}{\varpi_0},$$

and

$$|\mathbf{b}(t)|^2 = |\mathbf{b}(0)|^2 + 2\mathbf{b}(0) \cdot \left(\frac{A}{\varpi_0} \mathbf{M_0} + \mathbf{r_0} \zeta_0 \right) t + \left(\frac{A}{\varpi_0} \mathbf{M_0} + \mathbf{r_0} \zeta_0 \right)^2 t^2,$$

one obtains:

$$f = \frac{1}{\sqrt{1 + 2\zeta_0 t + \left(M_0^2 + \zeta_0^2 \right) t^2}}, \tag{11.15}$$

and, after differentiation,

$$\frac{df}{dt} = -f^3 \left(\zeta_0 + \left(M_0^2 + \zeta_0 \right) \right) t.$$

Then, we can determine the new parallax

$$\varpi(t) = \varpi_0 f. \tag{11.16}$$

Differentiating (11.15), one gets the proper motion vector at time t. After reduction, one simplifies the expression and gets

$$\mathbf{M_0}(t) = \left(\mathbf{M_0}(1 + \zeta_0 t) - \mathbf{r_0} M_0^2 t\right) f^3. \tag{11.17}$$

Finally, the sixth parameter ζ_0 becomes

$$\zeta_0(t) = V_r(t) \frac{\varpi(t)}{A} = \frac{d\,(b(t))}{dt} f \frac{\varpi_0}{A}.$$

The derivative of $b(t)$ is readily obtained knowing that $f^2 = \mathbf{b}(0)^2/\mathbf{b}(t)^2$. So finally, one has

$$\zeta(t) = \left[\zeta_0 + \left(M_0^2 + \zeta_0^2\right)t\right] f^3. \tag{11.18}$$

Formulae (11.15), (11.17) and (11.18) describe fully the transformation from the parameters at time T_0, $(\alpha_0^\star,\ \delta_0,\ \varpi_0,\ \mu_{\alpha_0^\star},\ \mu_{\delta_0},\ V_{r_0})$, and their value at time $T_0 + t$, $(\alpha^\star(t),\ \delta(t),\ \varpi,\ \mu_{\alpha^\star}(t),\ \mu_\delta(t),\ V_r(t))$. This transformation is reversible and the same formulae may be used backwards by replacing t by $-t$. Needless to say, this transformation includes the effect of secular acceleration.

Let us remark, however, that the origin of the proper motion vector $\mathbf{M_0}(t)$ is $\mathbf{u_0}(t)$, and it is, therefore, necessary to consider the local propagated triad $\mathbf{p}(t)$, $\mathbf{q}(t)$, $\mathbf{r}(t)$, where $\mathbf{r}(t)$ is the unit vector of $\mathbf{b}(t)$ so that one has:

$$\mathbf{M}(t) = \mu_{\alpha^\star}(t)\mathbf{p}(t) + \mu_\delta(t)\mathbf{q}(t).$$

11.3.5 Uncertainty propagation

The above transformation of astrometric parameters from time T_0 to time $T_0 + t$ is only part of the expected results. It is also necessary to determine r.m.s. uncertainties for these parameters, and also to see how correlations develop with time. Let us first consider the simplistic approach with formula (11.13), namely assuming correlation only between a coordinate and the corresponding component of the proper motion,

$$\alpha^\star(t) = \alpha_0^\star + t\mu_{\alpha_0^\star},$$
$$\delta(t) = \delta_0 + t\mu_{\delta 0}.$$

The derivation of uncertainties at time t is given in the example of Section 4.2.

$$\sigma_{\alpha^\star}^2(t) = \left[\sigma_{\alpha^\star}^2 + 2t\rho_{\mu_{\alpha^\star}}\sigma_{\alpha^\star}\sigma_{\mu_{\alpha^\star}} + t^2\sigma_{\mu_{\alpha^\star}}^2\right]_0,$$
$$\sigma_\delta^2(t) = \left[\sigma_\delta^2 + 2t\rho_{\mu_\delta}\sigma_\delta\sigma_{\mu_\delta} + t^2\sigma_{\mu_\delta}^2\right]_0. \tag{11.19}$$

This result can also be obtained using the generalized transformation of Section 4.2.3.

With the above stated assumption concerning the correlations, the variance–covariance matrix at time T_0 is

$$C = \begin{pmatrix} \sigma_{\alpha^*}^2 & 0 & \sigma_{\alpha^*}\sigma_{\mu_{\alpha^*}}\rho_{\alpha^*\mu_{\alpha^*}} & 0 \\ 0 & \sigma_\delta^2 & 0 & \sigma_\delta\sigma_{\mu_\delta}\rho_{\delta\mu_\delta} \\ \sigma_{\alpha^*}\sigma_{\mu_{\alpha^*}}\rho_{\alpha^*\mu_{\alpha^*}} & 0 & \sigma_{\mu_{\alpha^*}}^2 & 0 \\ 0 & \sigma_\delta\sigma_{\mu_\delta}\rho_{\delta\mu_\delta} & 0 & \sigma_{\mu_\delta}^2 \end{pmatrix}.$$

The Jacobian $\mathcal{J}$ of the transformation is

$$\mathcal{J} = \begin{pmatrix} 1 & 0 & t & 0 \\ 0 & 1 & 0 & t \\ 0 & 0 & 1 & 0 \\ 0 & 0 & 0 & 1 \end{pmatrix},$$

and if one computes $\mathcal{S} = \mathcal{J}C\mathcal{J}^{\mathsf{T}}$, one obtains the new $\sigma_{\alpha^*}^2(t)$ and $\sigma_\delta^2(t)$ as given by (11.19), and the two new correlation coefficients:

$$\rho_{\alpha^*\mu_{\alpha^*}}(t) = \frac{\sigma_{\alpha^*}\sigma_{\mu_{\alpha^*}}\rho_{\alpha^*\mu_\alpha^*} + \sigma_{\alpha^*}^2 t}{\sigma_{\alpha^*}(t)\sigma_{\mu_\alpha^*}},$$

$$\rho_{\delta\mu_\delta}(t) = \frac{\sigma_\delta\sigma_{\mu_\delta}\rho_{\delta\mu_\delta} + \sigma_{\mu_\delta}^2 t}{\sigma_\delta(t)\sigma_{\mu_\delta}}. \tag{11.20}$$

The other correlation coefficients remain null and the r.m.s. uncertainties in μ_{α^*} and μ_δ are not changed.

For the most general case of the complete transformation of the six parameters as presented in Section 11.3.4, one must take the more-general 6×6 variance–covariance matrix. The 6×6 terms of the Jacobian can be found in ESA (1997, Section 1.5.5).

11.4 Magnitudes

Although photometry is not an astrometric technique, the apparent intensity of a celestial source is a major parameter in assessing the quality of the astrometric observations and of the results that can be derived. For instance, it is evident that the precision with which a star position may be observed depends directly on its brightness. The fainter the celestial objects appear to be, the more difficult it is to set a micrometer on its image, or to measure it on a photographic plate or on a CCD. At some level of faintness, the object is no longer observable even if its image is exposed for some time on the detector. Alternatively, bright objects in the field of view saturate the detector, and cannot be measured as accurately as well-exposed images on CCDs. To take this fact into account in the astrometric reduction procedure, it is necessary to quantify the photometric properties of the source. It is therefore useful that the bases of this quantification be presented here.

11.4.1 Luminous intensity and flux

The intensity of a source characterizes the energy that is emitted by the source in a unit of time. It is generally frequency dependent, and one defines a specific intensity, I_λ, at a wavelength λ, or a frequency, $\nu = c/\lambda$, expressed in hertz. The flux is expressed in W/(m² Hz sr). In general, what is useful is the integral of I_ν over the medium and the instrumental filters. Let $T(\lambda)$ be the transmission, that is the proportion of light that is recorded by the instrument. The transmitted intensity is

$$ I_T = \int_0^\infty T(\lambda)\, I_\lambda d_\lambda. $$

What is actually observed is an energy collected by a detector. This is measured by the *flux density*, or flux which is the power of the radiation per unit area, expressed in W/(m² Hz), or W/m², if it is a global transmitted flux density. It is a very small quantity, and in radio astronomy, it is expressed in janskys (1 jansky = 10^{-26} W/(m² Hz)).

The energy collected by a receiver per second, or flux, is the power going through some surface and is expressed in watts.

11.4.2 Apparent magnitudes

In astronomical photometry, the flux observations are not absolute. The expression of flux in watts is not easily measured, and it is the ratio of fluxes, which is generally determined. This leads to the definition of magnitude scales, which all refer to some standard stars used to define a magnitude equal to zero. If F_0 is the flux recorded for this reference, and F is the flux recorded for another star in the same experimental conditions, one defines the magnitude m of the second star as

$$ m = -2.5 \log \frac{F}{F_0}, \tag{11.21} $$

where the logarithm is the decimal logarithm. The difference of magnitudes between two stars S_1 and S_2 whose fluxes are respectively F_1 and F_2 is

$$ m_1 - m_2 = \Delta m = -2.5 \log \frac{F_1}{F_2}. \tag{11.22} $$

A good approximation, although it is not exact, is that a difference of 5 magnitudes corresponds a ratio of fluxes equal to $(2.5)^5 \simeq 100$.

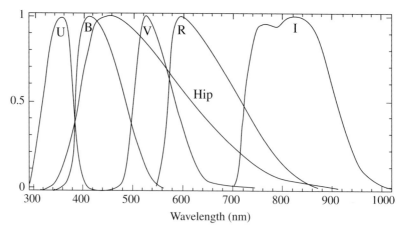

Fig. 11.6. Sensitivity curves for the *UBV* Johnson system, the *R* and *I* bands, and the wide-band Hipparcos system *Hip*. The medium wavelengths are respectively 0.365, 0.44, 0.55, 0.70, and 0.90 μm (Zombeck, 1990).

11.4.3 Magnitude scales

The flux measured depends on the instrument and on the type of receivers. This could lead to great inconsistencies in magnitudes determined by different instruments, using different filters. For this reason, there is a general consensus in defining magnitude systems or scales, with reference filters that are used to reproduce the transmissions required by the definition. Figure 11.6 shows the sensitivity curves of the most popular system called *UBV* (*U* = ultraviolet, *B* = blue, *V* = visual). It is defined by a certain number of standard stars which set the origin (mag = 0) and the scale. Another wide-band photometric scale was derived from the photometric observations by Hipparcos (see also Fig. 11.6), while the million stars of the *Tycho Catalogue* have magnitudes given in scales that are very close to *B* and *V*. Another photometric scale was commonly used in photographic astronomy and is called photographic magnitudes m_{pg}. It used to correspond to the response of early photographic plates. It is close to the blue Johnson filter, and the two scales are related by

$$m_{pg} = B - 0.11.$$

One should also mention the *I* (infrared scale), which is becoming increasingly useful in astrometry, and another important photometric system defined for the SLOAN Digital Sky Survey (SDSS). The photometric system consists of five filters whose characteristics are given in Table 11.1. Most of the astrometric instruments, as well as the atmospheric refraction, are sensitive to the spectral distribution of energy. As a consequence, the image of the star is spread along a small spectrum.

Table 11.1. *Wavelengths of the SLOAN photometric survey*

Filter name	Central wavelength	Wavelength width
u′	350	60
g′	480	140
r′	625	140
i′	770	150
z′	910	120

Depending on the color of the star, the position of its centroid is shifted. An indicator of this shift, to be corrected for astrometric purposes, is the color index c defined as

$$c = B - V.$$

For instruments that are sensitive to the red part of the spectrum, one uses also the color index $V - I$ rather than $B - V$.

11.4.4 Absolute magnitudes

The magnitudes defined in the preceding section are apparent magnitudes, that is the apparent luminosity as seen from the Earth after corrections for the absorption by the atmosphere and the instrument, so as to reproduce the sensitivity required by the definition of the chosen scale. But this does not give any information on the actual luminosity of the star. To do this, an absolute magnitude is defined as the magnitude the star would have if it were at a distance of 10 pc. Since the apparent luminosity decreases as the inverse square of the distance, the absolute magnitude M is defined as

$$m - M = 5\log\rho - 5 = -5(1 + \log\varpi).$$

The quantity $m - M$ is called the distance modulus. It is necessary to specify in which photometric system the absolute magnitude is reckoned (M_V, M_B, etc.).

Actually, the important astrophysical quantity is the total energy flux emitted by the star. It is not directly obtained by observation, and the corrections (for instance to M_V) imply modeling of the star as a function of its effective temperature, gravity, spectrum, etc., and correcting for interstellar extinction. The results are bolometric magnitudes which are not currently used in astrometry, although they are the correct parameters to be used in analyzing the mass–luminosity relationship and in discussing physical properties of stars.

11.4.5 Stellar spectral types

Often, in astrometric star catalogs, the color is not provided and, instead, there is an indication of their spectral type. Although various spectral types are defined by criteria in terms of the presence, or absence, of some spectral lines, their determination is partly a matter of interpretation, especially when one is led to consider a composite spectrum of two stars in a close binary system. It is not, therefore, a parameter as well quantified as a color index.

Let us recall the definition of the main spectral classes, designated by a letter followed by a number between 0 and 9, representing an evolution of the characteristics within a class. A small letter may follow indicating certain peculiarities of the spectral lines (emission lines, sharp or nebulous lines, etc.). The most frequent classes are as follows.

- **O**: Hot blue stars with strong helium HeII absorption lines and a strong ultraviolet continuum. Temperatures are above 30 000 K.
- **B**: Hot blue stars with HeI absorption lines and hydrogen appearing in the latest sub-classes (B5–B9). Temperatures range between 12 000 and 30 000 K.
- **A**: White stars with very strong hydrogen lines, and CaII lines appearing in the latest sub-classes. Temperatures are between 7500 and 12 000 K.
- **F**: White or yellowish stars with strong CaII lines, hydrogen lines are weaker, and other ionized metallic lines appear in the latest sub-classes. Temperatures range between 6000 and 7500 K.
- **G**: In these yellow stars, the CaII lines are still strong, and many iron and other metallic ionized and neutral lines appear, while hydrogen lines become weak. Temperatures range between 5300 and 6000 K. The Sun is a G2 star.
- **K**: Orange and reddish stars, with a large quantity of neutral metallic lines. Some CH and CN molecular bands are appearing. Temperatures are between 4000 and 5300 K.
- **M**: Very red stars, with many molecular bands. Temperatures may be as low as 2500 K.

In addition, some special classes have been defined among very hot stars (Q, P, W), cold stars (S, R and N depending upon the presence of some molecules), and, at the extreme edge, brown dwarfs, which are visible essentially in the infrared.

There is another important classification derived from the spectra and from the knowledge of the absolute magnitudes, the luminosity class.

- **I**: Supergiants
- **II**: Bright giants
- **III**: Giants
- **IV**: Sub-giants
- **V**: Main sequence
- **VI**: Sub-dwarfs
- **VII**: White dwarfs.

Table 11.2. *The relation between B–V and the main spectral types*

Spectrum	B–V	Spectrum	B–V
O 5	−0.32	G 0	0.60
O 9	−0.31	G 2	0.64
B 0	−0.30	G 5	0.68
B 2	−0.24	G 8	0.72
B 5	−0.16	K 0	0.81
B 8	−0.09	K 2	0.92
A 0	0.00	K 3	0.98
A 2	0.06	K 5	1.15
A 5	0.15	K 7	1.30
A 7	0.20	M 0	1.41
F 0	0.33	M 1	1.48
F 2	0.38	M 3	1.55
F 5	0.45	M 5	1.61
F 8	0.53	M 8	2.00

The theory of stellar evolution shows how stars may move from one luminosity class to another depending upon their age, mass, metal content, etc. The main-sequence stars correspond to a stable situation, while the star is burning hydrogen in its core. The first four categories correspond to stars that have hotter cores, and are burning helium, or heavier elements like carbon or oxygen. They have a much larger atmosphere and are brighter than main sequence stars. Sub-dwarfs are metal-poor stars, and white dwarfs are in a degenerate state at the end of stars' lives.

Generally, the spectral classes are not used as such in the reduction of astrometric observations or in determining the chromatic part of the refraction. One assumes that the star is a main-sequence star, and we give, in Table 11.2 the relation between B–V and the main spectral types. This table may be interpolated, but one must have in mind that important biases may be introduced if the luminosity class is not V, especially for K, G, and F type stars, as shown in Fig. 11.7, and, even more so, in the case of composite spectra of double stars.

But this is not sufficient for very accurate astrometry. In the case of the measurements of centroids of images of stars on CCDs, the spectra are necessary to characterize the differences in point spread functions of the stars, and determine accurate relative positions of different types of stars.

11.5 Pre-Hipparcos star catalogs

In our presentation of star catalogs, a special emphasis is given to the *Hipparcos Catalogue*, because it is the basis of the realization of the celestial reference system

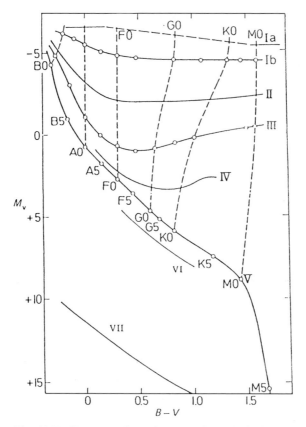

Fig. 11.7. Correspondence between the color index *B–V* and the absolute magnitude M_V, the spectral type and the luminosity class. (From Zombeck, M. V., 1990, *Handbook of Space Astronomy and Astrophysics*, 2nd Edn, p. 65. Reprinted with the permission of Cambridge University Press.)

in optical wavelengths and was the driving force in the adoption of the International Celestial Reference System (ICRS, see Section 7.1).

The common characteristic of star catalogs before Hipparcos is that they gave positions and proper motions in the dynamic reference frame, which meant that they had to be referred to the equinox and the equator as defined by the dynamical frame. Generally a star catalog contains the position (right ascension and declination) in a specified reference system reduced at a common date, called *epoch*. When a catalog is built from many years of observations, it also contains the proper motions. But it generally does not include parallaxes or photometric data which are presented in specific catalogs.

Depending upon the way they are obtained, one may distinguish between observational catalogs, fundamental catalogs, and reference catalogs, even if in some cases catalogs may have some characteristics of two of these classes.

11.5.1 Observational catalogs

These contain the results of the observations by a given instrument during a certain interval of time. They may simply be a list of observed positions, but they are often published after some reduction, which transforms the observed positions into a set of coordinates in a specified reference system for the epoch of the observations. Two possibilities exist.

(i) *Relative catalogs*. Among the observed stars, there is a certain proportion of stars whose positions are taken from a catalog that is already given in the specified reference system, like a fundamental or a reference catalog. The positions of these reference stars are used to determine the instrumental parameters, like the plate constants in photographic or CCD astrometry, or the constants of a transit instrument (collimation, inclination, decentering, etc.). Correcting the raw observations of the other stars with the values found for these parameters, usually for a limited area of the sky, provides their coordinates in the system of the reference stars.

(ii) *Absolute catalogs*. In these catalogs, the reduction to the reference system is made using observations of the Sun, planets, or minor planets, which allow us, by using a theory of their apparent motion in the reference system, to link the observed star positions to the equinox, the origin of the right ascensions, and to the equator, for the origin of declinations. Fundamental azimuth is also required. Actually, there were only a few meridian circles in the world that were able to observe the Sun, the planets, and stars, so that there were only a small number of absolute catalogs, which were the basis of the construction of a reference frame.

The construction of an observational absolute catalog is a lengthy and complex procedure. A detailed description may be found in chapter 17 of Woolard and Clemence (1966). The notion of absolute catalogs based upon the observation of the Sun has become obsolete with the new definition of the celestial reference system, ICRS, so we do not insist on this aspect of observational astrometry.

This is not the case of relative catalogs, which will now be the normal way to publish star-position observations. The methods are described in detail in Chapter 14. There is a very large number (several hundreds, possibly one thousand) of such catalogs built since the beginning of the nineteenth century. Each transit or meridian instrument in usage produced such a catalog, or list of observations, every few years. Many of them are quoted in Eichhorn (1974) who also gives references to existing lists of star catalogs prior to this date. Another source is the general index to the FK3, FK4 and FK5 catalogs.

11.5.2 Fundamental catalogs

A fundamental catalog is a catalog compiled from a large number of instrumental catalogs, in particular from all available absolute catalogs, to define the frame, and to build an ensemble of stellar positions and proper motions in the current reference system. It becomes the reference frame consistent, not only with the definition of the system (see Chapter 8), but also with the values of the fundamental astronomical constants that are adopted at that time (constants of precession, nutation, aberration, etc.). Again it is very complicated and lengthy work that was undertaken only rarely (every 20 years or so), and used to take about ten years of a team's work. Since it has now become essentially a historical interest, we refer the reader to chapter 18 of Woolard and Clemence (1966). Two series of fundamental catalogs were built during the twentieth century by German and American astronomical institutes respectively.

The German series of fundamental catalogs was started in the nineteenth century by Auwers, who compiled the *Fundamental Catalog for the Zone Observations* for the Astronomisches Gesellschaft and included 539 Northern and 83 Southern stars. It was followed by the *Neuer Fundamental Katalog* (NFK, by J. Peters, 1907) in the beginning of the twentieth century and then by the FK3 (Kopff, 1937), the FK4 (Fricke *et al.*, 1963) and finally the FK5 (Fricke *et al.*, 1988). The FK4 was the last catalog based uniquely on a dynamical reference system (see Chapter 8). The system of right ascensions was based upon 45 absolute catalogs and the same constant of precession as FK3 (Gliese, 1963). It included 1535 stars. Later, a list of supplementary stars was added, the FK4 supplement. The basic FK5 catalog includes the FK4 stars, and was derived as a correction to the FK4 based upon 85 absolute and 90 observational catalogs. An FK5 extension, including 3117 stars, added differentially to the system, was later issued (Fricke *et al.*, 1991). The FK5 system was obtained by applying a correction to the FK4 equinox (Fricke, 1982), the IAU expressions for general precession (Lieske *et al.*, 1977), and a discussion of the galactic rotation effect on the proper motions. The consequence of this particular novelty is that, although referred to a dynamical system, the FK5 is not strictly based upon motions of the Sun and planets, and is, therefore, a mixed stellar–dynamical reference frame, a choice fully justified by the announced objective to stick as well as possible to the observations, among which the observations of the Sun and the planets were the weakest part.

The American series of fundamental catalogs was mostly compiled at the US Naval Observatory. It originated with the *Catalogue of Fundamental Stars* published by S. Newcomb in 1899, which contained the positions of 1257 stars for the years 1875.0 and 1900.0, without distinguishing proper motions from the effect of the luni-solar precession. The reference system was purely dynamical, based upon Newcomb's theories of the motions in the Solar System. There was no direct connection between this catalog and its successors: the *Preliminary General Catalogue* (PGC) compiled by L. Boss in 1910 and the *General Catalogue* (GC) due to B. Boss *et al.* (1937), which comprised 33 342 star positions and proper motions. The latter were not very good, with uncertainties of the order of 20 mas/year. One of the reasons was that, for many of the stars, few earlier observations were available and thus, the accuracy of the catalog quickly degraded. For this reason, a less ambitious project based upon many newer observations, was undertaken and Morgan (1952) produced the N30 catalogue of 5268 stars from 60 observational and absolute catalogs. The reference system (N = Normal System) was based on observations of the Sun, Moon and Planets for the position of the equinox and the equator. The catalog gives positions for the epoch 1950.0 and proper motions obtained generally using normal N30 positions and older GC positions. The N30 catalog is much more accurate than the GC, but the mean uncertainty of proper motions is of the order of 3–7 mas/year, still larger than the FK3, where there was a smaller number of stars, but each of them with a longer history of observations in both absolute and relative catalogs. This, together with the continuity of the FK series, led the IAU to adopt the latter as the basic reference for astrometric observations. A very detailed description of the principles on which all these fundamental catalogs (except FK5) were built can be found in Eichhorn (1974).

11.5.3 Positional reference catalogs

Fundamental catalogs provide the positions of only a few stars, and there is a need to have the coordinates of as many stars as possible with an acceptable precision. This need was recognized around 1880 and a worldwide project, called *Carte du Ciel*, was organized with the cooperation of 20 observatories equipped with two refractive telescopes mounted on a single tube.

The reference stars were obtained by observations with meridian circles. The photographs covered fields of $2°2 \times 2°2$ with corner to center overlapping each surrounding plate. Most of the sky was covered in 35 years, except a few missing zones that were re-observed later. The series of the plates, reproduced on paper, represent as such the so-called *Carte du Ciel* and are a record of the sky to about magnitude 14–15. In conjunction with the *Carte du Ciel*, the *Astrographic Catalogue* plates were taken. These sought to reach 11 th magnitude and the x-y plate measurements were published. The measurements of the plates concerned on the average 200 stars. So, the coordinates of about 1.5 million stars were computed with an actual accuracy of the order of $1''$ to $2''$. This accuracy was primarily the result of the accuracy of reference stars and the resulting plate constants. New measurements of plates and their reductions have been undertaken in several places; better computing facilities and measuring machines, together with better coordinates of reference stars derived from modern catalogs, have provided better first epoch coordinates for subsequent proper motion determinations (Sections 11.5.5 and 11.6.2).

More recently, another need for a positional reference catalog arose, and the *Astrographic Catalogue* was too old and too small to cope with it. It was necessary to produce a dense catalog of star positions up to magnitude 15 in order to have reference stars everywhere in the sky that could be used as guide stars for the observations with the Hubble Space Telescope. The result was the *Guide Star Catalog* (Russell *et al.*, 1990). It is a catalog of approximately 18.5 million objects and includes stars from magnitude 9 up to magnitude 15 at galactic poles and 13.5 in the Milky Way. It was constructed from the reduction of about 1500 Schmidt plates taken by the Palomar Schmidt and the UK Schmidt in Australia. The reduction of the scanned plates was made using the AGK3 in the Northern Hemisphere and the SAO and the *Cape Photographic Catalogues* in the Southern Hemisphere (see Sections 11.5.4 and 11.5.5). The uncertainties range from $0\rlap{.}''5$ in the Northern Hemisphere to $2''$ in the Southern Hemisphere, with some larger peaks for stars near the edges of the plates. No attempt was made to make an average position for stars observed in two or more plates; all the derived positions are given. This enhances the inhomogeneity (already seen as a the result of the heterogeneity of the references) of the catalog. Nevertheless, it serves its original purpose in a satisfactory manner. An improved version is presented in Section 11.7.3.

11.5.4 Zone catalogs

Between the *Astrographic Catalog* and the observational or fundamental catalogs, there was space for large collections of precise star positions. Such catalogs are generally, but not always, referred to some uniquely defined reference system. This is the case of the last two *Astronomisches Gesellschaft* catalogs, AGK2 and AGK3. They were preceded by the AGK1, which was an ensemble of several zone catalogs within given limits in declination, referred to different epochs and different reference systems. This was similar to the *Astrographic Catalogue*, but it was more precise.

The first consistent catalog was the AGK2, published between 1951 and 1958, which includes about 180 000 stars with declinations between $-2°$ and $+90°$. The positions were referred to the FK3 system, average epoch 1960, using a compilation of transit instrument observations reduced to this system, called the AGK2A. It comprised about 13 000 stars with precisions of the order of $0\rlap{.}''20$ in δ and $0\rlap{.}''15$ in $\alpha \cos \delta$. No proper motions were computed. They were determined using a new series of photographic plates, taken in 1960–63, and the re-observation by transit circles of 21 000 stars in the FK4 system, and published under the name of AGK3-R. A repetition of the observations of stars was carried out around 1960, and, together with the AGK3-R positions, reduced to obtain proper motions (AGK3, Dieckvoss, 1971). Finally, the r.m.s. uncertainty of AGK3 positions at epoch was of the order of $0\rlap{.}''18$ and $0\rlap{.}''009$ per year for proper motions.

Two large-zone catalog programs were initiated in the 1920s. The Yale catalogs are an ensemble of declination zone catalogs that were designed to cover the full sky. All plates were taken, but not all the zones were reduced and, in particular in the Southern Hemisphere, the zone between 30° and 50°, and 60° to 90° are lacking, or were not published. The program was stopped and replaced by the Lick Northern Proper Motion Program, completed in 1994, and the Yale–Cordoba Southern Proper Motion Program, which has converted to CCD observations. The common objective was to determine directly the proper motions with respect to galaxies, assumed to be fixed. The drawback is that there is a zone of avoidance of ±20° around the galactic plane.

Another large program that came to completion is the *Cape Photographic Catalogue*, which extends from −30° to −90°, adjacent to the Yale equatorial zone catalog (−30° to +30°). It includes the positions of more than 68 000 stars at epoch 1950.0 with r.m.s. uncertainties of 0″.15 to 0″.22 in $\alpha \cos \delta$ and in δ, and proper motions are present everywhere except between −30° and −35°.

11.5.5 Compilation catalogs

The zone catalogs described in the preceding section cover a large part of the sky, but not completely, and often in a non-homogeneous manner. The objective of compilation catalogs is to have a full sky coverage in a single reference system. To construct them, fundamental, zone and observation catalogs are used, and a mean value for the positions at epoch and proper motions are computed from several sources.

The first such catalog was the *Smithsonian Astrophysical Observatory Star Catalog* (SAO, 1966) of 258 997 star positions and proper motions. The reference system is the FK4 system, which meant that positions in the GC and FK3 had to be put in the FK4 system using transformation formulae. The other stars were taken from AGK2-AGK1, Yale zones, the Cape photographic or astrographic zones, and transit observations performed in Melbourne. Despite the reduction to a common epoch and system, it is a heterogeneous catalog. However, it has been a very useful tool for the reduction of photographic observations of artificial satellites (its prime objective), and was actually very widely used as a reference-position catalog in all observatories, essentially because there was no other easily available catalog with a large number of stars.

Because of the source-dependent inconsistencies in the SAO it was necessary to compile a reference catalog from scratch to accurately extend the FK4 from the seventh to the ninth magnitudes. The *International Reference Stars Catalog* (IRS, Corbin 1992), based on the AGK3R and Southern Reference Stars (SRS) catalogs, with 36 027 stars, met this need for both the FK4 as well as the FK5 after a conversion to its system. Indeed, until the appearance of Hipparcos the IRS was the primary link between the fundamental catalogs for some 20 years in the Northern Hemisphere and almost 10 years in the Southern Hemisphere. So close was the connection of the IRS to the fundamental catalogs that it was chosen to provide the data for 2125 of the 3117 stars in Part II of the FK5 (Fricke, Schwan and Corbin, 1991).

In spite of the appearance of the IRS there was still a need for an SAO type catalog with higher precisions of individual stars and a system that conformed to the FK4 and FK5. The errors of the SAO were increasing with time, reaching mean uncertainties of about $1''$ by 1990. In fact, two catalogs, both based on the IRS, were produced in the late 1980s to correct this situation.

The *Positions and Proper Motions Catalogue* (PPM) was compiled at the Astronomisches Rechen Institut by Röser and Bastian (1991) for 181 731 stars North of $-2°5$, and by Bastian, Röser *et al.* (1993) for 197 179 stars South of that. In general the positions and proper motions were derived from positions in the AGK series, the *Astrographic Catalogue*, and other photographic and meridian circle catalogs. The system is that of the FK5 for 2000.0 and the epochs of the mean positions are 1930 in the North and 1960 in the South. The formal errors for the positions and motions are, respectively $0''27$ and 4.3 mas/year in the North and $0''11$ and 3.0 mas/year in the South.

The *Astrographic Catalog Reference Stars* (Corbin and Urban, 1988; Corbin, 1992) was compiled both to provide reference stars for a new reduction of the *Astrographic Catalogue* and, like the PPM, to serve as a replacement of the SAO. The PPM was compiled to give the best possible, individual, proper motions at the time, hence the inclusion of the *Astrographic Catalogue* data. On the other hand, the ACRS was constructed from 170 meridian circle and photographic catalogs, minus the AC, to provide a system of positions and motions closely tied to both the FK4 and FK5 for 320 211 stars. Thus, with a central epoch of 1949.9 and position and proper motion errors of $0''09$ and 4.6 mas/year, respectively, an ACRS position could be projected to either 1995 or 1905 (the average AC epoch) with a precision of $0''22$ for the purpose of data reduction to the FK4 or FK5.

Both the PPM and the ACRS served the astronomical community well for more than seven years, until the appearance of Hipparcos and Tycho. PPM, with its higher density and higher individual star precision was often preferred for reductions of data over small fields, including detailed reductions of Schmidt plates. ACRS, on the other hand, with its strong systematic ties to FK4 and FK5, was generally used for kinematic analysis and reductions of plates, such as the USNO TAC, over large areas. Since the ACRS was used to make new reductions of the AC (AC 2000), it lived on in the proper motions of the ACT (Urban *et al.*, 1988). The ACRS was revised and used to make an improved version of the AC which, along with the database of ACRS catalogs, allowed the determination of the Tycho 2 proper motions (Høg *et al.*, 2000b). See Section 11.6.2.

11.6 Hipparcos-based catalogs

The great turning-point of stellar astrometry is the success of the Hipparcos mission (Section 2.5.1). The resulting catalog (ESA, 1997) was released in 1996 for the members of the team having participated in its construction, and in 1997 for general distribution. On January 1st, 1998, it officially became the optical reference frame, representing the ICRS, and replacing the FK5 as the fundamental catalog. Now, the new star catalogs are to be referred to this system using the *Hipparcos Catalogue*, but since the star density is small, one often uses its extension, the *Tycho Catalogue*, as an intermediary link to the ICRS. Several catalogs have already appeared and they will be described after a presentation of the *Hipparcos* and *Tycho Catalogues*.

11.6.1 The Hipparcos Catalogue

The complete data, described in Section 2.5.1, were reduced by two independent international consortia, FAST and NDAC, by different though similar methods. They are presented in Kovalevsky *et al.* (1992) for FAST and Lindegren *et al.* (1992) for NDAC. Each constructed a complete independent catalog from the same photon counts, but the modeling used, including the rejection criteria, were not the same, so that the final correlation between the consortia results was on the average 0.7. This allowed a statistical improvement from a merging of the two catalogs. The scheme retained was to go back one step in the data reduction and estimate the star positions from the abscissae on reference great circles, taken together and considered as correlated observations in a new least-squares solution for the astrometric parameters (position, proper motion and parallax). This procedure resulted not only in optimally combined parameters, but also in the correct variance–covariance matrix for these data, a result that would not be possible by a direct weighted mean of the values of the parameters obtained by the consortia. A detailed description of the method used is given in chapter 17 of Volume 3 of the *Hipparcos Catalogue* (ESA, 1997).

The catalog so obtained was loosely linked to the FK5 system, because both consortia have constrained their catalogs to some FK5 stars. The next step was to change the reference system by a fixed, and a linearly time-dependent, spin, in order to link it to the ICRS, using the ICRF (see Section 7.2). This was possible only if some Hipparcos stars could be directly measured with respect to ICRF radio sources. Hipparcos radio stars were observed by radio interferometry with respect to these sources. In addition, if proper motions of Hipparcos stars were determined optically with respect to galaxies, one could assume that the latter being also very distant, these proper motions were also referred to the extragalactic system, ICRS. Actually, eleven different determinations of either both rotations, or only of the time-dependent spin, were obtained and a weighted mean of these results was needed to rotate the intermediary merged catalog (Kovalevsky *et al.*, 1997). The final result is that the *Hipparcos Catalogue* reference frame coincides with the ICRF with an uncertainty of 0.6 mas in position at epoch (1991.25) and 0.25 mas/year in time-dependent spin. Evidently, because of this latter uncertainty, the representation of the ICRS is slowly degrading with time.

The *Hipparcos Catalogue* includes astrometric data of 117 955 stars (about three stars per square degree) up to magnitude $V = 12.4$ mag, although it is complete only up to

$$-V \leq 7.9 + 1.1 \sin |b|,$$

for spectral types earlier or equal to G5 and

$$-V \leq 7.3 + 1.1 \sin |b|,$$

for spectral types later (redder) than G5, where b designates galactic latitudes. This corresponds to about 52 000 stars. The others were selected for their astrophysical or astrometric interest (Turon *et al.*, 1992b). All relevant astrophysical data, together with pre-Hipparcos astrometric parameters, were published for all program stars in the *Input Catalogue* (Turon *et al.*, 1992a).

The median uncertainties for bright stars (Hipparcos magnitude $H_p < 9$) are:

- 0.77 mas in $\alpha \cos \delta$
- 0.64 mas in δ
- 0.88 mas per year in $\mu_\alpha \cos \delta$
- 0.74 mas per year in μ_δ
- 0.97 mas in parallax.

Actually, the uncertainties depend on the Hipparcos magnitude, H_p, and on the ecliptic latitude, b, of the star. The latter effect is the result of the scanning law that was chosen in such a way, that the scan never approaches the Sun by more than $43°$. Figure 11.8 illustrates these dependencies.

For about 20 000 stars, the parallax is determined to better than 10% and for 30 000 others to better than 20%. Comparisons with parameters independently estimated (Magellanic clouds, distant clusters and supergiants) show that the systematic errors do not globally exceed 0.1 mas. However locally, they may be significantly larger because of existing correlations.

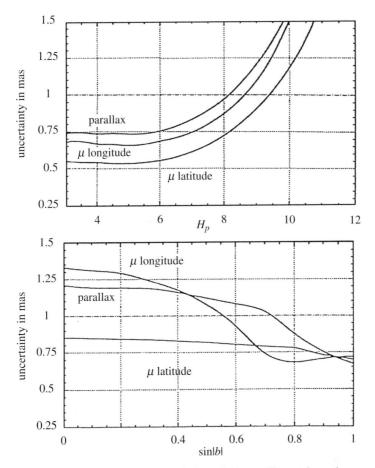

Fig. 11.8. Mean standard uncertainties of the parallax and yearly proper motions of all the single stars of the *Hipparcos Catalogue* as functions of the magnitude H_p and the ecliptic latitudes b (Kovalevsky, 2002). The uncertainties in position follow similar trends as proper motions.

Hipparcos data reduction allowed the determination of the parameters of double, and in some cases multiple, stars (see Chapter 12). These were determined for 13 425 systems out of which almost 3000 were new. Over 10 000 others were recognized as suspected doubles, but no solution could be found for their parameters. This was the case of very close systems, or systems with components with magnitude differences larger than 3.

In addition, the treatment of photon counts gave a very precise and consistent set of magnitudes for 118 217 stars in the Hipparcos magnitude system (H_p) which is described in Fig. 11.6. The median uncertainty for bright stars ($H_p < 9$) is 0.0015 mag derived from about 13 million separate observations. These data detected about 11 600 variable stars, of which over 8200 were new.

11.6.2 The Tycho Catalogues

In Section 2.5.1, it was mentioned that in addition to the main grid, there was a star-mapper which simultaneously scanned the sky with a primary objective of determining the orientation of the satellite as a function of time. In contrast with the main grid, in which only a very small part of the field was observed, all the stars crossing the star mapper grids provided signals that were recorded. Knowing the final orientation, as determined by the main reduction, and the time of crossing of the star-mapper grids by star images, it is possible to determine a function of the position at this time. From all such observations of a given star made during the mission, one may determine the positions and other astrometric parameters of these stars. The method is described in Høg *et al.* (1992) and in great detail in Volume 4 of the *Hipparcos* and *Tycho Catalogues* (ESA, 1997). A total of 1 052 031 star positions was obtained, with, in some cases, significant proper motions. This represents, with an additional 6301 Hipparcos stars not recognized by the Tycho data reduction, about 25 stars per square degree. The limiting magnitude is $V_T \simeq$ 11.5 and the completeness is estimated to be up to $V_T = 10.5$ mag. The mean astrometric uncertainty for all stars at epoch 1991.25 is 25 mas, but the median for bright stars ($V_T < 9$ mag) is 7 mas. Since through the orientation, the positions are in the *Hipparcos* system, the same rotation as for the main catalog was applied, so that the *Tycho Catalogue* is also on the ICRS.

As for the main mission, the analysis of photon counts provided photometric data. For this purpose, the light from the star mapper was split by a dichroic mirror in such a way that there were two photometric observations, called V_T and B_T very close to the Johnson V and B systems (Section 11.4.3). The median photometric uncertainty for all stars was 0.07, 0.06, and 0.10 mag for B_T, V_T and $B_T - V_T$, respectively. For bright stars ($V_T < 9$ mag) it is 0.014, 0.012 and 0.019 mag, respectively.

The precision of the Tycho observations is worse by a factor of about 10 than in the case of Hipparcos. This is more or less proportional to the actual time of observation per star in the programs. But this is largely compensated by the number of stars. This is by far the most precise observation catalog covering the whole sky in existence, remarkable for its large number of stars, exceeding all other catalogs except the low-precision *Astrometric Catalogue*, *Guide Star Catalog*, and the USNO *A2 Catalog*. It is bound to be the major reference in the construction of further compiled catalogs.

Later, a new reduction of the Tycho observations was performed using a more sophisticated procedure, and allowing less stringent conditions for accepting data. Høg *et al.* (2000a) constructed the *Tycho-2* catalogue containing positions and two color photometry of 2 538 913 stars for the epoch J2000.0. It is 99% complete to Tycho magnitude $V_T = 11.0$, and 95% to $V_T = 11.5$. Positions are accurate to 10–100 mas. Proper motion uncertainties are 1–3 mas/year, depending on magnitude.

Using over 140 astrometric catalogs, proper motions were derived, and positions were propagated to J2000.0. Two-color photometry from Tycho is included in the catalog. The catalog is referenced to the Hipparcos Catalogue Reference System (HCRS) and is for the epoch J2000.0 (Høg *et al.*, 2000a). One source of more information about astrometric catalogs is http://ad.usno.navy.mil.

11.7 Optical reference star densification

The Hipparcos and Tycho catalogs are not vey dense, and there is a need to add positions and proper motions of many more stars in the same reference system. This was the case for *Tycho-2*. But there are many more. Let us first mention the *AC 2000* catalog compiled by the US Naval Observatory (Urban and Corbin, 1998). It contains 4.62 million stars on the Hipparcos system for the epoch of observations published in the *Astrographic Catalogue* (Section 11.5.3). The positions of the *Astrographic Catalogue* were used in connection with modern positions reduced to the Hipparcos system to determine proper motions. In some sense, *AC 2000* is an update of the *Astrographic Catalogue*.

Another achievement is the US Naval Observatory *ACT Reference Catalog*, which contains 988 758 stars from the *Tycho Catalogue* and provides proper motions. This was obtained by combining Tycho positions, with those taken from the *AC 2000*. The large epoch span between the two catalogs yields proper motions with uncertainties of the order of 3–5 mas/year.

In addition, new catalogs were compiled using new observations.

11.7.1 USNO CCD Astrographic Catalog (UCAC)

The UCAC is a current astrometric observing program for a global sky coverage down to 16th magnitude in a red wavelength. It is expected to be a catalog of about 80 million stars. Positions should be accurate to 20 mas in the 10–14 magnitude range and 70 mas at 16th. Proper motions will have magnitude dependent errors of 1–12 mas/year, determined from catalogs as used for *Tycho-2* for brighter stars and USNO A2.0 for the fainter stars. The program started in February 1998 and is expected to be complete in 2004. Photometry will have errors of about 0.1 to 0.3 magnitude in a single, non-standard color. The reference system is the HCRS, via *Tycho-2*, and is for the epoch J2000.0.

UCAC1 is a preliminary catalog with stars in most of the Southern Hemisphere from observations taken at Cerro Tololo, Chile. UCAC2 is a preliminary catalog of about 40 million stars covering the sky from declination −90 to +24, and in some areas to +30. This catalog is available from the US Naval Observatory (nz@pisces.usno.navy.mil) (Zacharias *et al.*, 2000).

11.7.2 USNO A2.0, and derived catalogs

USNO A2.0 contains 526 230 881 stars, which were detected in the digitized images of three photographic sky surveys. The Palomar Optical Sky Survey (POSS-I) covered the Northern sky and the Southern sky down to −30° declination in blue- and red-sensitive emulsions. The rest of the Southern sky was covered by the Science Research Council (SRC)-J and the European Southern Observatory (ESO)-R surveys. Only stars appearing in both wavelengths were included in the catalog. USNO A2.0 is on 10 CD-ROMs with right ascensions and declinations for J2000.0 on the HCRS for the mean of the blue and red plates. Blue and red magnitudes are given for each star. Positions are thought to be accurate to about 250 mas, and no proper motions are given. The systematic errors due to the faintness of the stars could be as high as the random errors. The epoch of observations are declination dependent.

USNO SA2.0 is a subset of USNO A2.0 selected to fit on one CD-ROM giving uniform spatial distribution in an intermediate magnitude range that would be useful for reference stars for CCD observations.

USNO B will be the successor to USNO A2.0 and is expected to include proper motions and many more stars. The Northern and Southern proper motion survey plates will be included as part of the bases for this catalog.

11.7.3 HST Guide Star Catalog

The measures of the Hubble *Guide Star Catalog* described in Section 11.5.3, were re-processed in order to reduce the plate based systematic errors. The new version, HST 1.2, contains positions for 19 million stars down to 16th magnitude, but it is not complete to that magnitude. It is based on plate measures made at the Space Telescope Science Institute (STScI). There are no proper motions and positional errors are estimated to be about 500 mas at the plate epoch (Morrison *et al.*, 2001).

Hubble *Guide Star Catalog II*, expected to be issued in 2004, will be a combination of first and second epoch POSS and SES plates digitized at the STScI to produce positions, proper motions, magnitudes and colors to 18th magnitude for operational purposes. The epoch of observations is generally in the 1970 to 1980 period, and it is on HCRS via Tycho, but it is subject to systematic errors due to faintness. This has much smaller systematic errors than the GSC. Also, plate images are compressed and made available as scan data (McLean *et al.*, 1998).

11.7.4 2MASS catalog

The Two Micron All Sky Survey (2MASS) has imaged the entire sky in near-infrared J(1.25 μm), H(1.65 μm), and Ks(2.16 μm) bands from Mount Hopkins, Arizona and Cerro Tololo, Chile. The 10-sigma detection levels reached 15.8, 15.1, and 14.3 mag at the J, H, and Ks bands respectively. The 2MASS data produces an image atlas with 1″ pixels, a point source catalog of over 300 million objects, and an extended source catalog of 1 to 2 million sources, mostly galaxies. The positions are accurate to about 200 mas. Part of the catalog has been released, but the data are being reprocessed and the complete catalog was expected to be available in 2002. This will be improved by use of *Tycho-2* and internal overlapping plate techniques (Skrutskie, 2001). Information on 2MASS can be found at http://ipac.caltech.edu/2mass.

12

Double and multiple star systems

A very common feature in the sky is the presence of two or more stars very close one to another. The frequency of such configurations is larger by some orders of magnitude than would be the case if the distribution of stars on the sky were homogeneous. In some cases, indeed fortuitously, two stars may happen to be seen in almost the same direction. These are optical binary stars, and if one determines their parallaxes, it is found that they are actually very far from each other. But in most of the cases, it is found that they are at the same distance and are physically bound. Actually, it is believed that at least half of the stars are not single, like our Sun, but form gravitationally bound systems of two stars – these are called *binaries* or *double stars* – or more: these are *multiple stars*.

Let us consider the simplest case; two separate stars attract each other under the universal gravitational attraction (Newton's law). The theory that governs the resulting relative motion is the two-body problem.

12.1 The two-body problem

The Newtonian law of gravity states that two particles P_1 and P_2 of mass m_1 and m_2 attract each other with a force, F, directly proportional to the product of their masses, and inversely proportional to the square of the distance, r, between them:

$$F = k^2 \frac{m_1 m_2}{r^2}.$$ (12.1)

If the masses are expressed in solar masses and r in astronomical units, k is the Gaussian gravitational constant, $k = 0.017\,202\,098\,95$. It is, with the speed of light, the second defining constant of the system of astronomical constants.

In vectorial notation, (12.1) becomes

$$\mathbf{F} = -\frac{k^2 m_1 m_2}{|P_1 P_2|^3} \mathbf{P_1 P_2},$$ (12.2)

for the force acting on P_1, and with the opposite sign for the force acting on P_2.

It is sometimes useful to write this in a reference system of coordinates with an origin at the center of mass of the system, a point G such as

$$m_1\mathbf{GP_1} + m_2\mathbf{GP_2} = 0. \tag{12.3}$$

A consequence of the general theorems of mechanics is that G is not accelerated, so this reference system is inertial. From (12.3), one deduces

$$\mathbf{P_1P_2} = -\frac{m_1 + m_2}{m_2}\mathbf{GP_1} = \frac{m_1 + m_2}{m_1}\mathbf{GP_2}.$$

Substituting these expressions into (12.2), one gets the equations of motion of each component, dividing by m_1, or m_2, to get accelerations:

$$\frac{d^2\mathbf{GP_1}}{dt^2} = -\frac{k^2 m_2^3}{(m_1 + m_2)^2}\frac{\mathbf{GP_1}}{|GP_1|^3},$$

$$\frac{d^2\mathbf{GP_2}}{dt^2} = -\frac{k^2 m_1^3}{(m_1 + m_2)^2}\frac{\mathbf{GP_2}}{|GP_2|^3}. \tag{12.4}$$

So, each body P_i is attracted towards the center of mass as if a mass equal to $m_i^3/(m_1 + m_2)^2$ was concentrated in G. If we now eliminate $\mathbf{GP_1}$ and $\mathbf{GP_2}$ from (12.3) and (12.4), we get

$$\frac{d^2\mathbf{P_1P_2}}{dt^2} = -\frac{k^2(m_1 + m_2)\mathbf{P_1P_2}}{|P_1P_2|^3}. \tag{12.5}$$

In a system of non-rotating axes centered on one of the bodies, the motion is that which would be caused by the attraction exerted by a central mass, equal to the sum of the masses of the two bodies. Whether one studies relative or absolute motion, the form of the equations is the same and we shall write

$$\frac{d^2\mathbf{r}}{dt^2} = -\mu\frac{\mathbf{r}}{r^3}, \tag{12.6}$$

with $\mu = k^2 m$, where m is either $(m_1 + m_2)$ or one of the factors in Equations (12.4).

12.1.1 Shape of the trajectories

Let us apply the fundamental laws of mechanics to the motion of a point P of mass m defined by (12.6) around a point O. The theorem on angular momentum, with respect to the origin point O, states that the derivative of the angular momentum

$$\frac{d\mathbf{\Omega}}{dt} = m\mathbf{OP}, \quad \frac{d^2\mathbf{OP}}{dt^2} = 0.$$

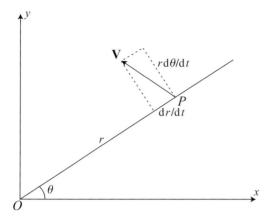

Fig. 12.1. Position and velocity of a point P in a polar reference coordinate system.

Since both vectors are colinear, then the angular momentum Ω is constant and the motion is planar. Let us take this plane as the reference and a system of polar coordinates r, θ (Fig. 12.1) in this plane. At time t, the velocity $\mathbf{V}$ can be split into two components:

(i) the radial velocity dr/dt,
(ii) the tangential velocity $r d\theta/dr$.

The constant angular momentum is called the constant of areas and is equal to

$$\Omega = r^2 d\theta/dt. \tag{12.7}$$

We may also apply the theorem of kinetic energy which states that, in the absence of external forces, it is constant,

$$\frac{1}{2}mV^2 - \int \frac{-\mu}{r^2}\frac{dr}{dt}dt = C,$$

or, dividing by m and introducing the energy constant h, one has

$$2h + \frac{2\mu}{r} = V^2 = r^2\left(\frac{d\theta}{dt}\right)^2 + \left(\frac{dr}{dt}\right)^2. \tag{12.8}$$

One can eliminate time in Equations (12.7) and (12.8) and obtain the differential equation of the trajectory,

$$\frac{1}{r^4}\left(\frac{dr}{d\theta}\right)^2 + \frac{1}{r^2} - \frac{2\mu}{\Omega^2 r} - \frac{2h}{\Omega^2} = 0,$$

which is easily integrated using a new variable

$$w = \frac{1}{r} - \frac{\mu}{\Omega^2}.$$

The final result is

$$\frac{1}{r} = \frac{\mu}{\Omega^2}\left[1 + \sqrt{1 + \frac{2\Omega^2 h}{\mu^2}}\cos(\theta - \theta_0)\right], \tag{12.9}$$

where θ_0 is a constant of integration. For $\theta = \theta_0$, r is at a minimum. Let us take this point as the origin of angles. The Ox axis, so defined, is also called the line of apsides. The angle $v = \theta - \theta_0$ is called the *true anomaly*. Introducing as new notations the constant e and $p = \Omega^2/\mu$:

$$e = \sqrt{\frac{1 + 2\Omega^2 h}{\mu^2}} = \sqrt{1 + \frac{2hp}{\mu}},$$

and consequently,

$$2h = \frac{\mu(e^2 - 1)}{p}.$$

Equation (12.9) becomes

$$\frac{1}{r} = \frac{1 + e\cos v}{p}, \tag{12.10}$$

which is the equation of a conic section of eccentricity, e, and parameter, p, and is referred to one of its foci. One can, hence, introduce the semi-major axis, a, such that

$$p = a(1 - e^2),$$

from which, one concludes that

$$2h = -\frac{\mu}{a}.$$

Depending upon the value of e, the conic is an ellipse ($e < 1$), a parabola ($e = 1$), or a hyperbola ($e > 1$). We shall consider from now on only the case of the elliptic motion, the only one that is periodic and relevant to the study of double stars.

12.1.2 Kepler's laws

A general description of elliptical motion, under Newtonian attraction by a central body, is given by the three laws established by Kepler from observations of planets. He stated them in terms of the Sun and planetary motions. In the case of double stars, the Sun should be replaced by one of the components of the binary or the center of mass, since both cases refer to the general Equation (12.6). The planet is the other component of the binary.

(i) The orbit of a planet lies on a plane that passes through the Sun, and the area swept over by the line joining the Sun and the planet is proportional to the time elapsed. This corresponds to the fact that Ω is constant.
(ii) The orbit of a planet is an ellipse of which the Sun occupies one focus. The equation of this ellipse is given by (12.10).
(iii) The ratio between the square of the period of revolution and the cube of the semi-major axis is the same for all planets.

This law is generalized to any system by stating that this ratio depends only on the sum of the masses of the binary system. Let us prove it.

The area of an ellipse is $A = \pi a^2 \sqrt{1 - e^2}$. It can be also expressed in terms of Ω, as the integral over a period P:

$$A = \frac{1}{2} \int_0^P \Omega \, dt = \Omega P / 2 = \pi a^2 \sqrt{1 - e^2}.$$

We have also the relation

$$p = a(1 - e^2) = \Omega^2 / \mu.$$

By the elimination of $1 - e^2$ between these two equations, we get

$$\frac{a^3}{P^2} = \frac{\mu}{4\pi^2} = \frac{k^2(m_1 + m_2)}{4\pi^2}. \tag{12.11}$$

Another useful form of this equation is obtained as we introduce the mean motion $n = 2\pi / P$. Then Kepler's third law becomes

$$n^2 a^3 = \mu = k^2(m_1 + m_2). \tag{12.12}$$

With this notation, one easily transforms the two integrals (12.8) and (12.7) into

$$\Omega = na^2 \sqrt{1 - e^2},$$
$$V^2 = \mu \, (2/r - 1/a).$$

12.1.3 Kepler's equation

To complete the description of the Keplerian motion, one needs to be able to know where the body is at any time t. To do this, let us consider the ellipse as the projection of a circle. This is illustrated by Fig. 12.2, where the circle C is projected on the plane of the ellipse and rotated on this plane. Let C, O, and A be respectively the center, the focus, and the perifocus of the ellipse. Let P be the position on the ellipse

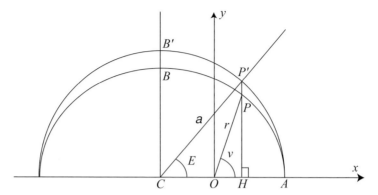

Fig. 12.2. Definition of the eccentric anomaly, E, and of the true anomaly, v.

at time t. It is the projection of the point, P', on the circle. The ratio HP/HP' is a constant equal to the ratio CB/CB', the semi-minor axis of the ellipse to the radius of the circle, that is $\sqrt{1 - e^2}$.

Note that, using (12.10), OA corresponds to the radius when $v = 0$, so that

$$OA = \frac{a(1 - e^2)}{1 + e} = a(1 - e),$$

and therefore $CO = ae$.

We define the *eccentric anomaly*, E, as the angle (CA, CP') expressed in radians.

Let us apply Kepler's first law to compute the surface swept by the radius-vector during the time $t - t_0$, t_0 being the time at which the body was in A, the perifocus, also called *periastron*

$$S = \text{Area}\,(AOP) = \frac{1}{2} \int_{t_0}^{t} \Omega\, dt = \frac{1}{2} na^2 \sqrt{1 - e^2}(t - t_0).$$

It is common practice to define the *mean anomaly* as the product $M = n(t - t_0)$, expressed in radians. So,

$$S = \frac{1}{2} a^2 \sqrt{1 - e^2}\, M.$$

The ratio between S and $S' = \text{Area}(AOP')$ is $\sqrt{1 - e^2}$, so that

$$S' = \frac{1}{2} a^2 M.$$

But $S' = \text{Area}(ACP') - \text{Area}(OCP')$. ACP' is a sector of a circle. Its area is $a^2 E/2$ (E is expressed in radians) and OCP' is a triangle whose area is

$$\frac{1}{2} \mathbf{CO} \times \mathbf{HP'} = \frac{1}{2} a^2 e \sin E.$$

Equating the two expressions for S', one gets

$$\frac{1}{2}a^2 M = \frac{1}{2}a^2 E - \frac{1}{2}a^2 e \sin E,$$

or, the Kepler equation, in its simple form:

$$E - e \sin E = M = n(t - t_0). \tag{12.13}$$

There are many numerical methods for solving this equation in order to obtain E knowing M. All involve an iterative procedure. With the use of computers, this is no longer a problem.

To return to r and the true anomaly v, it is sufficient to compute the coordinates of P. An examination of Fig. 12.2 gives immediately

$$x = r \cos v = a(\cos E - e),$$
$$y = r \sin v = a\sqrt{1 - e^2} \sin E,$$

from which we get

$$r = a(1 - e \cos E). \tag{12.14}$$

In addition, since

$$\tan^2 \frac{v}{2} = \frac{1 - \cos v}{1 + \cos v},$$

and considering that E and v are in the same half-plane, we deduce that

$$\tan^2 \frac{v}{2} = \frac{1 - e \cos E - \cos E + e}{1 - e \cos E + \cos E - e}$$

and hence,

$$\tan \frac{v}{2} = \sqrt{\frac{1 + e}{1 - e}} \tan \frac{E}{2}. \tag{12.15}$$

12.2 Apparent orbits of double stars

Physically, with the exception of very close binaries, whose actual separation is so small that their interaction involves deformations and exchanges of matter, one may consider that the two stars are behaving like independent bodies, except for their gravitational two-body-like behavior described in the preceding section. We shall consider only these separated systems, since the study of physically interactive binaries does not yet use astrometric techniques.

Actually, even among separated systems, it was customary to distinguish four different classes depending upon the way that they were observed. We shall keep this distinction, because the reduction procedures of the observed quantities are different. But in all the cases, the projection on the sky of the relative positions of the components is the same. For this we shall use differential coordinates (see Section 3.9.1). We shall take one of the components and call it the *primary* and center on it the local coordinate system. The coordinates of the *secondary* are the separation, ρ, expressed in arcseconds, and the position angle, θ, reckoned eastward from the northern direction of the local celestial meridian. The standard rectangular coordinates, expressed in seconds of arc, are

$$x = \rho \sin \theta,$$
$$y = \rho \cos \theta. \tag{12.16}$$

These formulae are sufficient for most practical cases. However, it may be advisable to apply a correction, noting that the ratio between a small angle, ρ^*, and its projection, $\tan \rho^*$, on the tangent plane to the celestial sphere is

$$\frac{\tan \rho^*}{\rho^*} = 1 + \frac{\rho^{*2}}{3},$$

where ρ^* is ρ expressed in radians. The correction is less than 0.05 mas for separations under $15''$.

12.2.1 Orbital elements

The actual orbit, not necessarily tangential to the line of sight, is projected on the sky and is seen as an ellipse, whose scale in seconds of arc is

$$a'' = a\varpi,$$

where ϖ is the parallax, and the semi-major axis, a, is expressed in astronomical units. It is to be noted that the projected ellipse is not a Keplerian ellipse, in the sense that the focus is not at the place of the primary, and its major axis is not along the projection of the diameter containing the periastron. But the areas on the apparent ellipse are projections of the areas on the actual orbital ellipse, so that the law of proportionality of the areas swept by the apparent radius vector holds. In addition, the center of the apparent ellipse is the projection of the center of the true orbit, so that the projection of the line of apsides is the line drawn from the center to the primary.

The position of the real ellipse with respect to the local coordinate system is defined by three angles, the orientation elements. The intersection of the plane of the ellipse with the tangent plane is the line of nodes (see Fig. 12.3). The position angle of the nodal point, Ω, corresponds generally to that node for which $0° \leq \Omega < 180°$. If radial velocity observations can determine at which point the secondary is approaching, then Ω corresponds to this point.

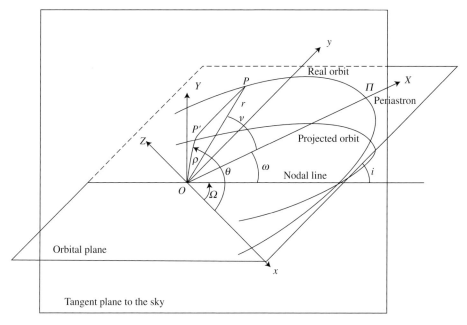

Fig. 12.3. Relations between the apparent and the true reduced orbits.

The angle of the orbital plane with the plane tangent to the celestial sphere is the *inclination, i*. Its value is chosen in such a way that if the position angle of the secondary increases with time, one takes $0° \leq i < 90°$. Otherwise, $90° \leq i < 180°$.

The third angle is the *longitude of the periastron, ω*. It is the angle in the true orbit from the nodal point to the periastron, Π, in the direction of the orbital motion.

The other elements characterizing the orbit are the dynamical elements pertaining to the true orbit and are the actual semi-major axis, a, obtainable only if the parallax is known, the period, P, which is the same for the apparent and the true ellipse, and the eccentricity, e, of the true orbit. Often a is replaced by the parameter $p = a(1 - e^2)$.

12.2.2 The Thiele–Innes constants

The problem is to find the relations that exist between the local apparent coordinates,

$$\mathbf{r} \begin{cases} x = \rho \sin \theta, \\ y = \rho \cos \theta, \end{cases}$$

and the reduced true coordinates, also expressed in seconds of arc,

$$\mathbf{R} \begin{cases} X = \varpi r \cos v, \\ Y = \varpi r \sin v. \end{cases}$$

To transform $\mathbf{R}$ into $\mathbf{r}$, one has to perform two transformations.

First, a rotation of $-\omega$ around the Z-axis perpendicular to the XY plane, followed by the projection, amounting to multiplying the new ordinates by $\cos i$. One gets

$$X_1 = X \cos \omega - Y \sin \omega,$$
$$Y_1 = \cos i (X \sin \omega + Y \cos \omega).$$

Then, a rotation of $-\Omega$ around the z-axis perpendicular to the tangent plane, and a permutation of abscissae and ordinates are performed to obtain the particular Oxy coordinate system in the tangent plane

$$x = X_1 \sin \Omega + Y_1 \cos \Omega,$$
$$y = X_1 \cos \Omega - Y_1 \sin \Omega.$$

One finally obtains

$$
\begin{aligned}
x = {} & X(\cos \omega \sin \Omega + \sin \omega \cos \Omega \cos i) \\
& + Y(-\sin \omega \sin \Omega + \cos \omega \cos \Omega \cos i), \\
y = {} & X(\cos \omega \cos \Omega - \sin \omega \sin \Omega \cos i) \\
& + Y(-\sin \omega \cos \Omega - \cos \omega \sin \Omega \cos i),
\end{aligned}
$$

which we write

$$
\left.
\begin{aligned}
x &= BX + GY, \\
y &= AX + FY.
\end{aligned}
\right\}
\tag{12.17}
$$

The coefficients A, B, F, and G are the Thiele–Innes constants:

$$
\begin{aligned}
A &= \cos \omega \cos \Omega - \sin \omega \sin \Omega \cos i, \\
B &= \cos \omega \sin \Omega + \sin \omega \cos \Omega \cos i, \\
F &= -\sin \omega \cos \Omega - \cos \omega \sin \Omega \cos i, \\
G &= -\sin \omega \sin \Omega + \cos \omega \cos \Omega \cos i.
\end{aligned}
\tag{12.18}
$$

The inverse formulae can be easily derived from these equations, giving

$$
\begin{aligned}
\sin(\omega + \Omega) &= \frac{B - F}{\sqrt{(B - F)^2 + (A + G)^2}}, \\
\cos(\omega + \Omega) &= \frac{A + G}{\sqrt{(B - F)^2 + (A + G)^2}}, \\
\sin(\omega - \Omega) &= \frac{-B - F}{\sqrt{(B + F)^2 + (A - G)^2}}, \\
\cos(\omega - \Omega) &= \frac{A - G}{\sqrt{(B + F)^2 + (A - G)^2}},
\end{aligned}
\tag{12.19}
$$

and

$$\tan^2 \frac{i}{2} = \frac{1 - \cos i}{1 + \cos i} = \frac{(B + F)\sin(\omega + \Omega)}{(B - F)\sin(\omega - \Omega)} = \frac{(A - G)\cos(\omega + \Omega)}{(A + G)\cos(\omega - \Omega)}.$$

$$(12.20)$$

Finally, the constant of areas on the projected ellipse is $\Gamma = \Omega^* \cos i$, where Ω^* is the constant of areas of the true ellipse and one can obtain it from

$$\Gamma = x\frac{dy}{dt} - y\frac{dx}{dt},$$

which is also equal, using formula (12.17), to

$$\Gamma = \left(X\frac{dY}{dt} - Y\frac{dX}{dt}\right)(AG - BF) = \frac{\Omega^*}{a''^2}(AG - BF).$$

These two expressions of Γ lead to

$$a''^2 = \frac{AG - BF}{\cos i},$$

while

$$\Gamma = \Omega^* \cos i = \frac{2\pi}{P}a''^2\sqrt{1 - e^2}\cos i,$$

provides the eccentricity e.

12.3 Resolved double stars

In this section, we deal with the classical case in which both components of the binary are observed, so that ρ and θ are directly determined. This is the case for visual, CCD, photographic, and speckle interferometry observations (for this reason, this category of double stars is called *visual*). The information obtained by Hipparcos for double, or multiple, stars had to undergo a complex reduction process before ρ and θ could be obtained. We will give the principle of the rather complex reduction process. Although Hipparcos produced only a relatively small fraction of double-star measurements, the method may influence the reduction of future space observations.

12.3.1 Hipparcos observations of double stars

As stated in Section 2.5.1 and illustrated by Fig. 2.6, the modulation curve of a double star is the combination of two periodic modulation curves, $S_1(t)$ and $S_2(t)$, produced by each of the components:

$$S_1(t) = I_1 + B + I_1 M_1 \cos(\omega t + \phi_1) + I_2 N_2 \cos 2(\omega t + \phi_1)$$
$$S_2(t) = I_2 + B + I_2 M_2 \cos(\omega t + \phi_2) + I_2 N_2 \cos 2(\omega t + \phi_2), \quad (12.21)$$

where B is the photon noise, M_i and N_i are the modulation coefficients of a single star, which are slightly color dependent, I_1 and I_2 are the intensities of each component, and ϕ_1 and ϕ_2 are the phases. They are linked to the projected separation of the components in the scanning direction by

$$\phi_1 - \phi_2 + 2k\pi = \frac{2\pi}{s}\rho\cos(\theta - \psi), \tag{12.22}$$

where θ is the position angle of the line joining the components, ψ is the known position angle of the scan, ρ is the separation, and $s = 1\rlap{.}{''}208$ is the grid step. The integer k takes into account the possibility that the projected separation is larger than the grid step.

The observed quantity is $S(t)$, which has the form

$$S(t) = I + B + IM\cos(\omega t + \Phi_1) + IN\cos 2(\omega t + \Phi_2),$$

where B is the photon noise. Identifying it with $S(t) = S_1(t) + S_2(t)$, one gets five equations

$$I = I_1 + I_2 + B,$$
$$IM\cos\Phi_1 = I_1 M_1 \cos\phi_1 + I_2 M_2 \cos\phi_2,$$
$$IM\sin\Phi_1 = I_1 M_1 \sin\phi_1 + I_2 M_2 \sin\phi_2,$$
$$IN\cos 2\Phi_2 = I_1 N_1 \cos 2\phi_1 + I_2 N_2 \cos 2\phi_2,$$
$$IN\sin 2\Phi_2 = I_1 N_1 \sin 2\phi_1 + I_2 N_2 \sin 2\phi_2.$$

The first equation is perturbed by the photon noise and is not useful. The next four equations are not independent and give only three relations between the three unknowns: $\phi_1 - \phi_2$, I_1 and I_2. This is not sufficient to solve (12.22) in which both ρ and θ are present. As a consequence, one needs to use simultaneously several observations made at different scanning angles, ψ, so that ρ and θ can be decorrelated. Therefore, Hipparcos provides mean values of ρ and θ over at least a few months, rather than a single observation. The advantage of this photometric observation is that one obtains, in addition, accurate magnitudes of the components derived from I_1 and I_2.

The procedures to get ρ and θ are complex, and several methods were used. Another difficulty is that in formula (12.22), $\phi_2 - \phi_1$ is defined modulo 2π. So, in addition, one has to determine the integer k. If the double star has been observed visually, one can derive it easily. If this is not the case, a special procedure is used to find a probable value of k. The different methods are described in Mignard *et al.* (1992) and Söderhjelm *et al.* (1992).

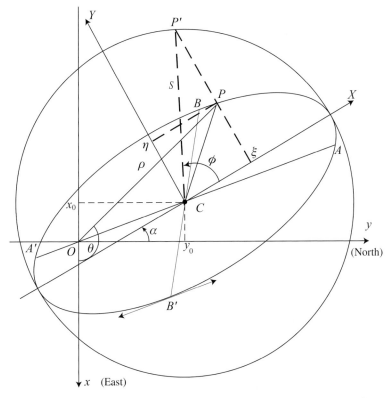

Fig. 12.4. Coordinates on the apparent orbit. In this figure, α is negative.

12.3.2 Double-star apparent-orbit determination

A first step is the determination of the apparent ellipse from observations. The observed vector **OP** is the sum of two vectors (Fig. 12.4)

$$\mathbf{OP} = \mathbf{OC} + \mathbf{CP},$$

where **OC**, with its two components x_0, y_0, represents the position of the center of the ellipse with respect to the coordinate origin at the primary. **CP** can be parametrized considering that it is an affine transform of a circle, the radius of which is the semi-major axis s of the ellipse, whose angle of position is α. The components of **CP** in the principal axes OXY of the ellipse are

$$\xi = s \cos \phi,$$
$$\eta = s\sqrt{1 - \varepsilon^2} \sin \phi,$$

where ε is the eccentricity of the apparent ellipse. To get the coordinates of **CP** in the relative xOy axes, a rotation of α has to be applied, followed by a permutation

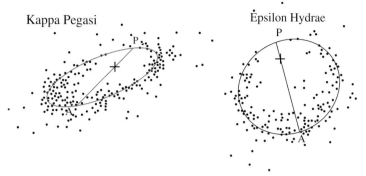

Fig. 12.5. Two examples of the determination of apparent orbits of κ Peg and ε Hyd as obtained using speckle astrometry observations. The widely dispersed points correspond to visual and photographic observations. Adapted from McAllister, 1996.

of abscissae and ordinates. Finally, one gets, with ϕ as a parameter:

$$y = y_0 + s \cos \phi \cos \alpha + s\sqrt{1 - \varepsilon^2} \sin \alpha \sin \phi = \rho \cos \theta,$$
$$x = x_0 - s \cos \phi \sin \alpha - s\sqrt{1 - \varepsilon^2} \cos \alpha \sin \phi = \rho \sin \theta. \tag{12.23}$$

The solution must be fitted in such a way that the integral of areas

$$\Gamma = \rho^2 \frac{d\theta}{d\rho} \tag{12.24}$$

is constant with time.

Another possibility is to eliminate the parameter ϕ, and replace (12.23) by a single equation of the ellipse:

$$(x - x_0)^2 (1 - \varepsilon^2 \sin^2 \alpha) + 2\varepsilon^2 (x - x_0)(y - y_0) \sin \alpha \cos \alpha$$
$$+ (y - y_0)^2 (1 - \varepsilon^2 \cos^2 \alpha) = s^2 (1 - s^2),$$

and determine the five unknowns x_0, y_0, s, ε, and α by least squares from the observations, keeping (12.24) as an external condition to be used to parametrize the path as a function of time and to determine the period.

The determination of a good apparent orbit is not an easy task. A first condition is that during the time span of observations one revolution took place, or at least that a large part of the orbit was scanned. Another condition is that the uncertainties of the observations remain rather small in comparison with the size of the orbit. Figure 12.5 illustrates this. It shows how good speckle interferometry observations could achieve a good apparent orbit and how dispersed the previous visual or photographic observations of close binaries were (in this case the separations are smaller than $0''\!.3$). It is to be noted, that the Hipparcos results agree very well with the speckle interferometry data; they have about the same uncertainties.

Once the apparent orbit and the period are determined, one has to place the actual orbit in space or, in other terms, to determine the three orientation angles, the eccentricity, and the reduced semi-major axis a''.

Since the center, C, of the ellipse was determined, CO is the projection of the line of apsides; the projections of the periastron, A, and apoastron, A', are therefore the intersections of CO with the apparent ellipse (Fig. 12.4). Similarly, the conservation of tangents to an ellipse in the projection, implies that the projections B and B' of the minor axis are the conjugate of AA'. These points correspond to values of the true anomaly of $0°$, $180°$, $90°$ and $270°$ respectively.

Equations (12.16), provide, from the equation of the apparent ellipse, the values of x and y for these four points. The corresponding values of X and Y on the actual ellipse are $(a''; 0)$, $(-a''; 0)$, $(0; a''\sqrt{1 - e^2})$, and $(0; -a''\sqrt{1 - e^2})$ respectively. With these values, one solves (12.17) for the four Thiele–Innes constants and a''. The inverse formulae (12.19) and (12.20) solve completely the problem and provide the three angular parameters.

Note that the real motion is not fully determined, and that one cannot distinguish between an orbit and the symmetric orbit with respect to the plane perpendicular to the line of sight. Nor is it possible to recognize whether the secondary is approaching, or receding, when passing the line of nodes.

If, in addition to the knowledge of the true orbit, even with these indeterminacies, the parallax ϖ is known, the physical semi-major axis $a = a''/\varpi$ is determined, and, applying (12.11), the sum of the masses of the binary is

$$(m_1 + m_2) = \frac{4\pi^2 a^3}{k^2 P^2}.$$

Further consequences on mass determination are given in Section 12.8.

12.4 Astrometric double stars

Not all the binary stars are observable visually, photographically, with Hipparcos, or by speckle interferometry. The limitations of the existing techniques are a small separation between the components and/or a large difference in luminosity. In the latter case, the secondary may be well separated from the primary, but too faint to be recorded. However, in some cases, an observable effect may be produced.

12.4.1 Absolute motion of binary components

As we have seen in Section 12.1, each component of a double star moves around the center of mass of the system. The paths are strictly similar. If m_1 and m_2 are the masses of the components P_1 and P_2, the forces are respectively proportional to m_2

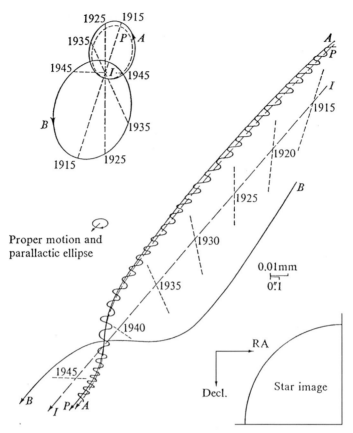

Fig. 12.6. Absolute mean paths of the components *A* and *B* of 99 Herculis and of the center of mass *I* (*G* in the discussion). The effect of parallax is given only for *A*. In addition the mean path of the photocenter *P* is also shown. From *Principles of Astrometry* by Peter van de Kamp, © 1967 by W. H. Freeman and Company. Used with permission.

and m_1 as shown by Equations (12.4). The dimensions of the orbits are constrained by the condition (12.3):

$$m_1\mathbf{GP}_1 + m_2\mathbf{GP}_2 = 0.$$

The ratio of the dimensions of the orbits is m_1/m_2. The center of mass *G* is the common focus of the two ellipses described by the components. In an inertial frame of reference, *G* has a proper motion and a parallactic ellipse common to P_1 and P_2. Figure 12.6, due to van de Kamp (1967), illustrates these motions for the double star 99 Herculis, both in a reference frame linked to the center of gravity, *I*, and in an inertial celestial frame showing the additional effect of proper motion. In the case of component *A*, the parallax is shown as well. Actually, it must be understood that the same parallactic motion exists also in the apparent displacement of the component, B, of the barycenter, I, and of the photocenter, P. The latter are omitted to simplify the drawing. In this example, both stars were observed and the difference of magnitudes was 3.4 mag. To illustrate the difficulty of the astrometric reduction of the plates, and the accuracy of the measurements, the apparent size of the star image is given in the lower right corner. Now, if B were much fainter, and would not be observable, only the motion of A would be seen, practically identical to the photocenter P. It would be found that the apparent motion of A is not linear, and one would interpret this by the presence of an unseen companion, B. In such cases, one says that the star is an *astrometric binary*. The invisible companion may be a star, a brown dwarf, or a planet.

12.4.2 *Orbit determination*

The apparent orbit of the visible component around the center of mass of the system is similar to the relative orbit studied in Section 12.3, but the new constant μ' is deduced from formulae (12.4) so that, calling m_1 the mass of the primary, and m_2 the mass of the unseen companion,

$$\mu' = \frac{m_2}{m_1 + m_2} \mu,$$

where μ is the corresponding gravitational constant for the relative motion. The difference in the technique of orbit determination is that the center of mass, G, is not an observable point, and, therefore, one has to model simultaneously the elliptic motion of the star P and the position and proper motion of G. Let us assume that the common effect of the parallax has been previously removed. At this point, since the apparent ellipse moves rigidly with the same proper motion $\vec{\mu}_G$ as the center of mass, it is sufficient to relate only G to a local fixed reference frame centered in O. One obtains

$$\mathbf{OP}(t) = \mathbf{OG} + \vec{\mu}_G(t - t_0) + \mathbf{GP}(t).$$

This leads to a modification of Equations (12.23) to allow them to be expressed in standard rectangular coordinates:

$$\left.\begin{aligned}
y &= y_0 + \mu_y(t - t_0) + S \cos\phi \sin\alpha + S\sqrt{1 - \varepsilon^2} \cos\alpha \sin\phi, \\
x &= x_0 + \mu_x(t - t_0) + S \cos\phi \cos\alpha + S\sqrt{1 - \varepsilon^2} \sin\alpha \sin\phi.
\end{aligned}\right\} \quad (12.25)$$

In this new set of equations there are two additional unknowns, the components, μ_x and μ_y of the proper motion. The procedure is very similar to the one described for resolved double stars except that, in the absence of positions for the companion, it is not possible to use the integral of areas, Γ, for additional fitting equations, but one should verify that the motion of G is linear within the uncertainties of the observations.

Once the apparent orbit is determined, one can compute the Thiele–Innes constants and the real orbit around the center of mass, which is the focus of the orbit and the position of the center of mass is therefore obtained.

12.5 Spectroscopic binaries

Spectroscopic observations may provide additional information on a double star; the radial velocities of each component vary with time, while describing elliptic orbits around the center of mass. The compound spectrum obtained is the addition of the spectra of the two components and one may determine the two radial velocities. We have seen (Section 12.2) that this knowledge allows us to determine the actual ascending node, and consequently, removes the indetermination of the inclination. However, spectroscopic observations are much more complex than the visual, photographic, CCD, or speckle, observations of stars, so a visual double star is generally not observed spectroscopically for this purpose. However, if one of the stars is faint with respect to the other, or if they are so close that they cannot be optically separated, spectroscopy is the only method of observation (except the special case of eclipsing binaries described in Section 12.6). Such double stars are called *spectroscopic binaries*.

12.5.1 Radial velocity of a component

Let us assume that the spectral lines of one of the components have been well identified, and that the radial velocity has been determined at a number of times, and corrected each time for the projection of the component of the Earth's velocity along the radius vector. Each corrected radial velocity is then the one that would be observed from the barycenter of the Solar System at that particular time. One must also correct the times for the difference of light times between barycentric and Earth-based observations. Figure 12.7 shows the theoretical shape of the radial velocity curve for one of the components as a function of time. Two cases must be considered.

(i) Measuring the difference of the radial velocities of the two components, the relative radial velocity is determined, and, hence, information about their relative orbit.

(ii) Measuring the absolute radial velocity, which is the sum of the constant radial velocity of the barycenter of the system and of the radial velocity of the component moving around the barycenter (see formula (12.5)).

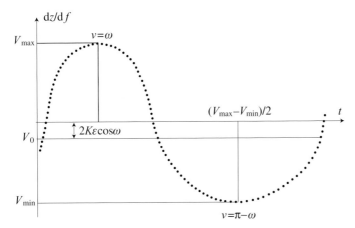

Fig. 12.7. The radial velocity of a component of a double star as a function of time.

In what follows, we shall consider the latter case, having in mind that the shapes of the absolute and relative orbits are the same, with a scale factor of $m_1/(m_1 + m_2)$.

The only information obtained spectroscopically is the time variation of the projection of the radius-vector on the line of sight Oz. From Fig. 12.3, we get

$$z = r \sin(\omega + v) \sin i,$$

from which we deduce

$$\frac{dz}{dt} = \left(\frac{dr}{dt} \sin(\omega + v) + r \cos(\omega + v) \frac{dv}{dt} \right) \sin i. \tag{12.26}$$

Let us compute dr/dt and dv/dt. From the value of the constant of areas, Ω, (12.7) computed in Section 12.1.1, we obtain

$$\Omega = r^2 \frac{dv}{dt} = na^2 \sqrt{1 - e^2},$$

and, using $1/r$ as given by (12.10)

$$r \frac{dv}{dt} = \frac{na^2 \sqrt{1 - e^2}(1 + e \cos v)}{a(1 - e^2)} = \frac{na(1 + e \cos v)}{\sqrt{1 - e^2}}. \tag{12.27}$$

From (12.14), one gets

$$r = a(1 - e \cos E)$$

$$\frac{dr}{dt} = e \sin E \frac{dE}{dt}.$$

Differentiating (12.13), one has

$$\frac{dE}{dt}(1 - e \cos E) = \frac{a}{r} \frac{dE}{dt} = n,$$

and using

$$r \sin v = a\sqrt{1 - e^2} \sin E,$$

we obtain

$$\frac{dr}{dt} = \frac{nae \sin v}{\sqrt{1 - e^2}}. \tag{12.28}$$

Replacing dr/dt and $r dv/dt$ in expression (12.26) one finally transforms it into

$$\frac{dz}{dt} = \frac{na \sin i}{\sqrt{1 - e^2}} (e \cos \omega + \cos(\omega + v)) = K (e \cos \omega + \cos(\omega + v)), \tag{12.29}$$

where we have set

$$K = \frac{na \sin i}{\sqrt{1 - e^2}},$$

which is a combination of the elements directly determinable from the observations.

12.5.2 Orbital elements of a spectroscopic binary

Let us assume that there is a sufficient number of observations. From the values of radial velocities, the periodic character of the variation is recognized and the period P can be determined by Fourier analysis. It is related to n by $n = 2\pi/P$. The trigonometric series derived from this analysis represent the velocity curve as given in Fig. 12.7.

Let us first remark that the maximum and the minimum of the curve, obtained by differentiating (12.29), correspond to $\sin(\omega + v) = 0$:

$$v = -\omega \quad \text{and} \quad v = \pi - \omega.$$

The difference between the values of the velocities at these points is

$$V_{\max} - V_{\min} = 2K. \tag{12.30}$$

But half of the sum of these velocities is

$$\frac{1}{2}(V_{\max} + V_{\min}) = \frac{1}{2}\left[\left(\frac{dz}{dt}\right)_{\max} + \left(\frac{dz}{dt}\right)_{\min}\right] + V_0 = 2Ke \cos \omega + V_0, \tag{12.31}$$

where V_0 is the radial velocity of the barycenter, which is not known *a priori*. It is also necessary to have an analytical expression for dz/dt. It can be shown (see for instance, Kovalevsky, 1967) that as a function of the mean anomaly, $M = n(t - t_0)$, one has:

$$\cos v = -e + \sum_{p=1}^{\infty} \frac{2(1-e^2)}{e} J_p(pe) \cos pM,$$

$$\sin v = \sqrt{1-e^2} \sum_{p=1}^{\infty} \frac{2}{p} \frac{dJ_p(pe)}{de} \sin pM,$$

where t_0 is the time at periastron and J_p is the Bessel function of pth order. We shall assume – and this is by far the most general case – that the eccentricity, e, is small, so that one can develop the coefficients in power series of e and keep only the first terms. For instance, up to the third order in e, one has:

$$\cos v = -e + \left(1 - \frac{9}{8}e^2\right)\cos M + \left(e - \frac{4}{3}e^3\right)\cos 2M$$

$$+ \frac{9}{8}e^2 \cos 3M + \frac{4}{3}e^3 \cos 4M,$$

$$\sin v = \left(1 - \frac{7}{8}e^2\right)\sin M + \left(e - \frac{7}{6}e^3\right)\sin 2M$$

$$+ \frac{9}{8}e^2 \sin 3M + \frac{4}{3}e^3 \cos 4M. \tag{12.32}$$

This is used to compute dz/dt. We obtain after some simple algebra

$$\frac{dz}{dt} = K[e \cos\omega + \cos\omega\cos v - \sin\omega\sin v]$$

$$= K\left[-\frac{1}{12}e^3 \cos(2M-\omega) - \frac{e^2}{8}\cos(M-\omega) + (1-e^2)\cos(M+\omega)\right.$$

$$+ \left(e - \frac{5}{4}e^3\right)\cos(2M+\omega) + \frac{9}{8}e^2 \cos(3M+\omega) + \frac{4}{3}e^3$$

$$\times \cos(4M+\omega) + \cdots\Bigg]. \tag{12.33}$$

The $e\cos\omega$ term cancels out. The mean value of each term over the period 2π of M is zero, and the mean value of dz/dt is zero, so that the mean value of the radial velocity is V_0. Consequently, (12.31) provides $2Ke\cos\omega$, as illustrated by Fig. 12.7, and hence $e\cos\omega$ with K given by (12.30). One can now also identify the trigonometric series obtained by the Fourier analysis of the observations,

$$V_r - V_0 = \frac{dz}{dt} = K\sum_j (C_j \cos jnt + S_j \sin jnt),$$

with the similar expression deduced from (12.32) and substituted into (12.29). One gets a set of equations, the first of which are, neglecting e^3 and $e^2/8$:

$$C_1 = (1 - e^2)\cos(\omega + nt_0),$$
$$S_1 = (1 - e^2)\sin(\omega + nt_0),$$
$$C_2 = e\cos(\omega + 2nt_0),$$
$$= e\cos(\omega + nt_0)\cos nt_0 - e\sin(\omega + nt_0)\sin nt_0,$$
$$S_2 = e\sin(\omega + 2nt_0),$$
$$= e\sin(\omega + nt_0)\cos nt_0 + e\cos(\omega + nt_0)\sin nt_0.$$

Substituting C_1 and S_1 into the second forms of C_2 and S_2 gives

$$C_2 = \frac{eC_1}{1 - e^2}\cos nt_0 - \frac{eS_1}{1 - e^2}\sin nt_0,$$

$$S_2 = \frac{eC_1}{1 - e^2}\cos nt_0 - \frac{eS_1}{1 - e^2}\sin nt_0.$$

Solving these equations, one obtains

$$C_2 = \frac{e\cos nt_0}{1 - e^2} \quad \text{and} \quad S_2 = \frac{e\sin nt_0}{1 - e^2},$$

from which one easily derives nt_0 and e with an accuracy that is most generally sufficient. Otherwise, one should use also C_3 and S_3 with complete e^2 terms. This can be avoided by an iterative method starting with the values of e and nt_0 obtained as explained above and introducing their variations Δe and $n\Delta t_0$ as unknowns in the original equations.

Finally, we have shown how one determines successively:

- the period P and hence the mean motion n,
- the quantities $K = na\sin i/\sqrt{1 - e^2}$ and $Ke\cos\omega$, hence $e\cos\omega$,
- the eccentricity e from which one derives ω and $a\sin i$,
- nt_0, the time t_0 of transit at the periastron.

There remains the longitude of the node Ω, which does not enter the equations, and the quantities a and i are not separated. If the binary is also observed as an optical binary, all the three parameters a, i, and Ω can also be determined.

We have presented here the classical treatment of observations of spectroscopic binaries. It is generally sufficient. However, a relativistic approach may be necessary when very accurate spectroscopic measurements are performed. The post-Newtonian theory for precision Doppler determination of binary star orbits can be found in Kopeikin and Ozenoy (1999).

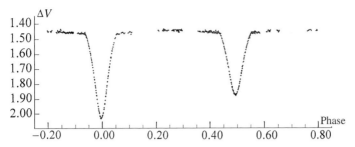

Fig. 12.8. Light-curve of V 451 Ophiuchi, observed at the Sierra Nevada 75 cm telescope, by Clausen, Gimenez and Scarfe (1986). The period is $2.196670 \pm 4 \times 10^{-6}$ days, the inclination is $85°\!\!.9 \pm 0°\!\!.5$, and the radii are respectively 2.64 ± 0.06 and 2.03 ± 0.05 Sun radii. (From Clausen, Giminez and Scarfe, 1986, *Astron. & Astrophys.*, **167**, 287–296. Reprinted by permission of *Astronomy and Astrophysics*.)

12.6 Photometric binaries

Photometric binaries, or eclipsing variables, are systems in which one of the components passes periodically, before and behind the other, so that the total intensity of light from the double star decreases. A typical light-curve of an eclipsing binary is given in Fig. 12.8.

12.6.1 Different types of eclipsing binaries

The problem is to determine the orbital elements from the observed light-curve. The great difficulty arises from the fact that the physical appearance of each star is not known *a priori* and this has a great impact on the shape of the curve. The simplest case corresponds to the Algol-type binaries, composed of two well-separated, spherical stars. For this reason, they are called *detached systems*. But, even in this case, the eclipse of the secondary may be complete or only partial, as in Fig. 12.9. The amplitude of the light variations depends upon the magnitudes of the components, and the shape of the light-curve depends, not only on the respective radii of the stars and the minimum apparent separation of the centers, but also on the limb darkening of each star, which may be quite large.

Other types of eclipsing binaries are more involved, because they correspond to very close binaries which strongly interact. In the case of β Lyrae-type stars, the gravitational forces are such that the tides distort the shapes of the stars which become elongated. The closer they are, the more difficult is the interpretation of the light-curve; there are exchanges of bright matter and the tidal energy heats the stars's surfaces facing each other. In some cases, such as in β Lyrae itself, one of the components fills its Roche limit. This class of binaries is called *semi-detached*. Even more complex systems are W UMa stars, or contact binaries, in which both

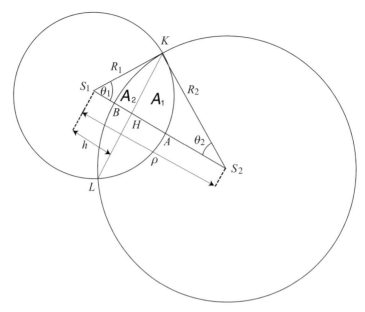

Fig. 12.9. Geometry of a partial occultation in an eclipsing binary.

components fill their Roche surfaces, and a continuous exchange of matter takes place between the two components.

In practice, each system is a particular case for which an astrophysical model must be proposed, and then modified as a function of the residuals of observations. These techniques are outside the scope of astrometry, and we give below only a simple example to show what astrometric parameters can be derived.

12.6.2 A simple model

Let us consider an Algol-type binary, consisting of two spherical stars that present no limb darkening. We assume that the radii of the stars are respectively R_1 and R_2, the distance between their centers, S_1 and S_2, is ρ, and the intensities of light are I_1 and I_2. When there is no eclipse, the total intensity received is

$$I = I_1 + I_2.$$

When there is an eclipse, one has to compute the eclipsed surface, S, the surface of the intersection of the projections of the two stars. Then, if star 1 occults star 2, the total intensity is

$$I_{1-2} = I_1 + \left(1 - \frac{S}{\pi R_2^2}\right) I_2, \tag{12.34}$$

and if star 2 occults star 1,

$$I_{2-1} = \left(1 - \frac{S}{\pi R_1^2}\right) I_1 + I_2. \tag{12.35}$$

The situation is symmetrical, so let us consider only the second case, assuming in addition that $R_2 > R_1$, and that the centers of the stars are outside the projections of the other star (Fig. 12.9). The surface S is composed of two parts: LAK and LBK. Let us take as a parameter the length $h = S_1 H$, where H is the intersection of the common chord LK. The area A_2 of LBK is, calling θ_2 the angle $(S_2 H, S_2 K)$, expressed in radians,

$$A_2 = R_2^2 \theta_2 - S_2 H \times HK, \tag{12.36}$$

which we express in terms of ρ and h using

$$S_2 H = \rho - h \quad \text{and} \quad HK^2 = R_2^2 - (\rho - h)^2$$

$$A_2 = R_2^2 \cos^{-1}\left(\frac{\rho - h}{R_2}\right) - (\rho - h)\sqrt{R_2^2 - (\rho - h)^2}. \tag{12.37}$$

One may express h as a function only of the two radii and ρ. Let us write

$$\rho^2 = (h + \rho - h)^2 = h^2 + (\rho - h)^2 + 2h(\rho - h). \tag{12.38}$$

In the triangle $S_2 H K$, one has

$$(\rho - h)^2 = R_2^2 - HK^2 = R_2^2 - \left(R_1^2 - h^2\right).$$

Introducing this into (12.37) one gets

$$\rho^2 = R_2^2 - R_1^2 + 2h^2 + 2h(\rho - h),$$

and finally,

$$h = \frac{\rho^2 - R_2^2 + R_1^2}{2\rho}.$$

Through Equations (12.36) and (12.37), one gets A_2 uniquely as a function of R_1, R_2 and ρ. Similarly one can express A_1, and finally S, as functions of only these parameters. All other dispositions of the two circles can be treated similarly and in all the cases, one gets a formula of the form

$$S = S(R_1, R_2, \rho), \tag{12.39}$$

which can be inserted into (12.34) or (12.35). In the formula, the units are arbitrary, since the expressions in parentheses are dimensionless. So it is possible to use arcseconds, and one may express ρ in terms of Thiele–Innes constants, and, therefore, in terms of orbital elements. Each observation of the ratio between the maximum intensity, when the stars are separated, and the intensity observed during an eclipse, provides an equation in terms of Thiele–Innes constants and the two radii. However, all parameters are not determined. The period is readily obtained from the observations, as well as the radii R_1 and R_2, and the inclination of the orbit i. The orbital eccentricity, e, and the longitude of periastron, ω, can be obtained only in the case when the light curve is very well determined and modeled.

12.6.3 Realistic case for detached binaries

The most important simplification that is made in Section 12.6.2 is to assume that the brightness of a star is the same over its apparent surface. This is not true, and one has to take into account the limb darkening. The uniformly bright disk of radius R is to be replaced by a brightness function depending on the distance to the center C,

$$B_r/B_c = F(r/R),$$

that is compatible with the theory of stellar atmospheres for the spectral type of the star. The ratio between the eclipsed and non-eclipsed intensities of a component, to be introduced in Equations (12.34) and (12.35), has to be computed numerically for each relative position of the components. A description of a method used to deal with this case can be found in Nelson and Davis (1972). Many papers have been published that present the reduction of light curves and discuss the resulting double-star parameters. For example, one may mention Popper (1984) and a series of other articles published by this author in the *Astronomical Journal*. A general review on the subject with the best results obtained to date is by Andersen (1991).

12.7 Particular objects

In this section, we shall present cases that either are subject to several observing techniques, or are not covered by any of them. We shall consider multiple stars, double pulsars, and finally discuss the matter of the search for extraterrestrial planetary systems.

12.7.1 Multiple stars

In many cases, stars are grouped in more than two components. If there are three or more components, the system is called a multiple star. A survey (Poveda, 1988), showed that 6–7% of non-single stars are multiples. This proportion is not very reliable, like the assumed proportion of double to single stars, because it is based on the discovered double or multiple stars, and there may be many more with invisible companions.

Among multiple stars, most are triple stars and almost all have a hierarchical structure. We mean by this that there is a close binary and a third component far away from the couple. Similarly many quadruple stars are actually a separated pair of close binaries. The motion of the separated pair can be studied in the same way as a widely separated resolved double star (Section 12.3), considering generally that the close binaries may be treated as a single stars in determining the relative orbit. For instance, in the Hipparcos data reduction many multiple stars were assigned a good double-star solution, by neglecting the faintest component or treating two close components as a single star. However, in some cases, a full triple-star model has fitted the data somewhat better. This latter case happened in hierarchical systems, and the improvement came from treating the close binary as sufficiently separated. In the general case, one would treat the binary either as a separate double system, or, more often, as a spectroscopic, or a photometric, binary (respectively, Sections 12.5 or 12.6).

When the distances between the components of a multiple star are of the same order of magnitude, such systems are called trapezia. None of the models described in this chapter can be applied. Generally the distances between the components are large (tens or hundreds of astronomical units), and only small parts of their relative motions have been observed. Therefore, the best that can be determined are the relative proper motions of the components. If spectroscopic data and parallaxes are available, the space velocities can be reconstructed. Orbital elements are not significant in those cases, and the theory of their motions is a particular case of the general three- or many-body problem in which every configuration is a special case to be investigated using numerical integration. Many such configurations are unstable, and it has been postulated that the final fate of trapezia is a hierarchical system, or a disruption with remaining close binaries and single stars. This hypothesis is supported by the fact that stars in trapezia are all young stars.

12.7.2 Invisible companions

A very popular application of astrometric binaries is the search for small invisible companions of stars, in particular the search for extra-solar planetary systems. Until recently, there was a gap between the smallest known stars, having a detectable luminosity in the visible spectrum, with a mass of about 0.08 solar mass, and the largest planet, Jupiter, with a mass of 0.001 solar mass. Theoretical considerations lead to the assumption that down to 0.08 solar mass, the body is sufficiently heavy to permit, after condensation, a start up of hydrogen burning; it is a star. At lower masses, the gravitational heating is insufficient to start and sustain nuclear fusion, but great enough to generate energy by gravitational compression. These objects essentially radiate in the infrared and their luminosity in visible light is very small. This justifies that they are known under the general term *brown dwarfs*. At much smaller masses, with a limit that is not well defined, around 0.010–0.015 solar mass, the objects are gas-giant planets which, like Jupiter, have central cores of rocks and ice, as inferred from their gravitational fields. Then, with masses of a few millionths of solar mass, we find the telluric, or terrestrial, type planets, the ones that are particularly important to find, since it is generally believed that only this type of object can shelter life.

The observing techniques are similar to the techniques used to observe double stars, that is essentially either astrometric or spectroscopic, as described in Sections 12.4 and 12.5, respectively. At present, the most effective method to find unseen companions, essentially brown dwarfs and some heavy gaseous planets, is the spectroscopic, or Doppler, method. However, the transit of a planet across a Sun-like star was detected photometrically (Charbonneau *et al.*, 2000). The application of these methods is presented in Chapter 14, together with a few others.

12.7.3 *Binary pulsars and pulsar companions*

The observation of pulsars and the data reduction is presented in Section 14.7. The data that are acquired through the observations of pulsars are described, for instance, in Backer (1993). Generally, the period of the pulses is constant. So, if there is a variation of the period, it is interpreted as Doppler shifts, or as modifications of the distance (an integral of the radial velocity). Therefore, the interpretation of the data in terms of orbital elements is analogous to the case of spectroscopic binaries, except that relativistic effects on the orbit have to be taken into account, because of the importance of the gravitational field in the vicinity of a neutron star. A review on the subject can be found in Taylor (1993). The first case was the double pulsar PSR 1913 + 6 in which a shrinking of the orbit was interpreted as an energy leakage due to the emission of gravitational waves (Taylor and Weisberg, 1989).

Similarly, planets have been discovered from the residuals of pulsar timing observations interpreted, as in the case of spectroscopic methods, as motions due to the perturbations by planets. Several of them have been discovered, either Jupiter-like, as in PSR 1828-11, or of a terrestrial type as in PSR 1257 + 12.

12.8 Determination of stellar masses

We have seen that masses of stars enter into the equations of the motion of double stars. Actually, unless one introduces some astrophysical models of stars, binaries are practically the only way to determine stellar masses, with the exception of the improbable use of light deflection by a star when another star passes behind it (see Section 6.4.4).

12.8.1 *Formulae for the astrometric method*

In previous sections, formula (12.11) provides a first relation between the masses m_1 and m_2 of the two components, the semi-major axis, a, of the relative orbit, and its period, P

$$\frac{a^3}{P^2} = \frac{k^2(m_1 + m_2)}{4\pi^2} = \frac{\mu}{4\pi^2}. \tag{12.40}$$

In this formula, a is expressed in astronomical units. In order to obtain the sum of the masses, it is necessary to know the parallax. Otherwise, if the orbit is an absolute orbit with respect to the center of mass I of the system, as shown in Fig. 12.6, the dimension of each orbit with respect to I is given respectively by

$$\frac{a_1^3}{P^2} = \frac{k^2 m_2}{4\pi^2} \quad \text{and} \quad \frac{a_2^3}{P^2} = \frac{k^2 m_1}{4\pi^2},$$

so that, since the periods are the same, one has

$$\frac{m_1}{m_2} = \left(\frac{a_1}{a_2}\right)^3. \tag{12.41}$$

12.8.2 *Masses of spectroscopic binaries*

Similarly, if the pair is a spectroscopic binary, and the spectral lines of each component can be identified, one gets for each star from formula (12.30)

$$K_1 = \frac{na_1 \sin i}{\sqrt{1 - e^2}},$$

$$K_2 = \frac{na_2 \sin i}{\sqrt{1 - e^2}}. \tag{12.42}$$

The ratio of the masses is obtained from

$$\frac{K_1}{K_2} = \frac{a_1}{a_2} = \frac{m_2}{m_1}. \tag{12.43}$$

No information concerning the sum of the masses is provided. However, considering the basic Equation (12.40)

$$m_1 + m_2 = \frac{4\pi^2 a^3}{k^2 P^2},$$

and replacing $a = a_1 + a_2$ by their values in Equations (12.41), one gets

$$m_1 + m_2 = \frac{4\pi^2}{k^2 P^2} \left(\frac{K_1\sqrt{1 - e^2}}{n \sin i} + \frac{K_2\sqrt{1 - e^2}}{n \sin i} \right)^3,$$

and

$$(m_1 + m_2) \sin^3 i = \frac{4\pi^2 (1 - e^2)^{3/2} (K_1 + K_2)^3}{n^3 k^2 P^2}),$$

and, since $n = 2\pi/P$, this simplifies into

$$(m_1 + m_2) \sin^3 i = \frac{P}{2\pi k^2} (1 - e^2)^{3/2} (K_1 + K_2)^3. \tag{12.44}$$

Using now (12.42), one derives easily

$$m_1 \sin^3 i = \frac{P}{2\pi k^2} (1 - e^2)^{3/2} (K_1 + K_2)^2 K_2,$$

$$m_2 \sin^3 i = \frac{P}{2\pi k^2} (1 - e^2)^{3/2} (K_1 + K_2)^2 K_1. \tag{12.45}$$

The inclination is not determinate, as can be seen from Section 12.5.2, but only $a \sin i$, and therefore, $(m_1 + m_2) \sin i$ can be obtained. However, formulae (12.43) and (12.44) may be used in statistical studies assuming some distribution of inclinations.

If the system is also a photometric binary, one can obtain the inclination, i (Section 12.6.2). In this case, (12.44) provides the masses. Actually, the conjunction of these two types of observations is, at present, the method that provides the best values of stellar masses with 1–2% uncertainty (Andersen, 1991).

12.8.3 Uncertainty analyses for astrometric mass determinations

To get masses, one needs to know simultaneously $m_1 + m_2$ and m_1/m_2. As far as the sum of the masses is concerned, one is led to use (12.40). It is legitimate to assume that the error on the period, P, is negligible, but one has to express a in astronomical units to be consistent with the known value of k^2. This implies that the parallax, ϖ, of the star is known. If a'' is the value observed in arcseconds, one has

$$a = \frac{\varpi a''}{\sin 1''},$$

where ϖ is also expressed in arcseconds. The error budget is

$$\frac{\delta a}{a} = \frac{\delta \varpi}{\varpi} + \frac{\delta a''}{a''},$$

and, in view of Equation (12.40),

$$\frac{\delta(m_1 + m_2)}{m_1 + m_2} = \frac{3\delta a}{a} = \frac{3\delta \varpi}{\varpi} + \frac{3\delta a''}{a''}. \tag{12.46}$$

A simple examination of this expression shows that the uncertainty of $\delta(m_1 + m_2)$ is necessarily large. In order to produce masses with an uncertainty of 2%, and assuming, as a first approximation, that there is no error on m_1/m_2, one should

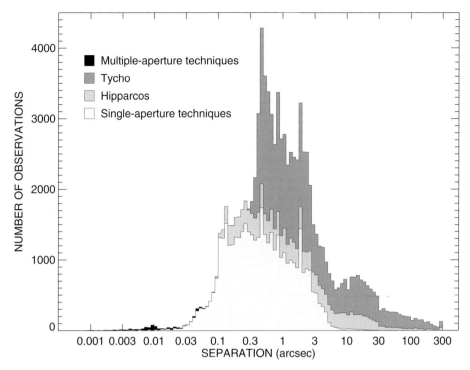

NUMBER OF OBSERVATIONS

Fig. 12.10. Distribution of double-star measures as a function of categories of techniques for 93 640 catalogued double stars (courtesy W. I. Hartkopf).

be able to get the parallax to better than 0.5%, and measure a'' to 0.2% at best. If the latter is possible with speckle interferometry techniques, the condition $\delta\varpi/\varpi$ to 0.5% (with parallaxes measured with an accuracy of 1 mas as in Hipparcos or the best long-focus observations) is satisfied only for parallaxes larger than $0''2$ (corresponding to distances smaller than 5 parsecs), that is for a minute sample of stars.

The objective of having masses to better than 1 or 2% is an important requirement of astrophysics, a requirement that could not be met by Hipparcos. One will have to do astrometry to a few microarcseconds in order to get this kind of accuracy for stars at distances up to 100 pc, so that most types of star are included in the sample.

12.9 Catalogs of double and multiple stars

In the past, double-star observers used to publish their observations in catalogs. Most of them are now given in various data bases, a list of which is available at the bibliographic service of the Centre de Données stellaires (CDS) in Strasbourg (cdsweb.u-strasbg.fr). The main data base of the CDS, called SIMBAD, provides an ensemble, as complete as possible, of data on stars: various catalog identifications, positions, kinematic, photometric and spectroscopic data. In the case of double and multiple stars, additional data are given, depending upon their category:

- astrometric, visual and interferometric binaries: separations and position angles of the secondaries;
- eclipsing binaries: luminosity curves, periods, phases and minimums;
- spectroscopic double stars: spectra, radial velocity curves of components, periods and phases.

There are a large number of double- and multiple-star catalogs and data bases in various observatories, generally for specific categories of double stars. They are generally linked to the CDS web site. Let us mention the most important in each domain.

- *Visual double stars*. The USNO Double Star CD in 2001 included the components of 84 489 systems (Mason *et al.*, 2001) with their epochs, position angles, separations, magnitudes, and spectral types. It is available as a CD issued by the US Naval Observatory. The WDS is continuously enriched. In 2002, 13 000 new couples were added. The 2001 CD also includes the Fifth Catalog of Orbits of Visual Binary Stars (Hartkopf *et al.*, 2001a), which presents 1465 orbits of 1430 systems. This number is also growing, and the sixth catalogue will include 1608 systems.
- *Interferometric double stars*. The third CHARA (2000) data base presents 73 776 high-accuracy observations obtained by different techniques (speckle or Michelson interferometry, space observations, occultations), including resolved or non-resolved pairs. The distribution of measurements as a function of separation is given in Fig. 12.10. The catalog is accessible at http://ad.usno.navy.mil/dsl/. In the USNO Double Star CD-ROM, one also finds the Third Catalog of Interferometric Measurements of Binary Stars (Hartkopf *et al.*, 2001b), which includes 64 779 resolved, and 9420 unresolved measures of 28 699 systems plus an additional 9425 observations reporting photometric information only. Many measurements are in common with the CHARA data base. The number of entries is also continuously growing and is available at http://ad.usno.navy.mil./wds/int4.html.
- *Eclipsing binaries*. A catalog of eclipsing binaries is currently updated at the Cracow Observatory. It contains measurements of about 1200 objects and is available at http://www.oa.uj.edu.pl/ktt/. It also includes published, yearly ephemerides and elements of the light curves.

A Washington Multiplicity Catalog is under preparation, whose goal is to provide a single consistent nomenclature scheme for all types of binary and multiple systems. Let us also add that a general data base for all types of double and multiple stars is being constructed at the Besançon Observatory, and is already accessible for about one hundred thousand entries at http://bdb.obs-besancon.fr/.

13

Astronomical phenomena

13.1 Motions involved

The combinations of different motions of Solar System bodies, involving orbital motions with their respective eccentricities and inclinations, as well as rotations with different planes and periods, lead to a variety of phenomena. These range from the obvious, such as the rising and setting of the Moon and planets, seasons on the Earth, and the phases of the Moon, to the less-obvious phases of the planets and apparent stationary and retrograde motions of the planets. The ellipticities and inclinations of the orbital motions of the bodies also lead to some variations of these phenomena.

The motions of the Earth, Moon, planets, and satellites lead to objects obscuring other objects and the sunlight onto those objects. The lack of visibility due to the proximity to the Sun, the visibility due to elongation from the Sun, or the positioning opposite to the Sun, are also resulting phenomena.

The motion of the Earth in its orbit determines the seasons and, with the Moon's motion, is the basis for the different calendars. The seasons are defined from the times of equinoxes and solstices. While the equinox may not be used for the reference frame anymore, it will continue to be the basis of defining the seasons, and can be determined from the Solar System ephemerides being used. Likewise, the times of perihelion and aphelion for the Earth and all the planets are determined from the ephemerides.

With the introduction of the new reference system, the International Celestial Reference System (ICRS), based on an extragalactic reference frame and independent of the orbital motions of Solar System bodies, it is necessary to clarify the definitions of times and occurences of some phenomena. This chapter will try to specify these definitions for the new system and indicate where the definitions have changed from previous practices.

There is extensive information about these phenomena in the *Explanatory Supplement* (Seidelmann, 1992). The phenomena are tabulated in the annual publications *Astronomical Almanac* and *Astronomical Phenomena*, or the *Ephémérides astronomiques* in the Annuaire du Bureau des longitudes. The information may also be obtained from the web at www.usno.navy.mil. The software MICA contains the programs and data necessary to calculate the various phenomena for any location.

13.2 Periods

The motions of all the objects in the Solar System are periodic, so this leads to a periodic behavior in the phenomena presented above. Thus, the synodic periods for planets are the mean intervals of time between successive conjunctions of a pair of planets as observed from the stars. The sidereal period is the mean interval of time for orbital motion to return the body to the same position with respect to a fiducial point as seen from the Sun. In the FK5 system that fiducial point was the catalog equinox, in the ICRS system the fiducial point is Celestial Ephemeris Origin.

For the Earth, there are a variety of possibilities for definitions of the year.

- The tropical year has traditionally been defined as the period of one complete revolution of the mean longitude of the Sun with respect to the dynamical equinox.
- The Bessilian year was defined as the period of one complete revolution in right ascension of the fictitious mean Sun as defined by Newcomb.
- Thus, the tropical year is longer than the Besselian year by $0.148T$ seconds, where T is in centuries from 1900.
- The Gregorian calendar has a calendar year of 365.2425 days.
- The Julian year has a period of 365.25 days, which is a period used for astronomical calculations.

For the Moon, one introduces the following additional periods.

- The lunation, the period between consecutive new Moons (see Section 13.4).
- The month, which is the period of one complete synodic, or sidereal, revolution of the Moon around the Earth (the calendar unit approximates the period of revolution).
- The Saros is a Babylonian lunar cycle of 6585.32 days, or 223 lunations, after which the centers of the Sun, Earth and Moon return to the same relative positions such that the eclipses recur approximately as before, but in longitudes approximately $120°$ to the West on the Earth.

The rotational period of an object can be given with respect to the fixed stars (sidereal periods) or with respect to its primary, the object about which it revolves (for the Earth a solar period). In some instances, there are natural resonances between rotation periods. This is the case for the Moon, whose mean rotational and orbital motions are close to the same. The variations in the orientation of the Moon's surface with respect to an observer on Earth are called librations. Physical librations are variations in the rate at which the Moon rotates on its axis (Eckhardt, 1981). The much larger optical librations are the result of variations in the rate of the Moon's orbital motion.

13.3 Rising/setting

Owing to the rotation of a body with respect to its primary and to the fixed stars, an observer on that body will see celestial objects rise and set with respect to the visible horizon. These phenomena can be calculated assuming a uniform spherical body and the absence of a refracting atmosphere. Correcting for the deviations from a sphere in height, variations in the terrain, and refraction of the variable atmosphere adds complexity and uncertainty to the predicted times of the rising and setting phenomena. At high latitudes the phenomena occur in a more varied manner than in the more equatorial regions. In winter, the sunrise and sunset phenomena do not take place at all. During the extended summer twilight period, rises and sets take place over extended periods of time with larger uncertainties due to the gradual nature of the event and the variations in the atmospheric effects. Rises and sets of other bodies are similarly affected, but with the occurrences dependent upon the orbits of the bodies.

13.3.1 The Sun

Sunrise and sunset are defined as the times when the upper limb of the Sun is on the astronomical horizon, specifically when the true zenith distance, referred to the center of the Earth, of the central point of the disk is 90° 50′. This is based on adopted standard values of 34′ for horizontal refraction and 16′ for the Sun's semidiameter. Because of this definition, and allowance for the semidiameter of the Sun, there is not an equal period of sunlight and darkness at the time of the equinox, or sunrise and sunset are not 12 h apart at that time.

Irradiation is an optical effect due to contrast that makes bright objects viewed against a dark background appear to be larger than they really are. Thus, the Sun and Moon, viewed against a darker background will appear larger than they really are. As they rise or set, the upper limb will appear higher in the sky than it really is. The magnitude of this effect, possibly as large as 10″, is dependent on the individual person's eyes and will vary from person to person.

Twilight is the interval of time preceding sunrise and following sunset, when the sky is partially illuminated by sunlight scattered by the upper layers of the Earth's atmosphere. There are three specific types of twilight:

(i) civil twilight comprises the interval when the zenith distance, referred to the center of the Earth, of the central point of the Sun's disk is between 90° 50′ and 96°;

(ii) nautical twilight comprises the interval from 96° to 102°;

(iii) astronomical twilight comprises the interval from 102° to 108°.

The amount of actual illumination is dependent on the weather, the specific location, and the phase of the Moon or other illumination. In general, in good weather normal activities can be carried out during civil twilight without artificial illumination. During nautical twilight both the horizon and the navigational stars are visible, and at the beginning, or end, of astronomical twilight it is dark enough for astronomical observations using telescopes. This means that at that time the indirect illumination from the Sun is less than the contribution from starlight and is of the same order as that from the aurora, airglow, zodiacal light, and the gegenschein.

13.3.2 The Moon

For calculating the times of moonrise and moonset, the true altitude of the center of the Moon is

$$h = -34' - \text{lunar semidiameter} + \text{horizontal parallax}, \tag{13.1}$$

where 34′ is horizontal refraction; h is approximately 5′ to 11′ depending on the distance of the Moon. The upper limb of the Moon is on the horizon, and no allowance is made for phase. Since the Moon revolves around the Earth in a synodic month, relative to the Sun, of a mean length of 29.53 days, the Moon appears to lose one rising and setting in that period. There is no moonrise on a day near the last quarter, and no moonset on a day near the first quarter. In high latitudes the times of the phenomena change rapidly and may not occur for extended periods, the Moon being continuously above or below the horizon. It is possible to have two moonrises, or two moonsets, during the same day, and for the times of moonrise and moonset to decrease from day to day, instead of the usual increasing pattern.

13.3.3 General formulae

To calculate the times of rising and setting of a body at true altitude h, an iterative approach is preferred. The Greenwich hour angle (GHA), declination δ, and horizontal parallax π for the Moon are required to about an arcminute of accuracy as a function of Universal Time (UT). Polynomial expressions or formulae for GHA, δ and π are desirable for the iterative procedure. The true altitudes, when the upper limbs are on the horizon, at the times of rise and set are, expressed in degrees:

$$h = -50/60 - 0.0353\sqrt{H} \quad \text{at sunrise and sunset,}$$
$$h = -34/60 + 0.7275\pi - 0.0353\sqrt{H} \quad \text{at moonrise and moonset,}$$
$$h = -34/60 - 0.0353\sqrt{H} \quad \text{at rise and set of a star or a planet,} \quad (13.2)$$

where H is the height, in meters, of the observer above the horizon, and the apparent altitude of the upper limb on the horizon is zero. For twilights, the precision is not important, and one can take the values of $h = -6°$ for civil twilight, $-12°$ for nautical twilight, and $-18°$ for astronomical twilight.

Then the time of rise or set, UT, in hours, is found by solving iteratively the equation

$$UT = UT_0 - (GHA + \lambda \pm t)/15, \tag{13.3}$$

where the plus sign is for rise and the minus sign for set, and t is the hour angle of the body at UT, which is given by

$$\cos t = \frac{\sin h - \sin \phi \sin \delta}{\cos \phi \cos \delta}. \tag{13.4}$$

Here λ is the longitude (positive to the East) and ϕ is the latitude of the observer.

$$\text{if } \cos t > +1, \text{ set } t = 0°;$$
$$\text{if } \cos t < -1, \text{ set } t = 180°. \tag{13.5}$$

UT_0 may be set to 12^h as an initial guess, but any value between 0^h and 24^h will work. After each iteration, one should add multiples of 24^h to set UT in the range of 0^h to 24^h. Then replace UT_0 by UT until the difference between them is less than $0^h.008$. If for several iterations $\cos t > 1$, it is likely that there is no phenomenon and the body remains above the true value of h all day. If, on the other hand, $\cos t < -1$, the body remains below the true altitude all day. In each lunation, around the first quarter, there is always a day when the Moon does not set, and another, around the last quarter, when the Moon does not rise.

At latitudes above $60°$ the algorithm may fail, and it is necessary to use a more systematic approach. If h_0 is the adopted true altitude, the times when $h = h_0$ are the roots of

$$\sin h_0 = \sin \phi \sin \delta + \cos \phi \cos \delta \cos(GHA + \lambda), \tag{13.6}$$

and these are the times of rising and setting. These roots, if they exist, will lie between the true altitude at 0^h and 24^h and the maximum and minimum altitudes during the day. The maximum and minimum altitudes occur at, or near, upper and lower transit. These times are found by setting $t = 0°$ and $t = 180°$ in

Equation (13.3) and iterating. Only for the Moon at high latitudes is it necessary to calculate the maximum and minimum altitudes more precisely. This calculation is done by fitting a second-order polynomial to the true altitude at points around upper and lower transit and differentiating with respect to time to find the turning point.

13.4 Phases

Since the source of light in the Solar System is primarily the Sun, the other objects are only illuminated on the hemisphere of the body facing the Sun. Observations made from other bodies, such as the Earth, see a portion of the object's illuminated hemisphere dependent upon the geometry of the location of the Sun, the object, and the Earth. If the body and the Sun are on opposite sides of the Earth, a fully illuminated hemisphere will be observed. As the body moves, reducing the angle at the Earth between the Sun and the body, the portion of the illuminated hemisphere observed will be reduced. The fraction of the apparent disk that is illuminated by the Sun is the phase. It depends on the planetocentric elongation of the Earth from the Sun, called the phase angle. Assuming that the body's apparent disk is circular, the terminator is the orthogonal projection, onto a plane perpendicular to the line of sight, of the great circle that bounds the illuminated hemisphere of the body. The terminator is generally an ellipse that becomes a straight line at phase angle $90°$ and a circle at phase angles $0°$ and $180°$. The most noticeable phase effect is on the Moon, but the planets are subject to the same phenomena, depending on the possible geometric relations. The phases of the Moon are determined from the differences between the apparent geocentric ecliptic longitudes of the two bodies. When the longitudes of the Moon minus that of the Sun are $0°$, $90°$, $180°$, and $270°$, the phases are new, first quarter, full, and last quarter, respectively. These are independent of the location on the Earth. Owing to the rapid variations in the distance and velocity of the Moon, the intervals between successive phases are not constant and there is no simple prediction formula. The phases of the Moon do not recur on exactly the same dates in any regular cycle.

There are two cycles that give approximate dates of phases of the Moon. The Metonic cycle of 19 tropical years, which nearly equals 235 synodic months (new Moon to new Moon), means that the phases of the Moon will recur on dates that are the same or differ by one, or occasionally two days, depending on the intervening leap years and perturbations of the lunar orbit. The Saros, which consists of 233 synodic months and equals 19 passages of the Sun through the node of the Moon's orbit, means phases will recur and so will eclipses. The durations of eclipses repeat also, since the Moon's apse has made 2.037 revolutions during the same period.

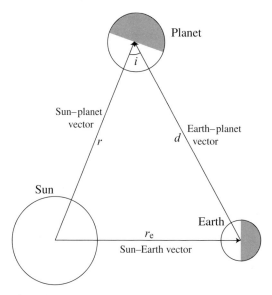

Fig. 13.1. Geometry of phase effect and solar phase angle.

The solar phase angle is the angular distance at the object between the Sun and the Earth. For the Moon, Mercury, and Venus the solar phase angle (i) in Fig. 13.1 can vary from $0°$ to $180°$. For Mars and its satellites i can be as large as $47°$, while for the other planets, their satellites and most asteroids, i can only reach a few degrees.

The phase modifies the observed magnitude, V, of an object. It is given by

$$V = V(1, 0) + 5\log_{10}(d) + \Delta m(i) \tag{13.7}$$

where $V(1, 0)$ is the magnitude of the object seen at a distance of 1 AU (astronomical unit) with a $0°$ phase angle, d is the Earth to object distance (in AU), and $\Delta m(i)$ is the observationally determined correction for variation in brilliancy of the planet with phase angle. The quantity $\Delta m(i)$ is measured empirically and is caused by two effects; one is the fraction of the illuminated disk seen from the Earth as it varies with i, and second is the result of the properties of diffuse reflection from the object's surface or atmosphere. $\Delta m(i)$ is expressed as a power series in i, including as many terms as are necessary. For the outer planets, a single term is usually adequate and is referred to as the *phase coefficient*.

The phases of the Moon are also used in determining lunar and luni-solar calendars. In some cases (Hebrew luni-solar calendar) an adopted mean value of the lunation cycle is used. In other cases, e.g. the Islamic lunar calendar, the months begin with the first visibility of the lunar crescent after conjunction. For civil purposes a tabulated calendar that approximates the lunar phase cycle may be used.

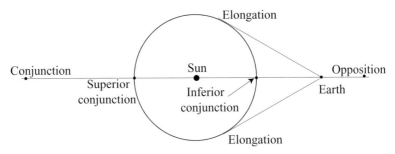

Fig. 13.2. Opposition, conjunctions, and elongations.

13.5 Oppositions, conjunctions, stationary points, elongations

As the Earth and the planets revolve around the Sun they reach various geometric extremes which are labeled opposition, conjunction, and elongations (Fig. 13.2). The times of these events are based on the apparent, geocentric, ecliptic longitudes and must be determined from ephemerides. The times of conjunctions and oppositions of planets are such that the differences between the longitudes of the planet and the Sun are 0° and 180°, respectively. For outer planets conjunctions take place when the planets are in the direction of the Sun as seen from the Earth, and oppositions take place when the planets are in the opposite direction from the Sun with respect to the Earth. For the interior planets, Mercury and Venus, superior conjunctions take place when the planets are more distant from the Earth than the Sun, and inferior conjunctions occur when the planets are closer to the Earth than the Sun.

In the apparent motions of the planets as observed from the moving Earth, the planets appear to move forward as their angular motion exceeds that of the Earth, then the planet will be stationary before it enters a period of apparent retrograde motion. This is followed by an apparent stationary status and the return to forward motion. The times when the planet appears to be stationary are called the stationary points. These times are tabulated in the *Astronomical Almanac*, or the *Ephémérides astronomiques* in the Annuaire du Bureau des longitudes, along with the times of oppositions and conjunctions.

The elongations of the planets, measured eastward or westward from the Sun, are the angle formed by the planet, the Earth, and the Sun, These are measured 0° to 180° East or West of the Sun. Because of orbital inclinations, the elongations do not necessarily pass through 0° or 180° as they change between East and West. For interior planets they cannot reach 180°, and thus reach greatest eastern and western elongations.

13.6 Eclipses

The motions of the various bodies are in orbital planes that are inclined to each other and moving and intersect at nodes that move, leading to different patterns

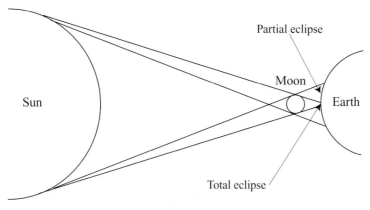

Fig. 13.3. Solar eclipse geometry.

of phenomena caused by bodies shadowing others and passing in front of other bodies. Thus, an eclipse takes place when a body passes into the shadow cast by another body. Specifically, when the Moon passes between the Sun and the Earth, it casts a shadow on the Earth (Fig. 13.3). Since the apparent sizes of the two bodies are approximately equal, the shadow is cast on a limited moving location on the rotating Earth. If the Moon is close to the Earth, its apparent size may be smaller than the Sun's and it leaves a visible ring of Sun around the Moon; this is an annular eclipse. The area covered by a complete shadowing of the Sun in an eclipse is called totality, while the regions of the Earth only experiencing part of the Sun obstructed by the Moon are in the partial phase. The portion of the shadow cone in which none of the light from an extended light source can be observed is called the umbra. The penumbra is the portion of a shadow in which light from an extended source is partially, but not completely, cut off by an intervening body, this is the area of partial shadow surrounding the umbra. Eclipses are excellent tests of the accuracies of the ephemerides of the Earth and the Moon and our knowledge of the rotation of the Earth, all on one reference system. Thus, historical records of eclipses, events that were generally dramatic and considered as omens and thus recorded in history, provide a means of determining the history of the Earth's rotation. The ephemerides can be calculated accurately into the past with some confidence. It should be noted also that at times of eclipses the sunlight is sufficiently reduced such that bright stars and planets can be observed around the Sun in the daytime. Thus, the prediction by Einstein's Theory of Relativity of the bending of the star's light by the mass of the Sun was confirmed by observations of the 1919 eclipse.

13.7 Occultations

When an apparently larger body passes in front of an apparently smaller body so that it cannot be seen by the observer, it is called an occultation. Thus, the Moon, planets, satellites, and minor planets can pass in front of stars and occult the light from the stars, or the Moon can pass in front of other Solar System objects. Observations of such events can be used to determine the size, shape, atmosphere, and possible duplicity of the occulting body as well as possible duplicity of the occulted body. The geometry of such observations is dependent on the specific location on the surface of the Earth. Multiple observations of the same event strengthen the results.

The calculation of the occultation by the Moon is based on a fundamental plane through the center of the Earth and perpendicular to the line between the star and the center of the Moon, which is the axis of the shadow. Taking the center of the Earth as the origin of the coordinates, the x-axis is the intersection of the Earth's equator and the fundamental plane, positive toward the East. The y-axis is perpendicular to the x-axis and positive to the North of the fundamental plane. Because of the great distance to the star, the fundamental plane is essentially perpendicular to the line between the center of the Earth and the star, and the Moon's shadow is a cylinder of invariable size, the diameter being that of the Moon. The coordinates of the center of the circle, where the shadow intersects the fundamental plane, or the axis of the shadow, are denoted x and y, and are measured in units of Earth radius.

The Besselian elements at the time of conjunction of the star and Moon in right ascension, when x is zero, describe the geometric situation at the time T_0 in UT of the conjunction in right ascension. They are the following:

- H is the Greenwich hour angle of the star at T_0,
- $Y = y$ at T_0,
- x', y' are the hourly rates of change in x and y,
- and α_s, δ_s are the right ascension and declination of the star.

Then, the coordinates of the center of the shadow in the fundamental plane are

$$x \sin \pi = \cos \delta \sin(\alpha - \alpha_s),$$
$$y \sin \pi = \sin \delta \cos \delta_s - \cos \delta \sin \delta_s \cos(\alpha - \alpha_s), \tag{13.8}$$

where α and δ are the right ascension and declination of the Moon at a particular Universal Time and π is the horizontal parallax. For prediction purposes these reduce to

$$x = \frac{\alpha - \alpha_s}{\pi} \cos \delta, \qquad y = \frac{\delta - \delta_s}{\pi} + 0.0087x(\alpha - \alpha_s) \sin \delta_s, \tag{13.9}$$

where α, δ and π are in degrees. At conjunction,

$$x = 0, \qquad y = Y = \frac{\delta - \delta_s}{\pi},$$
$$x' = \frac{\alpha' \cos \delta}{\pi}, \qquad y' = \frac{\delta'}{\pi} - Y\frac{\pi'}{\pi}. \tag{13.10}$$

The primed quantities are hourly variations that arise from differentiating the Moon's daily polynomial coefficients for α, δ and π. The rectangular coordinates of the center of the shadow on the surface of the Earth at Universal Time t from conjunction are

$$
\begin{aligned}
x &= x't, \\
y &= y't + Y, \\
z &= \sqrt{(1 - x^2 - y^2)},
\end{aligned}
\tag{13.11}
$$

where z is measured along the shadow axis from the fundamental plane toward the Moon. The longitude and latitude (λ, ϕ) of this point on the surface of the Earth are

$$
\begin{aligned}
\phi &= \sin^{-1}(y \cos \delta + z \sin \delta), \\
h &= \tan^{-1} \frac{x}{-y \sin \delta + z \cos \delta}, \\
\lambda &= h - (H + 15 \times 1.002\,738t).
\end{aligned}
\tag{13.12}
$$

The rectangular coordinates of the edge of the shadow on the surface of the Earth (ξ, η, ζ) are

$$
\begin{aligned}
\xi &= x - k \sin Q, \\
\eta &= y - k \cos Q, \\
\zeta &= \sqrt{(1 - \xi^2 - \eta^2)},
\end{aligned}
\tag{13.13}
$$

where $k = 0.2725$ is the radius of the Moon in Earth radii, and Q is varied through the range of $0°$ to $360°$. The longitude and latitude (λ, ϕ) of a point on the edge of the shadow are obtained from (13.12) by inserting (ξ, η, ζ) instead of (x, y, z).

The prediction of occultations of stars, or minor planets, by planets, or minor planets, depends on the search for conjunctions in right ascension within the limits of the difference of declination that permit an occultation. The limits are very small, approximately the sum of the horizontal parallax and the semidiameter of the planet. The prediction follows the principles of a lunar occultation, but the methods can be less formal and more direct, because the angles are smaller and the prediction less precise. An occultation will take place as seen from some path on the Earth's surface, if the difference, $\delta - \delta_s$, in apparent declination at the time of true conjunction in right ascension satisfies the condition

$$
|\delta - \delta_s| < (\pi' + s)/|\sin \rho|,
\tag{13.14}
$$

where π' and s are respectively the equatorial horizontal parallax and the semidiameter of the planet, and ρ is the position angle of its direction of motion. The methods of prediction are described by Taylor (1955).

13.8 Transits

When a relatively small body passes in front of a larger body, it is called a transit. Mercury and Venus can pass in front of the Sun as seen from the Earth. These events can take place only when the Earth and the planet are close to the same node of their orbit on the ecliptic. For Venus the events are rather rare, occurring twice within eight years and then not reoccurring for a hundred years. For Mercury there are May and November events, approximately fourteen transits per century. Based on knowledge of the orbital motions of the planets, observations of such events were used to determine the size of the Solar System, or the distance of the Earth from the Sun. The timing of these events is complicated by the human eye's responses to different levels of brightness of light. Photometric observations of stars are now being made in an effort to detect possible transits of extrasolar planets in front of stars. Such an event causes a temporary reduction in the apparent magnitude of the star (Charbonneau *et al.*, 2000).

A shadow transit is when the shadow of a smaller body passes in front of a larger body. The Galilean satellites of Jupiter experience the phenomena of eclipses, occultations, transits, and shadow transits. Observations of eclipses and occultations by teams of people deployed in a pattern perpendicular to the paths of the phenomena have been used to investigate the apparent diameter and shapes of the Sun and Moon and the lunar topography.

13.9 Mutual events

The satellites around Jupiter and Saturn are sufficiently coplanar and numerous that they have periods when they eclipse and occult each other as seen from the Earth. The prediction and observation of these events can be used for the improvement of the knowledge of the orbits and physical characteristics of the satellites. Predictions are prepared for these events and observation campaigns are organized. It has been recognized that CCD observations of satellites, when they appear to be close together, can be more accurate than widely separated configurations. This is because of the reduction in the variations in the atmospheric effect on the observations.

13.10 Physical ephemerides

For observations of Solar System objects, information about the orientation, illumination, appearance, and cartographic coordinates of points on the surface are needed. Thus, rotational elements, shapes and sizes of the bodies, and prime meridians must be adopted. For cases such as the Sun, where the surface cannot be seen, conventional values must be selected.

Planetocentric coordinates are generally used. The z-axis is the mean axis of rotation, the x-axis is the intersection of the planetary equator and an adopted prime meridian. Longitude of a surface point is positive to the East from the prime meridian as defined by the rotational elements. Latitude is the angle between the planetary equator and a line to the center of mass. The radial distance is measured from the center of mass to the surface point.

Planetographic coordinates are used for cartographic purposes and are defined based on an equipotential surface, or an adopted ellipsoid, as a reference surface. Longitude of a point is measured in the direction opposite to the rotation (positive to the West for direct rotation) from the cartographic position of the prime meridian. Latitude is the angle between the planetary equator and the normal to the reference surface at that point. The height of a point is specified as the distance above a point on the reference surface with the same longitude and latitude.

The physical ephemerides of the Sun, Moon, and planets are published in the *Astronomical Almanac*. The standard defining values of the rotational elements and cartographic coordinates of the Solar System bodies are published triennially by the IAU/IAG Working Group on Cartographic Coordinates and Rotational Elements of the Planets and Satellites (Seidelmann *et al.*, 2002).

13.11 Calendar

The periods of the motions of the Sun and Moon and the daily rotations of the Earth have been the natural bases for the development of calendars. However, some civilizations have based calendars on other Solar System bodies, such as the Mayan use of Venus. Unfortunately, none of the natural resulting periods, year, month, or day, is an exact multiple of the others. Thus, calendars based on lunar months require adjustments to remain close to the year defined by the apparent motion of the Sun. The exact motion of the Sun is not an integral number of days, so adjustments are required to keep the seasons or, more significantly, the religious holidays, at the same times in the calendar. The seasons and religious holidays are defined in terms of the vernal equinox, which strictly follows the motion of the Earth. There is also an ecclesiastical equinox defined for the determination of Easter without variations due to the true motion.

The tropical year is defined as the mean interval between vernal equinoxes. The expresssion, from the orbital elements of Laskar (1986), is used in calculating the length of the tropical year:

$$365.242\,189\,6698 - 0.000\,006\,153\,59T - 7.29 \times 10^{-10}T^2 + 2.64 \times 10^{-10}T^3$$

$$(13.15)$$

where

$$T = (JD - 2\,451\,545.0)/36\,525$$

and JD is the Julian Day number. The true interval from a particular vernal equinox to the next may vary from this mean by several minutes. The exact length of the tropical year is dependent on the general planetary theory used to determine the orbital elements. If a numerical integration is used, then the length of the integration used to form the mean values also becomes a factor. Thus, while the length of the tropical year does change, it is very slow.

The current calendar used internationally, is the Gregorian calendar. This was introduced by Pope Gregory in 1582 in some countries, and as late as 1927 for others. The Gregorian calendar has 365 days per year except for leap years of 366 days. Leap years are years evenly divisible by 4, except for centurial years which must also be evenly divisible by 400. Thus, 1800, 1900, 2100, and 2200 are not leap years, but 2000 is a leap year. The Gregorian calendar year is 365.2425 days long compared with the current tropical year of 365.242 189 days, a difference of 0.000 311 days, or 1 day per 3000 years. The IAU has adopted the Julian year of 365.25 days as a time unit for many computations such as precession and orbital elements, to avoid the variability and uncertainty of the value of the tropical year.

There have been statements of modifications of the Gregorian calendar by some countries to correct for this remaining discrepancy, but these statements have not been confirmed. Also there are frequent proposals for simplifying or improving the Gregorian calendar. The current total international acceptance of the Gregorian calendar will require a very significant international benefit before a new calendar will be adopted.

In many cases calendars have been based on the lunar cycle. The synodic month, the mean interval between conjunctions of the Moon and the Sun, corresponds to the cycle of lunar phases. The expression for the synodic month is based on the lunar theory of Chapront-Touzé and Chapront (1988),

$$29.530\,588\,8531 + 0.000\,000\,216\,21T - 3.64 \times 10^{-10}T^2, \qquad (13.16)$$

where $T = (JD - 2\,451\,545.0)/36\,525$ and JD is the Julian Day number. Any lunar phase cycle can deviate from the mean by up to seven hours.

Some lunar calendars (e.g. Islamic) begin a new month based on observations of the lunar crescent after new Moon. In these cases the beginning of the month depends on weather conditions, atmospheric clarity, and the location on Earth, since the Moon can generally only be observed about fifteen hours after new Moon and about $10°$ above the horizon. These conditions can generally be accurately predicted with about a $30°$ uncertainty in Earth longitude for actual observability. In some cases, predictions are not accepted and observations are required for the determination of the beginning of the month.

Since there is not an integral number of lunar months in a solar year, some calendars (e.g. Hebrew) have introduced intercalated months to adjust the months, such that religious holidays occur in the same season each year.

There are non-astronomical features to calendars, such as the seven day week. The origin of the seven day week is uncertain, with explanations ranging from weather cycles to Biblical and Talmudic texts. There is also an unproven assumption that the cyclic continuity of the week has been maintained without interruption from its origin in Biblical times. More about the calendar may be found in chapter 12 of the *Explanatory Supplement to the Astronomical Almanac* (Seidelmann, 1992).

14

Applications to observations

The previous chapters have provided the bases for the applications to observations. This chapter will provide examples of the reduction and analysis of observational data. Software for coordinate conversions and individual effects, such as precession, nutation, proper motions, aberration, refraction, etc., is available as NOVAS at the US Naval Observatory website. A software program that performs on-line the full reduction of a CCD plate, called PRIAM, is available at the IMCCE web site of the Paris Observatory.

14.1 Observing differences, classical and new systems

In Chapters 7, 8, and 10, we introduced fixed and moving reference systems. At present, we are in a transition period, when old and new systems may be used, although it is to be expected that the new system will progressively remain the only one used, because they have important advantages. However, it is still necessary to be acquainted with both.

Referring positions to the fixed reference frame at epoch J2000.0, namely the International Celestial Reference Frame (ICRF), instead of the reference frame and positions referred to the equinox and the ecliptic and based on the FK5, improves the accuracies of the positions. The ephemerides are now on the ICRF. Astrometric observations made with respect to ICRF positions only require the corrections for proper motions of the reference stars and appropriate differential corrections to reduce the magnitudes of the plate solution parameters (Section 14.5). Thus, for the general astronomer the ICRF has been introduced in place of the FK5 with an improvement in accuracy, but without an operational or systematic change. For the coordinate system of date, there are two classical systems and one new system (Section 8.4).

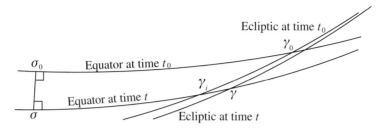

Fig. 14.1. Between times t_0 and t, the CEO, σ, moves perpendicularly to the two equators, while the equinox, γ, moves by $\gamma_0 - \gamma_i$ on the ecliptic at time t_0 and is shifted to γ because of the small motion of the ecliptic.

(i) The old uncorrected classical system based on the equinox, old precession values, old nutation theory, Celestial Ephemeris Pole (CEP), Aoki *et al.* (1982) expression for GMST in terms of UT1, and old constants.

(ii) The improved classical system uses the equinox, Celestial Intermediate Pole (CIP), new precession values, new nutation model, revised GMST expression, and new constants.

(iii) The new system introduces the Celestial Ephemeris Origin (CEO) in place of the equinox, CIP, precession–nutation model, stellar angle, and new constants. The precession–nutation model includes geodesic precession, sub-daily terms, and observationally determined accuracy improvements.

The primary differences between the first two systems and the new one, arise in the right ascensions due to the CEO–equinox difference. For declinations, the differences are in milliarcseconds owing to the differences between the CEP and the CIP based on the differences in the nutation values. This means there are small differences in the location of the equator of date. The differences between the classical systems and the new system along the plane of the equator correspond essentially to the precessional motion of the equinox that does not affect the CEO (Fig. 14.1). For the rotation angle of the Earth the difference between the two systems can be determined by the difference between the stellar angle and the Greenwich sidereal time (GST). For a given time, it can be computed from formulae given in Section 10.6.2. For the difference between the right ascensions from the CEO and the equinox, they can be determined from the same quantity, the stellar angle minus the GST. The recommended observation reduction procedure for both systems is the application of the matrix method, the differences being the quantities used in the precession–nutation and Earth rotation matrices. Examples are given in Appendix A.

14.2 Determinations of positions

Observations can be made at many different wavelengths. At present, owing to the combination of accurate VLBI positions of extragalactic sources, accurate optical positions of stars, resulting from accurate ties between the optical and the radio sources, and a sufficient density of reference frame stars, it is possible to tie together the observations in the different wavelengths at about the 100 mas level to establish source identifications. However, care must be exercised to ensure that sufficient reference stars are being used and that the magnitude and color effects are taken into account.

When dealing with individual detector fields-of-view of stars, the reference stars in that field must be sufficient to determine the "plate constants", possible distortions, and differential corrections over the field. The reference stars must be corrected in position for their proper motions during the period from epoch to date of observation. In addition, the reference stars should be corrected for color and magnitude effects, to the extent that it is possible.

The reference star positions should be taken from the most accurate catalog covering the magnitude range of the exposure, and then including less-accurate reference stars in the same reference frame, as necessary to obtain sufficient coverage.

The star observed positions depend on time and can be written for time T_j as

$$\left.\begin{aligned} \alpha_i &= \alpha_{i0} + \mu_{\alpha i} t_j + \varpi_i P_j + L_{ij} + \eta_\alpha(t_j), \\ \delta_i &= \delta_{i0} + \mu_{\delta i} t_j + \varpi_i Q_j + M_{ij} + \eta_\delta(t_j), \end{aligned}\right\} \tag{14.1}$$

where α_{i0} and δ_{i0} are the standard coordinates of the ith star in the chosen system at some time T_0; $\mu_{\alpha i}$ and $\mu_{\delta i}$ are the star's proper motion components; $tj = T_j - T_0$; ϖ_i is the annual parallax ; P_j and Q_j are the parallax factors in α and δ respectively (see Section 6.2) at the epoch t_j for the position of the star; L_{ij} and M_{ij} are those nonlinear displacements from the star's epoch position, owing to components of a binary- or multiple-star system caused by orbital motion, accelerations of the proper motion, source structure changes, or other real or apparent displacements that may be present. The $\eta_\alpha(t_j)$ and $\eta_\delta(t_j)$ are the observational errors, which are assumed to be random.

14.3 CCD observations

Charge coupled device (CCD) observations are very similar to photographic plates, except that the CCDs are physically smaller, are much more efficient, are linear in their response, and have a larger dynamic range. CCDs also have the advantage that the exposures are immediately recorded digitally, so there is no need for the use of a measuring machine, and the images can be digitally processed, which allows co-adding exposures, stretching, removing background, etc. The CCDs, owing to their read-out flexibility, can be operated in several different ways. A CCD can be exposed in the stare mode, similar to a plate, and in the time delay integration (TDI) mode, where the CCD is read out at the same rate as the motion of the star across the detector. Thus, for example, if the telescope is held stationary or driven to compensate for declination change, and the rotation of the Earth is causing the stars to cross the field of view, the CCD could be read out continuously, observing a strip across the sky. This is basically how the SLOAN Digital Sky Survey observes. It is also possible to observe in a combination of the two modes such that, for example, a moving object is observed by clocking the CCD at the object's rate, then the stars are observed in the stare mode, followed by another exposure of the moving object. This is a means of obtaining accurate observations of a moving object with respect to the stellar background.

There are some characteristics of CCDs that must be considered and included in reduction procedures. The pixels do not all respond identically, so it is necessary to take flat fields at the same wavelength as the exposure and determine the response of each of the individual pixels. Also, it is necessary to obtain a bias frame to establish the zero level, or bias, of each of the pixels. In addition, dark frames should be taken to determine the reading that would be achieved by the CCD over time, if it were not exposed to any photons. It is also necessary to determine the geometric characteristics of the CCD, how large are the pixels, are they uniform in size and spacing, and are there gaps without response to photons? A detailed description of the applications of CCDs can be found in Martinez and Klotz (1999).

When CCDs are operated in space, they are exposed to radiation effects that must be considered. Cosmic rays appear in the exposures and must be removed as false readings, or non-existent stars. Long exposures can be split into two, such that cosmic rays will not appear in both exposures, but real objects will appear in both. When exposures are being made in multiple filters, or several exposures of moving objects are being made, the cosmic rays can be detected from their presence on individual frames. A more difficult problem is the charge transfer efficiency (CTE), or the corresponding charge transfer inefficiency (CTI). This is caused by the development over time of traps in the CCD silicon due to radiation. Thus, electrons that were generated by exposure to photons get delayed in the traps as the CCD is read out, and appear delayed and offset in position from the original detected position, due to the time constants of the traps. Techniques for dealing with CTI are currently under investigation. One approach is to build a notch in the bottom of the column channel and reduce the geometric space for traps. Another method is to expose the entire chip to a low level of illumination and fill all the traps prior to an exposure. For TDI mode, a line of charge can be injected before each expected star, so that the traps are filled before the charges associated with the star reach the traps. However, the trap release times are variable and currently not well understood.

The accuracies of CCD observations have been investigated for varying conditions. Astrometric survey observations were investigated by Zacharias (1996), who found that the base-level noise, more than the atmosphere, limits the accuracy achievable. The astrometric and photometric calibration of the Wide Field Planetary

Camera 2 on the Hubble Space Telescope has been investigated by Anderson and King (1999). The astrometric accuracies of the TDI mode of operation have been investigated for FAME by Triebes *et al.* (2000) and for GAIA by Gai *et al.* (2001).

14.4 Plate reduction

The plate reduction consists in determining the celestial coordinates of objects on a photographic plate or on a CCD. This implies a detailed modeling of the transformation between the sky and the plate, which includes the geometric gnomonic transformation (Section 3.7), the deformation of the image due to refraction (Section 6.1) and distortions introduced by the instrument.

14.4.1 Distortions

The distortions introduced into the data by the optical system and the detector need to be removed before determining the positions of the observed objects. These distortions can be known from previous investigations, or they may be determined from multiple observations of known fields, or dense fields, shifted with respect to one another.

Schmidt telescopes cause distortion patterns on the bent plates. These can be characterized for individual telescope systems and corrections introduced (Taff *et al.*, 1990a,b; Monet *et al.*, 1998). Many optical systems will have either barrel or pin cushion distortion that can be greatly reduced by use of third-order correction expressions when observations are made in the stare mode. In the TDI mode the distortions present a more complex problem, because the distortions cause a smear of the image in the read-out direction owing to differences in the apparent motion of the image on the detector, and a smear in the cross scan direction owing to an apparent shift of the image from a straight path down the CCD column. Vignetting and coma cause distortions of the point spread function (psf) which can be corrected by using different psfs, as a function of location in the optical system, in the reduction method. The methods of correcting for distortions can require large numbers of reference stars.

Corrections for the geometric effects of the CCD can be introduced in the point spread functions and in the determined centroid positions.

14.4.2 Differential corrections

When individual observations are made on a small field of view, the target object's position can be determined with respect to the reference stars directly by correcting the reference star positions according to Section 14.2 from their catalog positions at epoch to the time of observation, and corrections for differential refraction and

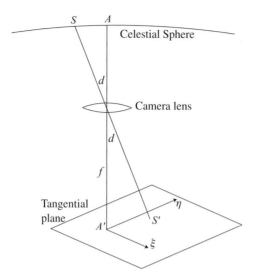

Fig. 14.2. The tangential plane concept.

aberration effects across the field. After differential reduction, this will give an observed position with respect to the reference catalog, equivalent to what was known as an astrometric position.

14.4.3 Plate solutions

The same procedure that has been traditionally used to determine the astrometric positions of measured coordinates of images on plates is applied to CCDs and is called plate solutions. While the fields-of-view are generally smaller on CCDs and might lead to lower-order expressions, the accuracies of CCDs are greater, and thus justify higher-order expressions. The technique is to use the reference stars to determine the coefficients of the plate solution expressions, and to apply the expressions to determine the astronomical coordinates of the images on the frame (Zacharias, 2001).

The projection of the sky on a plate, or a CCD, can be considered as shown in Fig. 14.2. Let ξ, η be the rectangular Cartesian coordinates, measured from the intersection of the optical axis and the tangent plane (A'), ξ toward the East and η to the North. The gnomonic projection gives the tangential coordinates (see Section 3.9.1):

$$\xi = \frac{\cos \delta \sin(\alpha - \alpha_0)}{\sin \delta_0 \sin \delta + \cos\delta_0 \cos \delta \cos(\alpha - \alpha_0)},$$

$$\eta = \frac{\cos \delta_0 \sin \delta - \sin \delta_0 \cos \delta \cos(\alpha - \alpha_0)}{\sin \delta_0 \sin \delta + \cos\delta_0 \cos \delta \cos(\alpha - \alpha_0)}. \tag{14.2}$$

where α, δ are the right ascension and declination of a star S and α_0, δ_0 are those of A. The units for ξ, η are radians.

The measured coordinates x, y, if exactly parallel to ξ, η, are

$$x = f\xi \quad \text{and} \quad y = f\eta \tag{14.3}$$

with f being the focal length and x, y, and f in the same units. This is the simplest model for the measurables. A more realistic plate model would be

$$\xi = P_1 x + P_2 y + P_3 + P_5 x + P_6 y + P_7 x^2 + P_8 xy + P_9 xr^2,$$
$$\eta = -P_2 x + P_1 y + P_4 + P_6 x - P_5 y + P_7 xy + P_8 y^2 + P_9 yr^2, \tag{14.4}$$

with the P_i values to be estimated in the adjustment process, using the reference stars with known positions (ξ, η) and $r^2 = x^2 + y^2$. An orthogonal model is represented by parameters 1 to 4. This allows for a zero point, scale and rotation between x, y and ξ, η. Adding parameters 5 and 6 gives a full linear model. Parameters 7 and 8 represent a tilt of the focal plane with respect to the ideal tangent plane, and parameter 9 is the third-order optical distortion term, which can be significant. Once the model is determined, then the values of ξ, η can be determined from the measured values of x and y for any other stars in the field. The spherical coordinates α, δ are obtained by solving

$$\cot \delta \sin(\alpha - \alpha_0) = \frac{\xi}{\sin \delta_0 + \eta \cos \delta_0},$$
$$\cot \delta \cos(\alpha - \alpha_0) = \frac{\cos \delta_0 - \eta \cos \delta_0}{\sin \delta_0 + \eta \cos \delta_0}. \tag{14.5}$$

14.5 Overlapping plate solutions

When doing a sky survey, it is highly desirable to tie together individual exposures, either on plates or CCDs, by having some overlap of the adjacent fields. The overlap percentage can vary depending on the purpose and density of the images. The Palomar Sky Survey (PSS) had the purpose of getting deep exposures covering the entire sky for the images thus achieved. It was not to determine accurate astrometric positions. Thus, it sought an overlap just to achieve complete coverage. An astrometric sky survey, seeking accurate positions of all objects and the ability to tie the positions together, will seek an overlap of 100%. This is achieved by the corners of a plate being at the center of each adjacent plate, so the plate overlaps 25% of each of four other plates. In this case, it is also desirable to have the overlapping plates taken close together in time. This is the procedure for UCAC (Section 11.7.1).

The overlapping plate method has been developed, similar to plate solutions, but using multiple plates to extend the sky area covered and to improve the accuracies of the resulting positions. Using reference stars to determine plate parameters for multiple plates at the same time, the positions of field stars are then determined. The sizes of the resulting matrices lead to the need to solve large matrices. The method has been described by Eichhorn (1960), Eichhorn, Googe and Gatewood (1967), Jefferys (1963), Eichhorn and Williams (1963), Eichhorn and Russell (1976), and Jefferys (1979).

One case of overlapping plates is the central overlapping plate case as proposed by Eichhorn and Jefferys (1971). In this case the plates are repeated exposures of the same field-of-view, so there are n frames of m stars with an image of each star on each frame. The solution of the resulting matrix of normal equations is given in Eichhorn (1988). This case is particularly applicable to parallax observations, where repeated observations of the parallax target stars are to be taken and measured with respect to the same reference stars.

Another application of the overlapping plate technique is for minor planets observed at crossing points in their orbits. The combination of common reference stars and different times and positions in the minor planet's orbit provides strong differential position measurements (Moller and Kristensen, 1976; Hemenway, 1980; Owen and Yeomans, 1994).

The overlapping plate method performs roughly a simultaneous fit of plate measurements to the reference star system and plate parameters. The basic idea is that each star is in a unique position. The observations are estimates of positions of the reference stars and measured rectangular coordinates of all stellar images on a number of detector images. So the unknowns are of two types, the plate parameters and all stellar positions. The field stars will greatly outnumber the reference stars. The proper choice of the relationship, or plate parameters, is critical to achieve high systematic accuracy with the overlapping plate method. The following development will apply equally to either a plate or CCD exposure.

A relatively simple derivation of the method will be presented to illustrate the method, more detailed developments are presented in references cited in this section. Using an imaging model such as

$$\mathbf{R} = \mathcal{M} \times \mathbf{P} + \mathbf{E}, \tag{14.6}$$

where

$$\mathbf{R} = \begin{pmatrix} x \\ y \end{pmatrix} \tag{14.7}$$

is the vector of the measured coordinates x and y.

$$\mathcal{M} = \begin{pmatrix} 1 & \xi & \eta & 0 & 0 & 0 & \cdots \\ 0 & 0 & 0 & 1 & \xi & \eta & \cdots \end{pmatrix} \tag{14.8}$$

is the model matrix, with the quantities ξ and η being the standard coordinates of the star with respect to the plate center of the plate in question.

$$\mathbf{P}^\top = (p_1 \quad p_2 \quad \cdots \quad p_J),$$

where $\mathbf{P}$ is the vector of the J ($j = 1, 2, \ldots J$) plate constants, written here as a column matrix (see Section 3.4), but represented here for editorial reasons as its transpose.

$$\mathbf{E} = \begin{pmatrix} e_x \\ e_y \end{pmatrix}$$

is the uncertainty vector. The model matrix must be developed to represent all effects present and its completeness is critical to the achievement of accuracy being sought. Thus, higher-order terms and terms dependent on magnitude and color indexes of stars may be included.

The relation between the actual position (α, δ) of a reference star and its observed position (α_0, δ_0) is

$$\left. \begin{array}{l} \alpha_0 \cos \delta = \alpha \cos \delta + e_\alpha, \\ \delta_0 = \delta + e_\delta, \end{array} \right\} \tag{14.9}$$

where e_α and e_δ are the uncertainties. Here the random uncertainties will be assumed independent with zero means and standard deviations σ_α, σ_δ, σ_α/w, and σ_δ/w respectively. The reference star weight w^2 is assumed known from the formal catalog errors. A more complicated model could be introduced for effects such as the errors depending on the positions on the plate, their magnitude, their color, and correlation between the errors for different stars. This establishes a standard nonlinear, least squares, adjustment problem. Now, the next steps are to obtain a solution from individual observation equations and a system of normal equations containing plate constants.

Assuming reasonable estimates of all unknowns (celestial coordinates of reference stars and reduction parameters for all plates), linearized plate measurement equations have the form

$$\Delta \mathbf{R} = \mathcal{U} \times \Delta \mathbf{S} + \mathcal{M} \times \Delta \mathbf{P}. \tag{14.10}$$

$\Delta \mathbf{P}$ is the unknown vector of corrections to the assumed vector of plate constants $\mathbf{P}$.

$$\Delta \mathbf{P}^\top = (p_1 - p_{1a} \quad p_2 - p_{2a} \quad \cdots \quad p_J - p_{Ja}),$$

where p_{ja} are the assumed values of the plate parameters and p_j, the actual values.

The unknown vector of corrections to the assumed celestial coordinates is $\Delta \mathbf{S}$. To avoid problems induced by the convergence of the meridians, it is written as

$$\Delta \mathbf{S} = \begin{pmatrix} \Delta\alpha\cos\delta \\ \Delta\delta \end{pmatrix} = \begin{pmatrix} (\alpha_0 - \alpha_a)\cos\delta \\ \delta_0 - \delta_a \end{pmatrix}, \tag{14.11}$$

where $\Delta\alpha$ and $\Delta\delta$ are corrections to the assumed values α_a, δ_a.

The vector of residuals is

$$\Delta \mathbf{R} = \begin{pmatrix} x_0 - x_c \\ y_0 - y_c \end{pmatrix},$$

where x_0, y_0 are the measured coordinates and x_c, y_c are the computed coordinates:

$$\begin{pmatrix} x_c \\ y_c \end{pmatrix} = \mathcal{M}\left(\xi(\alpha_a, \delta_a), \eta(\alpha_a, \delta_a)\right) \times \mathbf{P}_a. \tag{14.12}$$

Here, ξ and η are modeled by (14.2), α_a, δ_a are the assumed celestial coordinates of the star, and $\mathbf{P}_a$, the assumed vector of plate parameters.

$\mathcal{U}$ is the transformation matrix from celestial coordinates to plate measurements:

$$\mathcal{U} = \left(\frac{\partial(x, y)}{\partial(\alpha, \delta)}\right) = \left(\frac{\partial(x, y)}{\partial(\xi, \eta)}\right) \times \left(\frac{\partial(\xi, \eta)}{\partial(\alpha, \delta)}\right). \tag{14.13}$$

The first factor in this expression can be derived once the model is specified, and the second can be determined from the relations (14.2) between standard coordinates (ξ, η) and celestial coordinates (α, δ). All partial derivatives are evaluated at the assumed values (α_a, δ_a), and $\mathbf{P}_a$ and, hence $\mathcal{U}$, is appropriately adjusted.

Finally, $\mathcal{M}$ is the matrix of partial derivatives of the measurements with respect to the plate constants $\mathbf{P}$, which, because of the linearity of (14.6), is the model matrix

$$\mathcal{M} = \mathcal{M}\left(\xi(\alpha_a, \delta_a), \eta(\alpha_a, \delta_a)\right). \tag{14.14}$$

For the reference stars the catalog values of the spherical coordinates α_a, δ_a are estimators of their actual coordinates. Equations (14.9) must be rewritten to get the unknowns in the same form as those used in the measurement equations (14.10), thus

$$(\alpha_0 - \alpha_a)\cos\delta = \Delta\alpha\cos\delta$$
$$\delta_0 - \delta_a = \Delta\delta. \tag{14.15}$$

For weighting the reference star equations, one writes

$$\Delta \mathbf{R}_r = \mathcal{W}_r \times \Delta \mathbf{S}, \tag{14.16}$$

with

$$\Delta \mathbf{R}_r = \begin{pmatrix} w(\alpha_0 - \alpha_r) \cos \delta \\ w(\delta_0 - \delta_r) \end{pmatrix},$$

and

$$W_r = \begin{pmatrix} w & 0 \\ 0 & w \end{pmatrix},$$

while $\Delta \mathbf{S}$ is defined by Equation (14.11).

Let us now assume that there are N plates ($n = 1, 2, \ldots N$). The set of observation equations of the type (14.10) for unknown stars and (14.16) for the reference stars, the ith star and the nth plate can be incorporated into a single equation of the form

$$\mathcal{A}_i \times \Delta \mathbf{S}_i + \sum_{j=1}^{J} \mathcal{B}_{ij} \times \Delta \mathbf{P}_j = \mathbf{C}_i. \tag{14.17}$$

This equation is the type that can be incorporated in a matrix equation (see Section 14.5.2). Let us clarify the structure of this equation.

14.5.1 Equations for one star

Let us assume that one has I ($i = 1, 2, \ldots I$) field stars in the program. Because the plates are overlapping, the same star i is observed on K plates, ($k = 1, 2, \ldots K$). So, there are K vectorial equations (14.10) for the star i that we shall write as follows:

$$\Delta \mathbf{R}_{ik} = \mathcal{U}_{ik} \times \Delta \mathbf{S}_i + \mathcal{M}_{ik} \times \Delta \mathbf{P}_{jk}, \tag{14.18}$$

where the index ik means that the star i is measured on plate k whose plate parameters are jk. In other terms, the subset $[jk] \subset [j]$ denotes the plates on which an image of the star i occurs. The equality (14.18) represents actually two equations with the two star i unknowns, the components of $\Delta \mathbf{S}_i$, and the J components of $\Delta \mathbf{P}_k$.

If, now, we write all the K vectorial equations (14.18), we obtain a system of $2K$ equations. The unknowns are still the two components of $\Delta \mathbf{S}_i$ and KJ plate parameters (J per plate). One may write the ensemble of these equations as a unique vectorial equation

$$\mathcal{A}_i \times \Delta \mathbf{S}_i + \sum_j \mathcal{B}_{ij} \times \Delta \mathbf{P}_{jk} = \mathbf{C}_i. \tag{14.19}$$

In this equation, $\mathcal{A}$ is a $2 \times K$ order matrix formed by superposing the 2×2 matrices $\mathcal{U}_{ik}$, $k = 1$ to K,

$$\mathcal{A}_i^\top = \begin{pmatrix} \mathcal{U}_{i1}^\top & \mathcal{U}_{i2}^\top & \cdots & \mathcal{U}_{iK}^\top \end{pmatrix}. \tag{14.20}$$

Similarly,

$$\mathbf{C}_i^\top = \begin{pmatrix} \Delta\mathbf{R}_{i1}^\top & \Delta\mathbf{R}_{i2}^\top & \Delta\mathbf{R}_{iK}^\top \end{pmatrix}. \tag{14.21}$$

Finally, taking into account that to each equation there corresponds different $\Delta\mathbf{P}_{jk}$ and $\mathcal{M}_{ik}$, and that only the product of terms corresponding to the same k are not zero, one has the summation over j of

$$\mathcal{B}_{ij}^\top = \begin{pmatrix} \mathcal{D}_{ij1}^\top & \mathcal{D}_{ij2}^\top & \cdots & \mathcal{D}_{ijK}^\top \end{pmatrix}. \tag{14.22}$$

with

$$\begin{aligned} \mathcal{D}_{lj} &= \mathcal{M}_{ik} \quad &\text{if } l = k \text{ and } j = j_k \\ &= 0 \quad &\text{otherwise.} \end{aligned}$$

If the star is a reference star, the terms of Equation (14.16) are appended so that:

$$\begin{aligned} \mathcal{A}_i^\top &= \begin{pmatrix} \mathcal{U}_{i1}^\top & \mathcal{U}_{i2}^\top & \cdots & \mathcal{U}_{iK}^\top & \mathcal{U}_{ir}^\top \end{pmatrix}, \\ \mathcal{B}_{ij}^\top &= \begin{pmatrix} \mathcal{D}_{ij1}^\top & \mathcal{D}_{ij2}^\top & \cdots & \mathcal{D}_{ijK}^\top & 0 \end{pmatrix}, \\ \mathbf{C}_i^\top &= \begin{pmatrix} \Delta\mathbf{R}_{i1}^\top & \Delta\mathbf{R}_{i2}^\top & \cdots & \Delta\mathbf{R}_{iK}^\top & \Delta\mathbf{R}_{ir}^\top \end{pmatrix}. \end{aligned} \tag{14.23}$$

14.5.2 Solution for all stars

Using the same convention for writing matrices as superposition of smaller matrices, the observational equations for all the I stars can now be written as

$$\mathcal{A} \times \Delta\mathbf{S} + \sum_j \mathcal{B}_j \times \Delta\mathbf{P}_j = \mathbf{C}, \tag{14.24}$$

with

$$\mathcal{A} = \begin{pmatrix} \mathcal{A}_1 & 0 & . & . & . & 0 \\ 0 & \mathcal{A}_2 & . & . & . & 0 \\ 0 & 0 & . & . & . & 0 \\ 0 & 0 & . & . & . & 0 \\ 0 & 0 & . & . & . & 0 \\ 0 & 0 & . & . & . & \mathcal{A}_I \end{pmatrix}$$

$$\begin{aligned} \mathcal{B}_j^\top &= \begin{pmatrix} \mathcal{B}_{1j}^\top & \mathcal{B}_{2j}^\top & \cdots & \mathcal{B}_{Ij}^\top \end{pmatrix}, \\ \mathbf{C}^\top &= \begin{pmatrix} \mathbf{C}_1^\top & \mathbf{C}_2^\top & \cdots & \mathbf{C}_I^\top \end{pmatrix}, \\ \Delta\mathbf{S}^\top &= \begin{pmatrix} \Delta\mathbf{S}_1^\top & \Delta\mathbf{S}_2^\top & \cdots & \Delta\mathbf{S}_I^\top \end{pmatrix}. \end{aligned}$$

Assuming there are I stars and M plates, the normal equations are

$$
\begin{pmatrix}
A^\top A & A^\top B_1 & A^\top B_2 & . & . & . & A^\top B_M \\
B_1^\top A & B_1^\top B_1 & 0 & . & . & . & 0 \\
B_2^\top A & 0 & B_2^\top B_2 & . & . & . & 0 \\
. & . & . & . & . & . & 0 \\
. & . & . & . & . & . & 0 \\
. & . & . & . & . & . & 0 \\
B_M^\top A & 0 & 0 & . & . & . & B_M^\top B_M
\end{pmatrix}
=
\begin{pmatrix}
\Delta S \\ \Delta P_1 \\ \Delta P_2 \\ . \\ . \\ . \\ \Delta P_M
\end{pmatrix}
\times
\begin{pmatrix}
A^\top C \\ B_1^\top C \\ B_2^\top C \\ . \\ . \\ . \\ B_M^\top C
\end{pmatrix}.
$$

$$(14.25)$$

Advantage is taken of the fact that $B_j^\top \times B_k = 0$, if $j \neq k$. If the unknown vector of star coordinates ΔS is now eliminated, the following system of reduced normal equations results:

$$
\begin{pmatrix}
\mathcal{F}_{11} & \mathcal{F}_{12} & . & . & . & \mathcal{F}_{1M} \\
\mathcal{F}_{12}^\top & \mathcal{F}_{22} & . & . & . & \mathcal{F}_{2M} \\
. & . & . & . & . & . \\
. & . & . & . & . & . \\
. & . & . & . & . & . \\
\mathcal{F}_{1M}^\top & \mathcal{F}_{2M}^\top & . & . & . & \mathcal{F}_{MM}
\end{pmatrix}
\times
\begin{pmatrix}
\Delta P_1 \\ \Delta P_2 \\ . \\ . \\ . \\ \Delta P_M
\end{pmatrix}
=
\begin{pmatrix}
G_1 \\ G_2 \\ . \\ . \\ . \\ G_M
\end{pmatrix},
\qquad (14.26)
$$

with

$$
\mathcal{F}_{jj} = B_j^\top B_j - B_j^\top A (A^\top A)^{-1} A^\top B_j,
$$
$$
\mathcal{F}_{jk} = -B_j^\top A (A^\top A)^{-1} A^\top B_k, \quad \text{for } j \neq k,
$$

and

$$
G_j = \left(B_j^\top - B_j^\top A (A^\top A)^{-1} A^\top \right) C.
$$

Because of the simple structure of A, one can write

$$
\mathcal{F}_{jj} = \sum_i \left(B_{ij}^\top B_{ij} - B_{ij}^\top A_i (A_i^\top A_i)^{-1} A_i^\top B_{ij} \right), \tag{14.27}
$$

$$
\mathcal{F}_{jk} = -\sum_i \left(B_{ij}^\top A_i (A_i^\top A_i)^{-1} A_i^\top B_{ik} \right), \quad \text{for } j \neq k, \tag{14.28}
$$

$$
G_j = \sum_i \left(\left(B_{ij}^\top - B_{ij}^\top A_i (A_i^\top A_i)^{-1} A_i^\top \right) C_i \right). \tag{14.29}
$$

The summations in these formulae are over all stars (i.e. the index i), but B_{ij} is zero if star i does not appear on plate j. Thus, the summation in (14.28) is taken over all stars which are common on plates j and k, and $\mathcal{F}_{jk}$ is a zero matrix if there are no such stars. Since the matrices B_{ij} contain so many zeros, the matrix products in formulae (14.27) to (14.29) contain many trivial multiplications. These unnecessary computations can be avoided, if the following equations are used, as may be seen from Equations (14.19)–(14.23):

$$B_{ij}^\top B_{ij} = \sum_k M_{ik}^\top M_{ik},$$

$$A_i^\top B_{ij} = \sum_k U_{ik}^\top M_{ik},$$

$$B_{ij}^\top C_i = \sum_k M_{ik}^\top R_{ik}, \tag{14.30}$$

where the summations are over k, representing the measurements of the ith star made on the jth plate. In addition, one has:

$$A_i^\top A_i = \sum_k U_{ijk}^\top U_{ijk} \qquad \text{for field stars}$$

$$= \sum_k U_{ijk}^\top U_{ijk} + U_{ir}^\top U_{ir} \qquad \text{for reference stars,}$$

and

$$A_i^\top C_i = \sum_k U_{ijk}^\top \Delta R_{ijk} \qquad \text{for field stars}$$

$$= \sum_k U_{ijk}^\top \Delta R_{ijk} + U_{ir}^\top \Delta R_{ir} \qquad \text{for reference stars.}$$

Once these matrices are calculated all the combinations needed for the summations in Equations (14.27)–(14.29) can be found by using arithmetic operating on small matrices.

For the solution of the plate constant normal equations for the overlap pattern, the component matrices $\mathcal{F}_{jk}$ of the plate constant normal equations will be zero, whenever there are no stars common on the plates j and k, and there are no model constraints relating the constants of the two plates. If one neglects for the moment all constraints, then the plate constant normal equation coefficient matrix for a typical zonal pattern, in which every star appears on exactly two plates which overlap in right ascension only, assumes the form

$$\begin{pmatrix} \mathcal{F}_{11} & \mathcal{F}_{12} & 0 & . & . & . & \mathcal{F}_{1M} \\ \mathcal{F}_{12}^\top & \mathcal{F}_{22} & \mathcal{F}_{23} & . & . & . & 0 \\ 0 & \mathcal{F}_{23}^\top & \mathcal{F}_{33} & . & . & . & 0 \\ . & . & . & . & . & . & . \\ . & . & . & . & . & . & . \\ . & . & . & . & . & . & . \\ \mathcal{F}_{1M}^\top & 0 & 0 & . & . & . & \mathcal{F}_{MM} \end{pmatrix}. \tag{14.31}$$

Only three non-zero submatrices appear in any row. Because of the symmetry, for computational purposes it is only necessary to store two sub-matrices per row. However, even with so many sub-matrices, the number of elements is considerable. For example, for 90 plates and 24 plate constants per plate, more than 100 000 elements require storage. So it is advantageous not to introduce any more non-trivial elements than absolutely necessary. Thus, it is desirable not to introduce constraints on plate constants of pairs of non-overlapping plates. Methods of introducing constraints on the plate constants are given in Googe *et al.* (1970). Once the corrections to the plate constants have been found, the corrections to the celestial coordinates are given by

$$\Delta \mathbf{S}_i = \left(\mathcal{A}_i^\top \mathcal{A}_i \right)^{-1} \left(\mathcal{A}_i^\top \mathbf{C}_i - \sum_j \mathcal{A}_i^\top \mathcal{B}_{ij} \Delta \mathbf{P}_j \right). \tag{14.32}$$

Then $\mathcal{A}_i$, $\mathcal{B}_i$, and $\mathbf{C}_i$ should be evaluated from Equations (14.19) and (14.21) for reference stars as well as field stars.

Modifications or improvements on the overlapping plate method have been given in terms of quaternions (Jefferys, 1987), orthogonal functions (Brosche *et al.*, 1989), and a new principle of statistical estimation (Taff, 1988). The overlapping plate technique has also been suggested as a means of developing star catalogs from existing catalog material (Taff *et al.*, 1990a,b).

The existence of many Schmidt plates taken as all sky surveys reaching as faint as 21st magnitude provided the opportunity to develop very large astrometric positional catalogs for such faint magnitudes (Monet, 1998; Russell *et al.*, 1990). The challenge was how to obtain the best accuracy in the presence of the distortions of Schmidt plates. A number of techniques were developed and tested as part of the preparation of the Space Telescope *Guide Star Catalog* and the US Naval Observatory's massive A and B, and smaller SA and SB catalogs. These methods might be called the filtering, mask, plate modeling, and infinitely overlapping circles approaches (Taff *et al.*, 1990a,b; Lattanzi and Bucciarelli, 1991). The filtering and mask methods remove small-scale systematics, but the mask technique does not handle plate-to-plate variations. It also requires a large collection of plates that have been taken on the same telescope, same filter, same exposure and emulsion, and measured on the same measuring machine. Thus, the mask technique is best suited for reduction of large plate data bases. The filter technique requires a dense reference catalog. The plate model and mask technique give an indication of what is being modeled, and the plate model gives an estimate of the errors remaining. The combination of the mask and filter techniques can remove the systematics that are common to all the plates, while the mask method and the filter technique remove the smaller individual plate systematics (Morrison *et al.*, 1998). It has been found that there is a coma-like term, which is a function of magnitude and radial distance from the plate center, in Schmidt plates (Morrison *et al.*, 1996).

14.5.3 Global solutions

Global solutions, or block adjustments, techniques can be used to achieve wide-angle astrometric results from differential narrow-field observations. Overlapping fields are required for this approach. Instead of solving for individual plate parameters, the reduction scheme solves for all plate parameters and star positions simultaneously using the fact that each star has one unique position in the sky as observed on all CCDs or plates.

This technique was developed in the 1960s, but not really applied until the 1970s owing to the computer capabilities required. The method is very sensitive to any systematic effects not included in the model, and can thus lead to results that are worse than traditional methods, if the technique is not properly applied.

The Hipparcos sphere reconstruction process is an example of global solutions. Hipparcos has the advantage of providing many observations of each star as well as simultaneous observations of widely separated fields. This leads to a better conditioned global solution than overlapping fields.

14.6 Radio observations

At the present time, Very Long Baseline Interferometry (VLBI) observations of distant radio sources give the most accurate astrometric positions. These positions determine the International Celestial Reference Frame (ICRF). Since the sources are very distant, there are no apparent proper motions of the sources. However, owing to the physics of the generating sources, they are subject to changes in the source structure which has the characteristic of temporal changes in the positions. The VLBI observations are made of many sources from multiple sites over a period of time. They are made from the rotating Earth, so they are subject to all the kinematics of the Earth including the moving equator. Solutions of radio-source positions must be made in conjunction with solutions for many other parameters, such as station locations, Earth orientation parameters, precession–nutation, atmospheric effects, etc. For details, see chapter 12 of Thompson *et al.* (1986).

14.7 Millisecond pulsars

Millisecond pulsars require precise reduction procedures in order to achieve the full accuracy possible from their observations. They provide, in turn, a variety of information and tests on the accuracies of astrometry, relativity, and time. If the millisecond pulsars maintained constant rates, they could serve as independent sources of accurate time, and thus verify the accuracies of atomic time standards on Earth. The reduction of millisecond pulsars to the barycenter of the Solar System requires accurate ephemerides and masses of the bodies of the Solar System. Periodic variations in the pulsar data can indicate the inaccuracies in the reduction procedure. The determinations of the positions of the pulsars can be used as calibrators of the accuracy of the reference system to which the positions have been referred. Also, the millisecond pulsars provide tests of the theory of relativity. So, in general, pulsars that have been "recycled" through a process of mass exchange in an evolving binary star are believed to be rapidly spinning, strongly magnetized neutron stars with energetics dominated by their spindown luminosities. Their evolution has been modified by an accretion episode in which mass and angular momentum were transferred from an orbiting companion star. While recycled pulsars comprise only a small fraction of pulsars, they are older, spin faster, and have smaller magnetic fields than ordinary pulsars. Thus, they are the most demanding on astrometric reduction procedures, but also the most fruitful for results (Backer and Hellings, 1986; Rawley *et al.*, 1988; Taylor, 1992).

The timing of the arrival of a pulse requires proper handling of the polarization of the signal, accurate prediction of the true period and Doppler offset, averaging of the pulse profile, determination of a fiducial point of the pulse, and cross correlation of each average with a template of a high-sensitivity averages to produce the UTC time of arrival of a pulse at the observatory with an accuracy of about 0.001 period. The observatory UTC scale requires correction to a national standard UTC, then to TAI, and then to TCB (see Section 5.5), to achieve a uniform time as seen by an external observer. A correction for dispersion in the interstellar plasma is made by computing the total path delay at the barycentric observing frequency. The TCB times of arrival are related to the TCB times of emission by a model that includes the motion of the Earth, the motion of the pulsar, and the time of flight of the photons from the pulsar to the Earth.

The emission times T_N are modeled by the "pulsar spin equation", which contains polynomial terms for the rotational phase N_0 at time T_0, the spin rate Ω, its acceleration $\dot{\Omega}$, as well as, in some cases, the second derivative $\ddot{\Omega}$. The equation is

$$N(T) = N_0 + \Omega(T - T_0) + \frac{1}{2}\dot{\Omega}(T - T_0)^2 + \frac{1}{6}\ddot{\Omega}(T - T_0)^3. \qquad (14.33)$$

The seven pulsar parameters are pulse epoch, rotation rate, rotation acceleration or spinup, right ascension, declination, proper motion in right ascension, and proper motion in declination. Additional parameters are required for binary pulsars. Pulsar spins are decelerated by electro-magnetic torque. The deterministic spin parameters are determined by iterative least squares analysis of arrival time data, leaving residuals characterized by random jitter, discontinuities, and slow variations, or noise. Some pulsars have better characteristics than others.

14.7.1 Times used

In the following treatment of the pulsar timing observations, we shall restrict ourselves to the following terms of the general relativity metric (5.22) reduced to its second-order terms in c^{-2}:

$$g_{00} = -1 + \frac{2U}{c^2} + O(c^{-4})$$

$$g_{ii} = 1 + 2\frac{U}{c^2} + O(c^{-4}) \qquad (14.34)$$

$$g_{0i} = O(c^{-3}),$$

where c is the speed of light and U is the gravitational potential, that we write assuming that the Sun and planets are point-like objects (Equation (5.17)):

$$U = \sum_p \frac{Gm_p}{|\mathbf{r} - \mathbf{r}_p|} \qquad (14.35)$$

where $\mathbf{r}_p$ is the coordinate of the pth body, $\mathbf{r}$ is the coordinate of the field point, and the sum is over the Sun and planets of the Solar System. The origin of the coordinate system, chosen to be the Solar System barycenter, must be approximately at rest relative to the mean rest frame of its gravitational sources, for the higher-order terms of (14.34) to remain small. More terms are required in (14.34) to describe the dynamics of the Solar System to modern accuracies of observation (postlinear terms in g_{00} and g_{0i} are needed), but the form is sufficient to discuss the principle of relativistic timing effects. Other terms could be added at will, and actually are necessary for precise pulsar timing.

Equation (14.34) defines the physical significance of the time coordinate. It is the time of a clock infinitely far from the Solar System and at rest with respect to the Solar System. In practice, among possible equivalent time scales, the present choice is the Barycentric Coordinate Time (TCB), which describes the dynamics of pulsars, planets, and photons. The observations are made in a local time scale, which is linked to the TAI or the Terrestrial Time (TT), called here τ, which is kept by terrestrial clocks. These two times are related by Equation (5.25), the "proper time equation",

$$\frac{d\tau}{dt} = \sqrt{1 - \frac{2U}{c^2} - \frac{v^2}{c^2}}, \qquad (14.36)$$

where $\mathbf{r}(t)$ is the location of the clock on the surface of the Earth, $\mathbf{v} = \dot{\mathbf{r}}$ is the coordinate velocity of the clock, and $U(\mathbf{r})$ is the instantaneous total gravitational potential at the clock's location. The relation between TCB and TT is given by Equations (5.37) and (5.39).

For PSR1937 + 21, time transformation accuracy of the order of 100 ns over decades of observations is required. In (14.34) higher-order terms arising from the c^{-4} term in g_{00} and g_{ii}, as well as from postlinear terms in g_{0i} as given by (5.22) are of order 10^{-16} for terrestrial clocks and should be considered. It was developed by the BIPM/IAU Joint Committee on Relativity (Petit, 2000).

Millisecond and binary pulsars exhibit very stable spinning and/or orbital periods, so their rotational, or orbital, phases might be used as time references, generally named pulsar time scale (PT) and binary pulsar time scale (BPT). Theoretical and observational investigations of the stabilities of PT and BPT have been considered (Matsakis and Foster, 1996; Ilyasov *et al.*, 1998). The PT is not ideally uniform, owing to variations of the phase of rotation, stochastic variations in the delay of signal through the interstellar medium, and errors in determining the relative position of the observer and the pulsar. In certain situations the scale of BPT is far more stable than PT; however, the accuracies are limited by the presence of noise (Deshpande *et al.*, 1996). While pulsar timing data are improving by increased time periods of data and more accurate observations, terrestrial atomic time standards are also improving in accuracy. Pulsar data from the 1980s were probably degraded by the instabilities of the terrestrial time scales used at that time. Comparisons of real pulsar data and terrestrial time scales show that these two kinds of data currently have comparable stabilities (Matsakis *et al.*, 1997; Kaspi *et al.*, 1994).

Pulsar rates of rotation are determined by evolutionary circumstances, not by fundamental constants of nature such as control atomic clocks. Therefore, the second should not be defined by the pulsar frequencies. If a number of pulsars are observed regularly, an ensemble average pulsar time scale should be possible. It might have better long-term stability than existing time standards, and could establish relative instabilities of different realizations of terrestrial time.

14.7.2 Time arrival model

Once the terrestrial observations of pulse arrival times are transformed into the coordinate time of reception, a model is needed for determining the coordinate time of reception as a function of the coordinate time of emission of the pulse. Models of pulse arrival times have been published by Manchester and Taylor (1977), Downs and Reichley (1983), and Hellings (1986).

Assuming the pulsar is at $\mathbf{R}_0$ at some time T_0 and has a velocity $\mathbf{V}_0$, its position at the time of emission of the Nth pulse is

$$\mathbf{R}_N = \mathbf{R}_0 + \mathbf{V}_0(T_N - T_0). \tag{14.37}$$

The vector $\mathbf{V}_0$ includes both radial and transverse, or proper, motion components. The position of the receiving station on the Earth is r_N at time t_N when the Nth pulse is received. $\mathbf{r}_N$ is the sum of two terms, $\mathbf{q}_N$, the position of the center of the Earth at time $\mathbf{t}_N$ with respect to the barycenter of the Solar System, and $\mathbf{p}_N$, the intersection of the rotation axes of the antenna relative to the center of the Earth at the same time. $\mathbf{q}_N$ is found from numerical integration of the equations of motion in ICRS, and $\mathbf{p}_N$ is found from known station coordinates and a known Earth rotation model.

The photons from the pulsar will follow null trajectories of the geometry. The first-order coordinate system with metric of (14.34) and (14.35) is used. Neglecting second-order terms introduces an error of less than 1 ns in absolute time of flight for crossing the Solar System. To first order the line element is

$$d\sigma^2 = (-1 + 2U)c^2dt^2 + (1 + 2U)(dx^2 + dy^2 + dz^2) = 0. \tag{14.38}$$

Considering that the signal path is not deviated by the gravitational field of the Sun, as a first approximation, introduces a maximum error of 36 ns. Assuming that the Sun and planets are at rest during the signal passage, causes an error in the time delay by each body by the factor of the ratio of its speed to c. The expression for the time of flight of the Nth pulse from a pulsar at $\mathbf{R}_N$ to the Earth at $\mathbf{r}_N$ is determined by integrating (14.38) along the path from $(T_N, \mathbf{R}_N)$ to $(t_N, \mathbf{r}_N)$, with the result being

$$c(t_N - T_N) = |\mathbf{R}_N - \mathbf{r}_N| - 2\sum_p \frac{Gm_p}{c2} \ln\left[\frac{\mathbf{n} \cdot \mathbf{r}_{pN} + r_{pN}}{\mathbf{n} \cdot \mathbf{R}_{pN} + R_{pN}}\right], \tag{14.39}$$

where $\mathbf{r}_{pN}$ is the position of the receiver relative to the pth Solar System body at the time of closest approach of the photon to that body, and $\mathbf{R}_{pN}$ is the pulsar's position relative to body p at time T_N. Vector $\mathbf{n}$ is a unit vector in the direction of the pulsar. The sum in (14.39) is over all bodies along the photon path from the pulsar to the Earth, including interstellar objects. Unless the pulsar's line of sight lies close to some intervening object, the interstellar gravitational fields will be essentially constant over decades and need not be modeled. The delay due to the pulsar's gravitational field will also be constant. It may be necessary to include the effects of the planets, because Jupiter's gravitational field, for example, can cause a signal delay of as much as 200 ns.

Using (14.37) to expand the source position and assuming that the pulsar is very far away so that $V_0(T_N - T_0) \ll R_0$ and $\mathbf{n} \sim \mathbf{R}_0/R_0$, then (14.39) becomes

$$ct_N = cT_N + [(\mathbf{n} \cdot \mathbf{V}_0)\Delta t_N - (\mathbf{n} \cdot \mathbf{r}_N)]$$
$$- \frac{1}{R_0}[\mathbf{V}_0 \cdot \mathbf{r}_N - (\mathbf{n} \cdot \mathbf{V}_0)(\mathbf{n} \cdot \mathbf{r}_N)]\Delta t_N$$
$$+ \frac{1}{2R_0}[r_N^2 - (\mathbf{n} \cdot \mathbf{r}_N)^2] + \frac{1}{2R_0}[V_0^2 - (\mathbf{n} \cdot \mathbf{V}_0)^2]\Delta t_N^2$$
$$- 2\sum_p \frac{Gm_p}{c^2} \ln\left|\frac{\mathbf{n} \cdot \mathbf{r}_{pN} + r_{pN}}{2R_0}\right|, \tag{14.40}$$

where the definition

$$\Delta t_N = t_N - t_0 \sim T_N - T_0$$

is introduced. The combination

$$T_0 + \frac{R_0}{c} + 2\sum_p \frac{Gm_p}{c^2}\ln(2R_0) \tag{14.41}$$

is the zero-order time of arrival of the pulses at the Solar System. Defining this time as the Solar System time origin t_0, then (14.40) becomes

$$c(t_N - t_0) = c(T_N - T_0) + [(\mathbf{n}\cdot\mathbf{V}_0)\Delta t_N - (\mathbf{n}\cdot\mathbf{r}_N)]$$
$$-\frac{1}{R_0}[\mathbf{V}_0\cdot\mathbf{r}_N - (\mathbf{n}\cdot\mathbf{V}_0)(\mathbf{n}\cdot\mathbf{r}_N)]\Delta t_N$$
$$+\frac{1}{2R_0}[r_N^2 - (\mathbf{n}\cdot\mathbf{r}_N)^2] + \frac{1}{2R_0}[V_0^2 - (\mathbf{n}\cdot\mathbf{V}_0)^2]\Delta t_N^2$$
$$-2\sum_p \frac{Gm_p}{c^2}\ln|\mathbf{n}\cdot\mathbf{r}_{pN} + r_{pN}| + \Delta t_{DM}. \tag{14.42}$$

The first term on the right-hand side involves the model of the rotational dynamics of the pulsar, the second term is the first-order Doppler delay, the third term, which is proportional to Δt_N, comes from the pulsar's proper motion, and the fourth term gives the effects of annual parallax. The Δt_N^2 term arises from the increase in distance to the pulsar with time, if the pulsar has radial motion. The log term is the gravitational delay caused by space-time curvature and term Δt_{DM} represents the plasma delay.

The predicted time of arrival of the Nth pulse is explicitly a function of the constant parameters $\mathbf{n}$, $\mathbf{V}_0$, and $\mathbf{R}_0$, but implicitly a function of all parameters in the numerical integration of planetary positions determining $\mathbf{r}_N$. All of these parameters may be adjusted in fitting the model to the observed arrival times. The term $T_N - T_0$ is modeled by assuming that the pulsar rotation is a simple polynomial in time (14.33) with parameters to be adjusted to fit the data. The timing data are the only source of information about these parameters.

The vector $\mathbf{n}$ gives the right ascension and declination of the pulsar at time T_0. The arrival time data can be analyzed to determine positions (Manchester and Taylor, 1977). Independently, positions of pulsars can be determined by radio interferometric techniques. The two techniques can lead to differences in the determined positions. The differences can be due to uncertainties in Solar System ephemerides, reference frames, instabilities in the reference time standards, uncompensated changes in the propagation medium, rotational instabilities in the pulsars, and cosmic background of low frequency gravitational waves. The two components of proper motion contained in $\mathbf{V}_0$ are determined from the timing equation and by

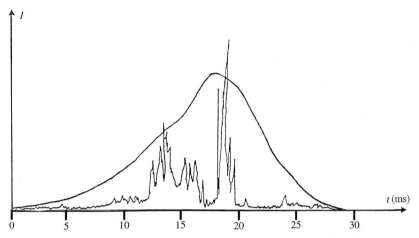

Fig. 14.3. Detail of a single impulse received from pulsar 0950 + 08 and the mean profile folded from 300 individual observations (adapted from Stinebring, 1982).

astrometry with interferometry. The accuracy of the proper motion solutions from pulsar arrival times is the result of slowly varying "noise" in pulsar arrival times. VLBI observations are able to determine parallaxes for nearby pulsars, and arrival time data can be used for determining parallaxes from (14.41) for some pulsars.

Individual radio profiles of a pulse are very irregular as shown in Fig. 14.3. Only after integration over a large number of periods can one obtain a smooth reproducible profile. The timing information is deduced from these integrated curves, which still show noise. There are many types of noise that may be present in the residuals. Taylor (1991) has proposed a dimensionless fractional instability, σ_z, which is an extension of the Allan variance (Section 4.5.3) This is a method to characterize the stability of pulsar time of arrivals.

However, the most common model is based on a mathematical description of several types of noise due to interstellar medium density fluctuations, the imperfections of planetary ephemerides and atomic clocks: random noise, white noise, frequency dependent noise, flicker noise, etc.

The usual premise for estimating pulsar parameters is that white noise dominates the time of arrival residuals. However, long-term monitoring of some pulsars has revealed the presence of a non-white component of noise that is of astrophysical origin (Kaspi *et al.*, 1994; Deshpande *et al.*, 1996). Such correlated noise is usually referred to as a colored Gaussian noise, or red noise, because its spectrum diverges at zero frequency. The red noise has a non-flat spectrum at low frequencies and can be represented by a single- or multiple-component power-law model. Although it is difficult to detect, the red noise contains information about the physical processes that take place in the neutron star interior, the interstellar medium, the early Universe, and terrestrial clocks. Thus, a rigorous method of treating red noise in the pulsar timing residuals has been of interest for a long while, and statistical procedures have been undertaken by Bertotti *et al.* (1983) and Blandford *et al.* (1984) to obtain unbiased estimates of pulsar parameters. There remains the need to improve statistical analysis of pulsar timing observational data in the presence of red noise as indicated by Bertotti *et al.* (1983), Blandford *et al.* (1984), and Kopeikin (1997, 1999).

The Solar System observations, including spacecraft, laser and radar ranging, VLBI, and optical, provide accurate means of determining the planetary ephemerides. However, the pulsar data become very precise tests of the motions of the planets and the Solar System barycenter, and should be used to indicate discrepancies in the Solar System ephemerides. In particular, the position of the Earth relative to the barycenter is not a well-determined quantity in the ephemerides. Variations in the barycenter in an absolute sense do not affect the relative motions of the planets, however, they can introduce periodicities in the Earth uncertainties. The effects of the uncertainties in masses of bodies in the asteroid belt and outer Solar System can affect the barycenter and pulsar timing data.

When the pulsar is a member of a binary system, then the pulse arrival time model is more complex. This is discussed, along with the relativistic description of the point mass binary, by Backer and Hellings (1986). Advanced developments of the motions of binary systems in general relativity are by Iyer and Will (1995), Jaranowski (1997), and Jaranowski and Shafer (1997).

Results of the applications of the above methods to pulsar observations are given, as examples, in Taylor and Weisberg (1989), Ryba and Taylor (1991a,b), and Kaspi *et al.* (1994).

14.8 Extrasolar planets

The discovery of planets around stars other than our Sun has been a goal for a long time. There have been many different methods pursued for this purpose. The efforts divide as a function of mass and period of the planet, with the size of the perturbations being a function of the planetary mass and the length of observations required being a function of the planetary period. The first planetary system discovered outside our Solar System was found around a neutron star, the 6.2 ms pulsar PSR 1257 + 12 (Wolszczan and Frail, 1992). Other planets have been detected by means of the measurement of variations in the radial velocities of stars. The astrometric measurement of periodic motions of stars has, to date, detected double stars, but not bodies small enough to be called planets. The photometric detection of the reduction of light from a star owing to the transit of a planet in front of the star is another means of detecting planets. Microlensing is a means of detecting more-distant planets. Another, more difficult, means of finding planets is the direct observation of the planet next to a star. The large difference in the light from the two bodies requires that some technique be used to reduce the light from the star, but not from the nearby planet. Techniques such as optical nulling are required.

After many false claims in the past, current surveys are indicating that 5% of main-sequence stars have companions of 0.5–8 Jupiter masses within 3 AU. The future for planetary discovery using the techniques below should be very bright.

14.8.1 Doppler technique

A star wobbles around the system barycenter as a result of the perturbations of planets in the system. For the Sun, the wobble due to Jupiter is 12.5 m/s, and that due to Saturn is 2.7 m/s. The general expression for the semi-amplitude, K, of the wobble of the stellar radial velocity due to an orbiting companion is given by Equation (12.29b):

$$K = \frac{na \sin i}{\sqrt{1 - e^2}}, \tag{14.43}$$

where a is the semi-major axis, n is the mean motion, e is the eccentricity, of the relative orbit of the companion about the primary, and i is the inclination of the orbit to the plane of measurement of K. Introducing the orbital period $P = 2\pi/n$ and Kepler's third law,

$$P^2 = \frac{4\pi^2 a^3}{G(M_\star + m_p)},$$

where m_p is the mass of the planet and $M_\star$ is the mass of the primary, one gets:

$$K = \left(\frac{2\pi G}{P}\right)^{1/3} \frac{m_p \sin i}{(M_\star + m_p)^{2/3}} \frac{1}{\sqrt{1 - e^2}}. \tag{14.44}$$

Thus, a Jupiter mass planet at 1 AU causes an amplitude of $28.0 \sin i$ m/s, which is now easily detectable. For a given amplitude, the detectability depends on the Doppler precision, the number of orbital cycles, and the number of observations per cycle. For orbital periods much shorter than the duration of observations, the use of Fourier analysis permits the detection of velocity amplitudes comparable to the Doppler errors. For orbital periods comparable to the observational history, detection requires amplitudes about four times the errors. As an example, Fig. 14.4 gives the graph of radial velocities that allowed Mayor and Queloz (1995) to report the first discovery of such an object around 51 Peg. The distance of 51 Peg as determined by Hipparcos is 13.5 pc. The discovered companion is a Jupiter-like planet. Because the inclination i could not be determined, one obtained only

$$M \cos i = 0.47 \text{ Jupiter masses.}$$

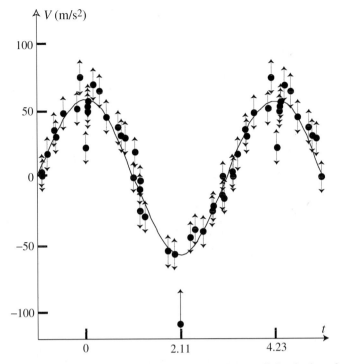

Fig. 14.4. Observations during one year of the radial velocity of 51 Peg with their uncertainties that led to the discovery of its planetary companion. Time in days is folded to the period of the system.

The period is 4.23 days, which means that the distance between the star and the planet is of the order of 0.09 AU, that is much closer than Mercury to the Sun. Now, every year, many new similar systems of a star accompanied by one or more close Jupiter-like planets are discovered by the Doppler method. In August 2000, at the General Assembly of the IAU, 48 extra-solar planets were known, and this number increases every month. In particular, three multi-planet systems had already been detected by that date.

The instrument designs, observational techniques, analysis methods, results, planet properties, and implications for Solar System formation are given in Serkowski (1976), Mayor and Queloz (1995), Marcy and Butler (1992, 1993, 1998, 1999, 2000), and Cummings, Marcy, and Butler (1999).

14.8.2 Astrometric technique

The angular wobble of a star due to a companion is proportional to both planet mass, m_p, and orbital radius, r in AU, and inversely proportional to the distance to the star, D in parsecs, and its mass, $M_\star$. The masses are in solar masses. The amplitude of the astrometric signal, θ, in arcseconds is given by

$$\theta = \left(\frac{m_p}{M_\star}\right)\left(\frac{r}{D}\right) = \left(\frac{m_p}{D}\right)\left(\frac{P}{M_\star}\right)^{2/3}.$$

(14.45)

Gatewood has formulated this, for stars of known parallax and mass, into a detection index, I, which is given as

$$I = \frac{A}{m\,P^{2/3}} = \frac{\varpi}{M^{2/3}},$$

(14.46)

where ϖ is the stellar parallax, m and M are masses of the companion and primary respectively, and A is the semi-major axis of the stellar motion.

The observational equation (14.1) is the basis for the astrometric technique. After the observations are corrected for observational effects, such as aberration, refraction, and nutation as discussed in Chapters 6 and 8, the position, proper motion and parallax constants can be determined by least squares. Then the problem is detecting the planetary induced perturbations in the presence of the noise of observational errors. There are some possible practical problems separating the linear perturbation effect from the proper motion term and the perturbations from the annual parallax terms, if the periods are close.

To achieve the accuracy required for planetary detection by the astrometric technique, photoelectric techniques have been introduced into astrometry. The approaches divide into images on the detector surface, modulating the light before it reaches the detector's surface, and interferometry. Gatewood (1987) has built the multichannel astrometric photometer with a Ronchi ruling and applied it on ground-based telescopes. Narrow-angle astrometry with long-baseline infrared interferometry can exploit the atmospheric turbulence properties over fields smaller than the baseline, and use phase referencing with a reference star in the isoplanatic patch. Using baselines > 100 m to minimize the photon-noise errors, and controlling systematic errors using laser metrology, accuracies of tens of microarcseconds per hour should be possible. The better approach is to observe from above the atmosphere, therefore missions such as FAME, SIM, and GAIA are well adapted for such investigation.

At present, only some heavy brown dwarfs have been discovered by this technique, precisions of the order of a milliarcsecond being insufficient to recognize slight nonlinearities of the proper motion of a star owing to the shift produced by the presence of small mass objects. Figure 14.5 illustrates this by showing what could be the apparent motion of the Sun as seen at 100 pc perpendicularly to the plane of the ecliptic. This motion is essentially governed by the motion of the outer planets and would be observable only if accuracies of the order of 10 μas were available. In addition, in order to detect the Earth or Venus, one should be able to recognize small periodic perturbations with periods of 1 and 0.6 years respectively

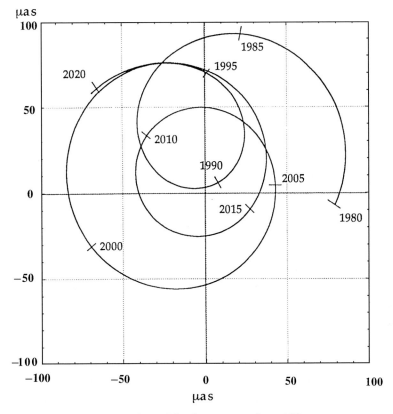

Fig. 14.5. Apparent motion of the Sun as seen from 100 pc.

and amplitudes smaller by a factor of about 2000 than the global amplitude shown in figure. So the recognition of terrestrial planets might be only marginally possible for the closest stars with microarcsecond astrometry.

Reviews of the astrometric approach to planet detection have been written by Black and Scargle (1982), Gatewood (1976, 1987), Gatewood *et al.* (1980), Colavita and Shao (1994), and Black (1995). The astrometric technique determines the mass and orbital inclination of a planet (see Sections 12.4.2 and 12.8).

14.8.3 Photometric technique

Planets may be detected as they transit in front of a star. The reduction in light from the star is the cross-sectional area of the planet multiplied by the surface brightness of the disk of the star. A Jupiter with a radius of 1 R_j will cause a solar type star to dim by 1% with a duration of hours, depending on the orbital radius (Hale and Doyle, 1994). Earth-size planets would cause 0.01% dimming. The dimming is distinguished in character as the transit includes a limb-darkened ingress and egress with a flat-bottomed mid-period. For a transit to occur, the orbital inclination, i, must be close to 90°, $\tan i > a/R$, where a is the planet semi-major axis and R is the star's radius. The probability, P, of transits, for randomly oriented orbital planes, is based on i between 90° and i', and is

$$P(i' \text{ to } 90°) = \cos i'.$$

If 4.7% of Jupiter-like satellites at 0.1 AU will transit and 4% of solar-type stars have such satellites (Marcy and Butler, 1999), then 0.19% of solar-type stars should exhibit transits. This estimate is subject to uncertainty owing to the small numbers involved. The detection of planetary transits can indicate the existence of planets which can be further observed. The radius of the planet can be determined from the dimming, the mass from Doppler measures, and thus the density can be computed. This technique can be pursued from the ground for Jupiter and larger-sized objects and from space missions, such as COROT and KEPLER, for Earth-size planets (Koch *et al.*, 1999; Léger *et al.*, 1999). There is now one example of a planet, previously discovered by Doppler technique, which was also recorded photometrically (Charbonneau *et al.*, 2000).

14.8.4 Microlensing observations

Gravitational microlensing of stars in the galactic bulge can reveal the presence of planets in orbit around the intervening lensing object (Section 6.4.4). Follow-up photometry of microlensing events can reveal the perturbations on the microlensing light curve by an attendant planet (Peale, 1997; Safizadeh *et al.*, 1999). Microlensing is most sensitive to planets that are about an Einstein radius from the lensing star. This is about 3–6 AU for a typical galactic bulge microlensing event. The duration of the perturbation on the light curve is proportional to the square root of the mass of the planet. The unique ability of microlensing is its ability for ground-based observations to detect Earth-mass planets with several astronomical units semi-major axis orbits around main-sequence stars. This procedure would yield statistics on such planets, but would be hard to follow-up because of the target's large distance and the uncertainties in identifying the lensing objects.

14.8.5 Direct detections

Direct detection of planets is difficult due to the small fluxes from the planets and the wings of the point spread functions (psf) of the star. In visible wavelengths, reflected light from a Jupiter will have a brightness of $\sim 10^{-9}$ of the star, while in the thermal infrared the contrast would be $\sim 10^{-4}$. The biggest challenges to detecting planets become the atmospheric seeing and exozodiacal light. Coronographic adaptive optics reduce seeing effects. The dark regions between speckles of a stellar image may enable planet detection (Labeyrie, 1995). Ground-based interferometric imaging in the infrared in methane absorption bands, which appear in Jovian atmospheres, may be able to reveal gas-giant planets. Earth-like planets will require space-borne interferometers that can eliminate the glare from the star by nulling, or destructive optical interference (Beichman, 1998; Angel and Woolf, 1997). Both ESA and NASA are developing missions for this purpose.

14.8.6 Conclusion

Comparing the methods presented, one may say that the spectroscopic method has the considerable advantage of being independent of the distance of the star. Its limitation lies in the uncertainties of the determination of radial velocities which cannot be indefinitely improved because of the intrinsic width and shape of the spectral lines. On the other hand, astrometry has still a large range of improvement as described in Section 2.6, and should provide in the future most of the data on extrasolar planetary systems. Finally, surveys using photometric techniques have good prospects for discovering new systems, but the quantitative investigations will have to be performed by spectroscopy and astrometry.

Appendix A

Examples

The examples in this appendix are given to provide specific examples of astronomical computations, to show the differences between old and new reference systems, and as verifications of computational methods. The sources of equations and detailed descriptions of the concepts from this book are cited. The textual material is not repeated in this appendix.

Example 1 Stellar angle and sidereal time

This is to illustrate the difference between the use of the equinox, the CEO and GMST, and the stellar angle. The stellar angle is described in Section 10.4 and the equation is (10.23).

For November 7, 2002, at 08.00 UT1, the Julian Date is 2 452 585.833 33.... The time difference T_u in Julian centuries from J2000.0 ($JD = 2\,451\,545.0$) is:

$$T_u = (2\,452\,585.833\,333\ldots - 2\,451\,545.0) = 1040.833\,333\ldots.$$

The stellar angle θ, expressed in radians, is, following (10.19):

$$\theta(T_u) = 2\pi(0.779\,057\,273\,2640 + 1.002\,737\,811\,911\,354\,48\,T_u)$$
$$= [2\pi(1044)] + 2\pi(0.461\,996\,504\,10)$$
$$= 166°.318\,6212$$
$$= 11^h\,5^m\,16\overset{s}{.}469.$$

GMST is discussed in Sections 10.5 and 8.7 and the equation is (10.21). The GMST is given by computing first $\tau_0 = $ GMST1 of 0^h UT1.

$$\tau_0 = 24\,110\overset{s}{.}548\,41 + 8\,640\,184\overset{s}{.}812\,866\,T'_u + 0\overset{s}{.}093\,104\,T'^2_u - 6.2 \times 10^{-6}T'^3_u\,s,$$

where,

$$T'_u = (2\,452\,585.5 - 2\,451\,545.0)/36\,525 = 0.028\,487\,3374,$$

so that,

$$\tau_0 = 270\,246\overset{s}{.}4288 = 3^d3^h4^m6\overset{s}{.}408\,79,$$

to which, one must add 8 hours at sidereal rate of $1.002\,737\,909\,350\,795 = 8^h1^m18\overset{s}{.}851\,77$.

Let us now apply the correction d$GMST$ given by the equation (10.22)

$$\mathrm{d}GMST = (-18330.266T'_u + 7.87T'^2_u + 1.27T'^4_u)\mu s.$$

One gets d$GMST = -522.17$ μs. So, GMST at $8\overset{h}{.}0$ UT1 on November 7, 2002, is $11^h5^m25\overset{s}{.}2606$.

Example 2 Apparent geocentric positions of planets from ICRF positions

This is to illustrate the calculation of apparent geocentric positions, the difference in planetary positions due to the difference between the equinox and the Geocentric Celestial Reference System, and the time transformation from TCB to TCG.

The ICRF position of Saturn for November 7, 2002, $JD = 2\,452\,585.5$, in equatorial rectangular coordinates in AU at 08.00 hour TCB is to be converted to apparent geocentric position in TCG in the old and new systems.

2a Time transformation

The time conversion to Geocentric Coordinate Time (TCG) from Barycentric Coordinate Time (TCB) is discussed in Section 5.5.1, Equations (5.40) and (5.41). There are secular and periodic terms, the principal periodic term is given below for the example. The difference, in seconds, is calculated from:

$$(\mathrm{TCB} - \mathrm{TCG})_{\text{secular}} = L_c(JD - 2\,443\,144.5) \times 86\,400$$
$$= 1.480\,826\,8457 \times 10^{-8}(2\,452\,585.833$$
$$-2\,451\,545.5) \times 86\,400$$
$$= 1.331\,038\,248\,\text{s},$$
$$(\mathrm{TCB} - \mathrm{TCG})_{\text{periodic}} = 0.001\,658\sin g + 0.000\,014\sin 2g,$$

where

$$g = 357\overset{\circ}{.}53 + 0\overset{\circ}{.}985\,003(JD - 2\,451\,545.0)$$
$$= 357\overset{\circ}{.}53 + 0\overset{\circ}{.}985\,003(2\,452\,585.833 - 2\,451\,545.0)$$
$$= 302\overset{\circ}{.}26.$$

So,

$$(\text{TCB} - \text{TCG})_{\text{periodic}} = -0.001\,415\,\text{s} \quad \text{and} \quad (\text{TCB} - \text{TCG})_{\text{total}} = 1.329\,624\,\text{s}$$

2b Barycentric positions

(1) Determine the barycentric vectors to the Earth and Saturn and the geocentric vector to Saturn. The ICRF position, $\mathbf{E}$, and velocity of the Earth, $\dot{\mathbf{E}}$, at $JD2\,452\,585.8333$ TCB in AU and AU per day are

$$\mathbf{E} = \begin{pmatrix} 0.705\,352\,335 \\ 0.635\,044\,686 \\ 0.275\,337\,634 \end{pmatrix} \quad \text{and} \quad \dot{\mathbf{E}} = \begin{pmatrix} -0.012\,381\,324 \\ 0.011\,164\,204 \\ 0.004\,840\,671 \end{pmatrix}.$$

The ICRF position, $\mathbf{Q}$, and velocity, $\dot{\mathbf{Q}}$, of Saturn at $JD2\,452\,585.8333$ TCB in AU and AU per day are

$$\mathbf{Q} = \begin{pmatrix} 0.937\,084\,026 \\ 8.317\,780\,581 \\ 3.395\,220\,616 \end{pmatrix} \quad \text{and} \quad \dot{\mathbf{Q}} = \begin{pmatrix} -0.005\,845\,312 \\ 0.000\,431\,021 \\ 0.000\,429\,677 \end{pmatrix}.$$

(2) Determine iteratively the barycentric position, $\mathbf{Q}$, and the geocentric position, $\mathbf{P}$, of Saturn at time $(t - \tau)$, where τ is the light time, so the light emitted from Saturn at $(t - \tau)$ arrives at the Earth at time t. After iteration, the positions of Saturn are:

$$\text{barycentric } \mathbf{Q} = \begin{pmatrix} 0.937\,364\,072 \\ 8.317\,759\,931 \\ 3.395\,200\,030 \end{pmatrix}, \quad \text{geocentric } \mathbf{P} = \begin{pmatrix} 0.232\,011\,737 \\ 7.682\,715\,245 \\ 3.119\,862\,396 \end{pmatrix},$$

where light time $= 4\,139.378\,81\,\text{s} = 0.047\,909\,477\,\text{d}$.

The unit vectors corresponding to the above vectors are:

$$\mathbf{e} = \begin{pmatrix} 0.713\,747\,382 \\ 0.642\,602\,937 \\ 0.278\,614\,680 \end{pmatrix} \quad \mathbf{q} = \begin{pmatrix} 0.103\,742\,588 \\ 0.920\,843\,875 \\ 0.375\,877\,685 \end{pmatrix} \quad \mathbf{p} = \begin{pmatrix} 0.027\,969\,164 \\ 0.926\,156\,275 \\ 0.376\,101\,423 \end{pmatrix}.$$

2c Correct for light deflection

Calculate the geocentric direction $\mathbf{p}_1$ of the planet, correcting for light deflection, $\Delta\mathbf{p}$, as discussed in Section 6.4.3, from

$$\Delta\mathbf{p} = \left(\frac{2\mu}{c^2 E}\right)\left(\frac{(\mathbf{p} \cdot \mathbf{q})\mathbf{e} - (\mathbf{e} \cdot \mathbf{p})\mathbf{q}}{1 + \mathbf{q} \cdot \mathbf{e}}\right).$$

The vector $\mathbf{p}_1 = \mathbf{p} + \Delta\mathbf{p}$ is a unit vector to order μ/c^2. The scalar products are

$$\mathbf{p} \cdot \mathbf{q} = \cos\zeta = 0.997\,115\,058, \quad \mathbf{e} \cdot \mathbf{p} = -\cos\psi = 0.719\,901\,038,$$
$$\text{and} \quad \mathbf{q} \cdot \mathbf{e} = \cos\lambda = 0.770\,508\,020.$$

In this case, the Earth and Saturn are in the same side with respect to the Sun, the deflection is very small. So, to allow comparisons, we give $\Delta\mathbf{p}$ with more decimals than necessary:

$$\Delta\mathbf{p} = \begin{pmatrix} 7.187 \times 10^{-9} \\ -0.250 \times 10^{-9} \\ 0.081 \times 10^{-9} \end{pmatrix} \quad \text{and} \quad \mathbf{p}_1 = \begin{pmatrix} 0.027\,969\,171 \\ 0.926\,153\,275 \\ 0.376\,101\,423 \end{pmatrix}.$$

2d Correct for aberration

Calculate the planetary direction in a geocentric fixed frame that is moving with the instantaneous velocity of the Earth as discussed in Section 6.3, from

$$\mathbf{p}_2 = \frac{\gamma^{-1}\mathbf{p}_1 + (1 + (\mathbf{p}_1 \mathbf{V}))\,\mathbf{V}/(1 + \gamma^{-1})}{1 + \mathbf{p}_1 \cdot \mathbf{V}}.$$

In this formulation, $\mathbf{V} = \dot{\mathbf{E}}/c$, $\gamma = (1 - V^2)^{-1/2}$; the velocity is in units of the velocity of light. So

$$\mathbf{V} = \begin{pmatrix} -7.150\,856\,36 \times 10^{-5} \\ 6.447\,906\,47 \times 10^{-5} \\ 2.795\,738\,27 \times 10^{-5} \end{pmatrix}$$

$$\mathbf{p}_1 \cdot \mathbf{V} = 6.823\,465 \times 10^{-5}$$

$$\text{and} \quad \mathbf{p}_2 = \begin{pmatrix} 0.027\,937\,465 \\ 0.926\,093\,292 \\ 0.376\,077\,142 \end{pmatrix}.$$

2e Precession–nutation correction (equinox system)

Apply precession and nutation as discussed in Section 8.4 by multiplying by the rotation matrix $\mathcal{R}$ as given in the *Astronomical Almanac* to obtain the apparent direction $\mathbf{p}_3$ from

$$\mathbf{p}_3 = \mathcal{R} \times \mathbf{p}_2.$$

$$\text{The matrix} \quad \mathcal{R} = \begin{pmatrix} 0.999\,999\,81 & -0.000\,559\,21 & -0.000\,243\,08 \\ 0.000\,559\,21 & 0.999\,999\,84 & -0.000\,015\,27 \\ 0.000\,243\,08 & 0.000\,015\,13 & 0.999\,999\,97 \end{pmatrix}.$$

Then,

$$\mathbf{p_3} = \mathcal{R} \times \mathbf{p_2} = \begin{pmatrix} 0.027\,328\,161 \\ 0.926\,103\,024 \\ 0.376\,097\,933 \end{pmatrix}.$$

2f Precession–nutation correction (new system)

Apply precession and nutation as discussed in Section 10.4.1 by multiplying by the precession–nutation matrix $\mathcal{PN}$ described in Section 8.7.2 to obtain the apparent direction $\mathbf{p_3'}$ from

$$\mathbf{p_3'} = \mathcal{PN} \times \mathbf{p_2}.$$

The matrix $\mathcal{PN} = \begin{pmatrix} 0.999\,999\,970\,483 & -0.000\,000\,014\,937 & 0.000\,242\,967\,915 \\ 0.000\,000\,011\,267 & 0.999\,999\,999\,886 & 0.000\,015\,105\,620 \\ -0.000\,242\,967\,915 & -0.000\,015\,105\,617 & 0.999\,999\,970\,369 \end{pmatrix}.$

Then,

$$\mathbf{p_3'} = \mathcal{PN} \times \mathbf{p_2} = \begin{pmatrix} 0.027\,328\,157 \\ 0.926\,103\,024 \\ 0.376\,118\,725 \end{pmatrix}.$$

2g Convert to spherical coordinates

Calculate the right ascensions and declinations from vectors $\mathbf{p_3}$. The results, for November 7, 2002, at $7^h59^m58\overset{s}{.}670\,376$ TCG, are

equinox system: $\alpha = 5^h53^m14\overset{s}{.}3430$, $\delta = 22° 5' 37\overset{''}{.}600$,

new system: $\alpha = 5^h53^m6\overset{s}{.}6525$, $\delta = 22° 5' 37\overset{''}{.}594$.

The difference in δ corresponds to the difference between the nutation theories used. The difference in α corresponds to the motion of the equinox with respect to the CEO.

Example 3 Astrometric positions of reference stars

This example illustrates how reference stars would be computed for a CCD observation of a program object at some observational date.

The observational program star is being observed on November 7, 2002, $JD2\,452\,585.5$, at 08.00 TT. Reference stars, three for the example, have been selected from the *Hipparcos Catalogue*, and their data are the following.

Number	Position	Parallax ϖ (mas)	Proper motion (mas/year) μ_α	μ_δ
84392	$\alpha_1 = 258°806\,244\,73$ $\delta_1 = +30°301\,251\,12$	2.96	1.84	13.98
84341	$\alpha_2 = 258°641\,307\,01$ $\delta_2 = +30°955\,792\,99$	9.84	−21.26	42.82
84733	$\alpha_3 = 259°765\,901\,84$ $\delta_3 = +30°905\,257\,83$	3.61	−16.19	5.59

The positions are on the ICRF, but their epochs are those of the *Hipparcos Catalogue*, 1991.25, $J\,D2\,448\,349.0625$ TT. For astrometric positions, the proper motions from epoch to observed time and the parallax corrections must be calculated and applied. The star positions dependent on time are given in Equation (14.1) as

$$\alpha_i = \alpha_{i0} + \mu_{\alpha i}t_j + \varpi_i P_j,$$
$$\delta_i = \delta_{i0} + \mu_{\delta i}t_j + \varpi_i Q_j,$$

where the terms for nonlinear motion and observational errors have been omitted.
    The proper motion corrections are to be computed with

$$t = 2\,452\,585.8333 - 2\,448\,349.0625 = 4236.7708 \text{ days} = 11.60 \text{ years},$$

$$
\begin{aligned}
\mu_{\alpha 1}t_j &= & 1.84 \times 11.60 & = 21.34 \text{ mas} \\
\mu_{\alpha 2}t_j &= & -21.26 \times 11.60 & = -246.62 \\
\mu_{\alpha 3}t_j &= & -16.19 \times 11.60 & = -187.80 \\
\mu_{\delta i}t_j &= & 13.98 \times 11.60 & = 162.17 \\
\mu_{\delta i}t_j &= & 42.82 \times 11.60 & = 496.71 \\
\mu_{\delta i}t_j &= & 5.59 \times 11.60 & = 64.84.
\end{aligned}
$$

The stellar parallax corrections are given in (6.23) as

$$\Delta\alpha \cos\delta = \varpi(x\sin\alpha - y\cos\alpha),$$
$$\Delta\delta = \varpi[(x\cos\alpha + y\sin\alpha)\sin\delta - z\cos\delta],$$

where x, y, z are the barycentric coordinates of the Earth which are for the date above

$$x = +0.709\,467\,924$$
$$y = +0.631\,312\,038$$
$$z = +0.273\,719\,198.$$

Then

$$\Delta\alpha_1 \cos \delta_1 = -1.70 \,\text{mas}, \quad \Delta\delta_1 = -1.83 \,\text{mas}$$
$$\Delta\alpha_2 \cos \delta_2 = -5.62 \,\text{mas}, \quad \Delta\delta_2 = -6.15 \,\text{mas}$$
$$\Delta\alpha_3 \cos \delta_3 = -2.16 \,\text{mas}, \quad \Delta\delta_3 = -2.23 \,\text{mas}.$$

The resulting positions for $JD2\,452\,585.833$ are:

for 84392 $\alpha_1 = 258°.806\,238\,26$ $\delta_1 = +30°.301\,205\,66$
84341 $\alpha_2 = 258°.641\,373\,69$ $\delta_2 = +30°.955\,653\,31$
84722 $\alpha_3 = 259°.765\,953\,32$ $\delta_3 = +30°.905\,239\,20.$

These are astrometric positions in the classical sense. Additional differential corrections for refraction, color, etc., can be introduced to reduce the size of the plate constants for the program object solution.

These positions can then be projected onto the plate, or CCD, by the gnomonic projection given in (3.52). Let (ξ, η) be rectangular Cartesian coordinates measured from the position of (α_0, δ_0). Then, setting $\Delta\alpha = \alpha_i - \alpha_0$,

$$\xi = \frac{\cos \delta \sin \Delta\alpha}{\sin \delta_0 \sin \delta + \cos \delta_0 \cos \delta \cos \Delta\alpha},$$

$$\eta = \frac{\cos \delta_0 \sin \delta - \sin \delta_0 \cos \delta \cos \Delta\alpha}{\sin \delta_0 \sin \delta + \cos \delta_0 \cos \delta \cos \Delta\alpha}.$$

If $\alpha_0 = 258°.7$ and $\delta_0 = 30°.5$, then

$$\xi_1 = 0.001\,600\,907 \quad \text{and} \quad \eta_1 = -0.003\,468\,882$$
$$\xi_2 = -0.000\,877\,504 \quad \eta_2 = 0.007\,953\,049$$
$$\xi_3 = 0.015\,964\,427 \quad \eta_3 = 0.007\,149\,155.$$

There is the alternate approach which is the use of space coordinates and motion matrices, as discussed in Section 7.8, rather than spherical coordinates. This is particularly applicable for positions near the pole, where accuracy can be lost.

Taking two stars from the *Hipparcos Catalogue* in the same region of the sky, but near the pole to which we have added hypothetical radial velocities.

Number	Position	Parallax π (mas)	Proper motion (mas/year)		Radial velocity (km/s)
			μ_α	μ_δ	
84525	$\alpha_1 = 259°.183\ 322\ 80$ $\delta_1 = +83°.700\ 273\ 94$	2.88	-7.85	22.31	30
84535	$\alpha_2 = 259°.238\ 585\ 72$ $\delta_2 = +89°.037\ 715\ 46$	3.72	-23.03	-3.07	40

Their position vectors, **r**, can be determined at the Hipparcos epoch, t, from Equation (7.7) as

$$\mathbf{r} = \begin{pmatrix} r\cos\delta\cos\alpha \\ r\cos\delta\sin\alpha \\ r\sin\delta \end{pmatrix}.$$

Note that r is the barycentric distance to the star, which can be computed from the parallax, ϖ, of the star. The additional problem here is the choice of the unit of length.

(1) If we take the astronomical unit (AU), then

$$r = 1/\sin\varpi,$$

by the more sophisticated approach described in Section 11.2.

So, for the stars above, the position vectors are

$$\mathbf{r}_1 = \begin{pmatrix} -1\,474\,839 \\ -7\,719\,170 \\ 71\,187\,247 \end{pmatrix} \quad \text{and} \quad \mathbf{r}_2 = \begin{pmatrix} -173\,873 \\ -914\,823 \\ 55\,439\,708 \end{pmatrix}.$$

The space motion vector, $\dot{\mathbf{r}}$ can be given by

$$\dot{\mathbf{r}} = \begin{pmatrix} p_x & q_x & r_x \\ p_y & q_y & r_y \\ p_z & q_z & r_z \end{pmatrix} = \begin{pmatrix} -\sin\alpha & -\sin\delta\cos\alpha & \cos\delta\cos\alpha \\ \cos\alpha & -\sin\delta\sin\alpha & \cos\delta\sin\alpha \\ 0 & \cos\delta & \sin\delta \end{pmatrix} \times \begin{pmatrix} 15sr\mu_\alpha \\ sr\mu_\delta \\ k\dot{r} \end{pmatrix}.$$

The factors $s = 2\pi/(360 \times 3\,600\,000 \times 365.25)$ converts from milliarcseconds per year to radians per day and $k = 86\,400/(1.495\,978\,70 \times 10^8)$ converts from km/s to AU/day.

The values of $\dot{\mathbf{r}}$ for the two stars are

$$\dot{\mathbf{r}}_1 = \begin{pmatrix} -003 & 730 & 584 \\ 0.020 & 236 & 249 \\ 0.019 & 549 & 060 \end{pmatrix} \quad \text{and} \quad \dot{\mathbf{r}}_2 = \begin{pmatrix} 0.017 & 015 & 406 \\ 0.001 & 329 & 326 \\ -0.023 & 136 & 621 \end{pmatrix}.$$

Then the positions of the stars for the epoch $t_1 = JD2\,452\,585.8333$, are determined from

$$\dot{\mathbf{r}}_i = \dot{\mathbf{r}}_{i1} + (t_1 - t)\ddot{\mathbf{r}}_i,$$

and the results are

$$\mathbf{r}_1 = \begin{pmatrix} -1\,474\,855 \\ -7\,719\,085 \\ 71\,187\,330 \end{pmatrix} \quad \text{and} \quad \mathbf{r}_2 = \begin{pmatrix} -173\,945 \\ -914\,816 \\ 55\,439\,610 \end{pmatrix},$$

giving

$$\alpha_1 = 259°\!.183\,092\,294 \quad \text{and} \quad \delta_1 = 83°\!.700\,345\,825$$
$$\alpha_2 = 259°\!.234\,163\,417 \quad \text{and} \quad \delta_2 = 89°\!.037\,705\,565.$$

(2) One may also choose parsecs. In this case, ϖ being expressed in milliarcseconds, one has

$$r = 1/1000\varpi.$$

The coefficient s remains the same, but k is to be divided by $A = 206\,264.806$, which is a parsec expressed in AU. The distances in parsecs are, respectively,

$$r_1 = 347.222\ldots \quad \text{and} \quad r_2 = 268.8172\ldots,$$

and the position vectors are

$$\mathbf{r}_1 = \begin{pmatrix} -7.150\,222\,76 \\ -37.423\,596 \\ 345.125\,514 \end{pmatrix} \quad \text{and} \quad \mathbf{r}_2 = \begin{pmatrix} -0.843\,311 \\ -4.435\,185 \\ 268.778\,817 \end{pmatrix}.$$

The resulting ccordinates at the observation time are of course the same.

Appendix B

Astronomical values

Preliminary remark. The numbers that are given in this appendix should not be considered as reference values. Although the present authors have tried to provide numbers that are as close as possible to what is estimated to be the best values, they do not guarantee either that they are the best possible, or that all are mutually consistent. In several cases, authorities that publish values of astronomical constants do not agree, and the choice of one or the other is necessarily subjective. Furthermore, as time goes on, better values will become available. For these reasons, we do not associate uncertainties to values, and one should consider these lists as providing orders of magnitude of the parameters and not as a basis for accurate and dependable calculations.

IAU system of astronomical constants, best estimates

SI units

The units meter (m), kilogram (kg), and second (s) are the units of length, mass and time in the International System of Units (SI).

Astronomical units

The astronomical unit of time is a time interval of one day (D) of 86 400 seconds. An interval of 36 525 days is one Julian century.

The astronomical unit of mass is the mass of the Sun (S).

The astronomical unit of length (AU) is that length (A) for which the Gaussian gravitational constant (k) takes the value 0.017 202 098 95 when the units of measurement are the astronomical units of length, mass and time. The dimensions of k are those of the constant of gravitation G: $L^3/(M\,T^2)$. The term "unit distance" is also used for the length A.

Defining constants

Gaussian gravitational constant: $k = 0.017\,202\,098\,95$

Speed of light: $c = 299\,792\,458$ m/s

Primary constants

Light-time for unit distance: $\tau_A = 499.004\,783\,8061$ s

Equatorial radius for Earth: $a_e = 6378\,140$ m

IUGG value: $a_e = 6378\,136.6$ m

Dynamical form-factor for Earth: $J_2 = 0.001\,082\,6359$

Geocentric gravitational constant: $GE = 3.986\,004\,418 \times 10^{14}\text{m}^3/\text{s}^2$

Constant of gravitation: $G = 6.673 \times 10^{-11}\text{m}^3/(\text{kg s}^2)$

Ratio of mass of Moon to that of Earth: $\mu = 0.012\,300\,0383$

General precession in longitude: $\rho = 5028''\!.790$

(per Julian century, at epoch J2000.0)

Obliquity of the ecliptic: $\varepsilon = 23° 26' 21''\!.4059$

(at standard epoch J2000.0)

Derived constants

Unit distance : $c\tau_A = A = 1.495\,978\,706\,91 \times 10^{11}$ m

Solar parallax: $\arcsin(a_e/A) = \pi_\odot = 8''\!.794\,144$

Constant of aberration $= \kappa_\odot = 20''\!.495\,52$

(for standard epoch J2000.0)

Flattening factor for the Earth: $f = 0.003\,352\,81 = 1/298.257$

Heliocentric gravitational constant:

$$A^3 k^2/D^2 = GS = 1.327\,124\,420\,76 \times 10^{20}\text{m}^3/\text{s}^2$$

Ratio of mass of Sun to that of the Earth:

$$(GS)/(GE) = S/E = 332\,946.050\,895$$

Ratio of mass of Sun to that of Earth + Moon:

$$(S/E)/(1 + \mu) = 328\,900.561\,400$$

Mass of the Sun: $(GS)/G = S = 1.9891 \times 10^{30}$ kg

Solar System values

System of planetary masses

Ratios of mass of Sun to masses of the planets

Mercury	6 023 600	Jupiter	1 047.3486
Venus	408 523.71	Saturn	3 497.898
Earth + Moon	328 900.561 400	Uranus	22 902.98
Mars	3 098 708	Neptune	19 412.24
		Pluto	135 200 000

Masses of minor planets

N° Minor planet	Mass in solar mass
(1) Ceres	4.7×10^{-10}
(2) Pallas	1.0×10^{-10}
(4) Vesta	1.3×10^{-10}

Masses of satellites

Planet	Satellite	Satellite / Planet
Jupiter	Io	4.70×10^{-5}
	Europa	2.56×10^{-5}
	Ganymede	7.84×10^{-5}
	Callisto	5.6×10^{-5}
Saturn	Titan	2.41×10^{-4}
Neptune	Triton	2×10^{-3}

Equatorial radii in km

Mercury	2 439.76	Jupiter	71 398
Venus	6 052.3	Saturn	60 000
Earth	6 378.137	Uranus	25 400
Mars	3 397.515	Neptune	24 300
		Pluto	2 500
Moon	1 738	Sun	696 000

Orbital elements of planets

Name	Semimajor axis (AU)	Eccentricity	Inclination on ecliptic	Sidereal orbital period in days
Mercury	0.3871	0.206	$7°00$	87.969
Venus	0.7233	0.007	$3°39$	227.701
Earth	1.0000	0.017	–	365.256
Mars	1.5237	0.093	$1°85$	686.980
Jupiter	5.2026	0.048	$1°31$	4332.59
Saturn	9.5547	0.056	$2°48$	10759.2
Uranus	19.2181	0.046	$0°77$	30688.5
Neptune	30.1096	0.009	$1°77$	60182.3
Pluto	39.4387	0.250	$17°14$	90469.7

Gravity fields of planets

Planet	J_2	J_3	J_4
Earth	$+0.001\,082\,6359$	-0.2533×10^{-5}	-0.1616×10^{-5}
Mars	$+0.001\,964$	$+0.36 \times 10^{-4}$	
Jupiter	$+0.014\,75$		-0.58×10^{-3}
Saturn	$+0.016\,45$		-0.10×10^{-2}
Uranus	$+0.012$		
Neptune	$+0.004$		

Mars: $C_{22} = -0.000\,055$, $S_{22} = +0.000\,031$, $S_{31} = +0.000\,026$

Gravity field of the Moon

$\gamma = (B - A)/C = 0.000\,2278$ $\qquad \beta = (C - A)/B = 0.000\,6313$

$C/MR^2 \quad = 0.392$ $\qquad I = 5552''7 = 1° \ 32' \ 32''7$

$C_{20} = -0.000\,2027$ $\quad C_{30} = -0.000\,006$ $\quad C_{32} = +0.000\,0049$

$C_{22} = +0.000\,0225$ $\quad C_{31} = +0.000\,0308$ $\quad S_{32} = +0.000\,00\,17$

$\qquad\qquad\qquad S_{31} = +0.000\,004$ $\qquad C_{33} = +0.000\,0018$

$\qquad\qquad\qquad\qquad\qquad\qquad\qquad S_{33} = -0.000\,0003$

Time scale factors

Average value of $d(TCB)/d((TT) - 1$,　　$L_B = 1.550\,519\,767\,72 \times 10^{-8}$

Average value of $d(TCB)/d(TCG) - 1$,　　$L_C = 1.480\,826\,867\,41 \times 10^{-8}$

Defining value of $d(TCG)/d(TT) - 1$,　　$L_G = 6.969\,290\,134 \times 10^{-10}$

Sun

Radius:	6.96×10^8 m
Semidiameter at mean distance:	$15'59\rlap{.}''63 = 959\rlap{.}''63$
Mass:	1.9891×10^{30} kg
Mean density:	1.41 g/cm^3
Surface gravity:	2.74×10^2 m/s$^2 = 27.9$ gal
Inclination of solar equator to ecliptic:	$7°\ 15'$
Longitude of ascending node: (T in centuries from J2000.0)	$75°\ 46' + 84'T$
Period of synodic rotation ($\phi =$ latitude):	$26.90 + 5.2\sin^2\phi$ days
Period of sidereal rotation adopted for heliographic longitudes:	25.38 days
Motion relative to nearby stars, apex:	$\alpha = 271°,\ \ \delta = +30°$
Motion relative to nearby stars, speed:	1.94×10^4 m/s $= 0.0112$ AU/day

Earth

Rotation of the Earth

Period with respect to fixed stars

Period in mean sidereal time:	$24^{\rm h}\ 00^{\rm m}\ 00\rlap{.}^{\rm s}0084$
Period in Terrestrial Time (TT):	$23^{\rm h}\ 56^{\rm m}\ 04\rlap{.}^{\rm s}0989$
Rate of rotation:	$15\rlap{.}''041\,067\,178\,669\,10$ s^{-1}
	$= 7.292\,115 \times 10^{-5}$ rads^{-1}

Annual rates of precession (T in centuries from J2000.0)

General precession in longitude:	$50\rlap{.}''287\,90 + 0\rlap{.}''022\,2226T$
Luni-solar precession in longitude:	$50\rlap{.}''387\,784 + 0\rlap{.}''004\,9263T$
Planetary precession:	$-0\rlap{.}''018\,8623 - 0\rlap{.}''047\,6128T$

Figure and gravity field of the Earth

Values in the IAU system and best estimates in 2001

	IAU system	Best estimate (2001)
Equatorial radius:	$a = 637\,8140$ m	$637\,8136.6$ m
Dynamical form factor:	$J_2 = 0.001\ 082\,63$	$0.001\ 081\ 6359$
		-2.8×10^{-11} year^{-1}
Flattening:	$f = 1/298.257$	
Polar radius:	$b = 6\,356\,755$ m	$6\,356\,752$ m
Mass of the Earth:	$M_\oplus = 5.9742 \times 10^{24}$kg	5.9742×10^{24} kg
Mean density:	5.52 g/cm	
Normal gravity:	gal $= 9.80621$	
	$-0.02593 \cos 2\phi$	
	$+0.00003 \cos 4\phi$ m/s^2	
Geocentric gravitational		
constant in 10^{14} m^3/s^2:	$GM_\oplus = 3.986\,005$	$3.986\,004\ 18$

Other constants (best values in 2001)

Mean equatorial gravity (g_e): $97.803\ 278$ m/s^2
Potential of the geoid (W_0): $62\,636\,856.0$ m^2/s^2
Mean angular velocity of the Earth (ω): $7.292\ 115 \times 10^{-5}$ rad/s
Geopotential scale factor ($R_0 = GM_\oplus/W_0$): $6\,363\,672.6$ m

For a point on the spheroid of the IAU system at geodetic latitude (ϕ):
1° of latitude: $110.575 + 1.110 \sin^2 \phi$ km
1° of longitude: $(111.320 + 0.373 \sin^2 \phi) \cos \phi$ km
Geodetic latitude (ϕ) – geocentric latitude (ϕ'): $692''\!.74 \sin 2\phi - 1''\!.16 \sin 4\phi$

Orbit of the Earth

	IAU System	Best estimate (2001)
Solar parallax:	$8''\!.794\ 148$	$8''\!.794\ 144$
Constant of aberration:	$20''.495\ 52(J2000.0)$	
Light-time for 1 AU:	$499.004\,782$s	$499.004\ 783\ 8061$s
Astronomical unit of length:	$1.495\ 978\ 70 \times 10^{11}$ m	$1.495\ 978\ 706\ 91 \times 10^{11}$ m
Mass ratio Sun/Earth:	$332\,946.0$	$332\,946.050\ 895$
Mass ratio Sun/Earth + Moon:	$328\,900.5$	$328\,900.561\ 400$
Mass ratio Moon/Earth:	$0.012\,3002$	$0.012\ 300\ 0383$
Mean eccentricity:	$0.016\ 708\ 617$	
Mean obliquity of the ecliptic:	$23°\ 26'\ 21''\!.448$	$23°\ 26'\ 21''\!.4119$
Annual rate of rotation on the ecliptic:	$0''\!.4704$	
Mean distance Earth–Sun:	$1.000\,001\,0178$AU	$1.000\,001\ 057\ 266\ 65$AU
Mean orbital speed:	29.7859 km/s	$29.784\ 766\ 966$ km/s
Mean centripetal acceleration:	$0.005\,94$ m/s^2	$5.930\ 113\ 4387 \times 10^{-3}$ m/s^2

Moon

Physical Moon

Mean radius	1738 km
Semi-diameter at mean distance	15'32''6
Mass	7.3483×10^{22} kg
Mean density	3.34 g/cm^3
Surface gravity	1.62 m/s^2 = 0.17 gal

Orbit of the Moon about the Earth

Sidereal mean motion of Moon:	$2.661\,699\,489 \times 10^{-6}$ rad/s[1]
Mean distance of Moon from Earth:	3.844×10^5 km = 60.27 Earth radii
	= 0.002 570 AU
Equatorial horizontal parallax at mean distance:	57'02''608 = 3422''608
Mean distance of center of Earth from	
Earth–Moon barycenter:	4.671×10^3 km
Mean eccentricity:	0.054 90
Mean inclination to ecliptic:	5°145 396
Mean inclination to lunar equator:	6°41'
Limits of geocentric declination:	±29°
Period of revolution of node:	6798 day
Period of revolution of perigee:	3232 day
Mean orbital speed:	1023 m/s = 0.000 591 AU/day
Mean centripetal acceleration:	0.002 72 m/s^2 = 0.0003 gal

Saros = 223 lunations = 19 passages of Sun through node
Saros = 6585 1/3 days

Glossary

aberration: the angular displacement of the observed position of a celestial object from its geometric position, caused by the finite velocity of light in combination with the motions of the observer and of the observed object.

aberration, annual: the component of stellar aberration resulting from the motion of the Earth about the Sun.

aberration, diurnal: the component of stellar aberration resulting from the observer's diurnal motion about the center of the Earth.

aberration, orbital: in space astrometry, the component of stellar aberration resulting from the spacecraft motion with respect to the center of the Earth.

aberration, planetary: the angular displacement of the observed position of a body of the Solar System produced by motion of the observer and the actual motion of the observed object in the Solar System.

aberration, secular: the component of stellar aberration resulting from the essentially uniform and rectilinear motion of the entire Solar System in space. Secular aberration is usually disregarded.

aberration, stellar: the angular displacement of the observed position of a celestial body resulting from the motion of the observer. Stellar aberration is divided into diurnal or orbital, annual, and secular components.

accuracy: evaluation of the uncertainty of a measurement due to the uncorrected systematic errors. With the precision, it is one of the two components of the uncertainty.

adaptive optics: active optics correctors that adjust the optics to compensate for the wavefront distortions due to changes in the inhomogeneity of the atmosphere as measured from close artificial or bright star measurements.

altitude: the angular distance of a celestial body above or below the horizon, measured along the great circle passing through the body and the zenith. Altitude is 90° minus zenith distance.

apparent place: the geocentric position in the CIRF (or GTRF) of a specific time determined by removing from the directly observed position of a celestial body the effects that depend on the topocentric location of the observer; i.e. refraction, diurnal aberration, geocentric (diurnal) parallax, polar motion, and Earth rotation. Thus, the position at which the object would actually be seen from the center of the Earth, displaced by the light bending and aberration, except the diurnal part (*see* aberration, planetary; aberration, stellar; aberration, diurnal), and referred to the true equator and the CEO or the equinox.

astrometric positions: observed positions with respect to GCRF reference stars at some epoch. Positions have not been corrected for annual or planetary aberration and light bending.

astronomical coordinates: the longitude and latitude of a point on Earth relative to the ITRF and the local vertical. These coordinates are influenced by local gravity anomalies.

astronomical unit (AU): the radius of a circular orbit in which a body of negligible mass, and free of perturbations, would revolve around the Sun in $2\pi/k$ days, where k is the Gaussian gravitational constant. This is slightly less than the semi-major axis of the Earth's orbit. Its best estimate is $1.495\,978\,706\,91 \times 10^8$ m.

azimuth: the angular distance measured clockwise along the horizon from a specified reference point (usually north) to the intersection with the great circle drawn from the zenith through a body on the celestial sphere.

barycenter: the center of mass of a system of bodies; e.g. the center of mass of the Solar System or the Earth–Moon system.

barycentric: with reference to, or pertaining to the barycenter of the Solar System. Otherwise, the mass system should be specified (e.g. the Earth–Moon system).

Barycentric Celestial Reference Frame (BCRF): the realization of the BCRS centered at the barycenter of the Solar System by extragalactic objects (kinematic BCRF) or coordinates and motions of planets (dynamical BCRF); specifically, a barycentric ICRF.

Barycentric Celestial Reference System (BCRS): a global space fixed coordinate system centered at the barycenter of the Solar System, defined by the metric tensor of the IAU 2000 resolutions. More specifically, it is a barycentric International Celestial Reference System (ICRS).

Barycentric Coordinate Time (TCB): the coordinate time for BCRS, a coordinate system at the barycenter of the Solar System, related to Geocentric Coordinate Time (TCG) and Terrestrial Time (TT) by relativistic transformations that include secular terms.

Barycentric Dynamical Time (TDB): an independent argument of ephemerides and equations of motion that are referred to the barycenter of the Solar System. A family of time scales results from the transformation by various theories and metrics of relativistic theories of Terrestrial Time (TT). TDB is a coordinate time, which differs from TT only by periodic variations. Since TT and TDB have the same rates, owing to the relativistic transformation of the space-time metrics, the implied units of distance are different.

Barycentric Ephemeris Time (T_{eph}): an independent time argument of barycentric ephemerides. This time scale is a linear function of TCB. Each ephemeris defines its own version of T_{eph}. The linear drift between T_{eph} and TCB is chosen so that the rates of T_{eph} and TT are as close as possible for the time span covered by the particular ephemeris.

catalog equinox: the intersection of the hour circle of zero right ascension of a star catalog with the celestial equator.

catalog positions: positions on the ICRF.

Celestial Ephemeris Origin (CEO): origin for right ascensions on the true equator of date. Its motion has no component along the true instantaneous equator. Also called Celestial Intermediate Origin (CIO).

Celestial Ephemeris Pole (CEP): the reference pole for nutation and polar motion; the axis of figure for the mean surface of a model Earth in which the free motion has zero amplitude. This pole has no nearly-diurnal nutation with respect to a space-fixed or Earth-fixed coordinate system. (Used from 1984 to 2003 with the IAU 1980 Theory of Nutation.)

celestial equator: the plane perpendicular to the Celestial Ephemeris Pole or the Celestial Intermediate Pole. Colloquially, the projection onto the celestial sphere of the Earth's equator.

celestial intermediate equator, or the true equator: equatorial plane through the center of the Earth and perpendicular to the Celestial Intermediate Pole (CIP) at some epoch.

Celestial Intermediate Origin (CIO): see CEO.

Celestial Intermediate Pole (CIP): Geocentric equatorial pole determined by the IAU 2000A precession–nutation model which provides the transformation from the ICRF to the GCRF.

Celestial Intermediate Reference Frame (CIRF): geocentric reference frame defined by the true equator, CIP, and CEO on a specific date. It is also called Geocentric True Reference Frame (GTRF).

celestial pole: either of the two points projected onto the celestial sphere by the extension of the Earth's axis of rotation to infinity.

celestial sphere: an imaginary sphere of arbitrary radius, upon which celestial bodies may be considered to be located. As circumstances require, the celestial sphere may be centered at the observer, at the Earth's center, or at any other location.

center of figure: that point so situated relative to the apparent figure of a body that any line drawn through it divides the figure into two parts having equal apparent areas. If the body is oddly shaped, the center of figure may lie outside the figure itself.

center of light: same as center of figure except referring to the illumination, each point being weighted by its intensity. It is also called photocenter.

collimation: alignment of an optical system.

coordinate time: the fourth component of the space-time metric in the Theory of Relativity. TCB and TCG are the coordinate times associated with the barycentric and the geocentric coordinate sytems.

Coordinated Universal Time (UTC): the time scale available from broadcast time signals. UTC differs from TAI (see International Atomic Time) by an integral number of seconds; it is maintained within ±0.90 second of UT1 (*see* Universal Time) by the introduction of one second steps (leap seconds).

correlation coefficient: a measure of the dependency between two variables.

covariance: the product of the standard deviations of two variables and their correlation coefficients.

datums: local geodetic coordinate systems.

declination: angular distance north or south of the celestial equator. It is measured along the hour circle passing through the celestial object. Declination is usually given in combination with right ascension or hour angle.

deflection of light: the angle by which the path of a photon is altered from a straight line by the gravitational field of a massive body. The path is deflected radially towards the mass. In the case of the Sun, it amounts up to $1''\!.75$ at the Sun's limb. Correction for this effect, which is independent of wavelength, is included in the reduction from CIRF to BCRF.

deflection of the vertical: the angle between the astronomical vertical, perpendicular to the local equipotential surface, and the geodetic vertical, perpendicular to the reference terrestrial spheroid.

delta T, ΔT: the difference between Terrestrial Time and Universal Time; specifically the difference between Terrestrial Time TT and UT1:

$$\Delta T = \text{TT} - \text{UT1}.$$

delta UT1, ΔUT1 (or ΔUT): the predicted value of the difference between UT1 (UT) and UTC, transmitted in code on broadcast time signals:

$$\Delta \text{UT1} = \text{UT1} - \text{UTC}.$$

diurnal motion: the apparent daily motion caused by the Earth's rotation, of celestial bodies across the sky from East to West.

dynamical equinox: the ascending node of the Earth's mean orbit on the Earth's true equator; i.e. the intersection of the ecliptic with the celestial equator at which the Sun's declination is changing from South to North.

dynamical time: the family of time scales introduced in 1984 to replace ephemeris time as the independent argument of dynamical theories and ephemerides. Examples: Barycentric Dynamical Time and Terrestrial Time.

Earth orientation: the Earth's orientation with respect to ICRF. It is the combination of the Earth's rotation around the polar axis, the polar motion, and the precession–nutation.

Earth potential model: list of values of spherical harmonics that specify the potential of the Earth.

Earth rotation angle: another name for the stellar angle.

eclipse: the obscuration of a celestial body caused by its passage through the shadow cast by another body.

ecliptic: mean plane of the Earth's orbit around the Sun as determined from and designated by an ephemeris of the Earth (e.g. ecliptic of DE 405).

ecliptic longitudes and latitudes: coordinates measured from the equinox with respect to the ecliptic, which can be of date, mean, or conventionally fixed. They can be apparent, true, or mean, geocentric or barycentric.

Einstein circle or ring: the circle or ring image of a star formed when a star and a gravitational lens are aligned so the light is deviated by the same amount in all directions.

Einstein distance: the angle between the true direction to the source of the light and the direction where the light is seen owing to the deviation by a gravitational lens.

Einstein radius: radius of the Einstein ring.

ephemeris hour angle: an hour angle referred to the ephemeris meridian.

ephemeris longitude: terrestrial longitude measured eastward from the ephemeris meridian.

ephemeris meridian: a fictitious meridian that rotates independently of the Earth at the uniform rate implicitly defined by Terrestrial Time (TT). The ephemeris meridian is $1.002\,738\Delta T$ east of the International (Greenwich) meridian, where $\Delta T = \mathrm{TT} - \mathrm{UT1}$.

Ephemeris Time (ET): the time scale used prior to 1984 as the independent variable in gravitational theories of the Solar System. In 1984, ET was replaced by dynamical time.

ephemeris transit: the passage of a celestial body or point across the ephemeris meridian.

epoch: an arbitrary fixed instant of time or date used as a chronological reference datum for calendars, celestial reference systems, star catalogs, or orbital motions.

equation of the equinoxes: the right ascension of the mean equinox referred to the true equator and equinox; alternatively apparent sidereal time minus mean sidereal time.

equation of time: the hour angle of the true Sun minus the hour angle of the fictitious mean sun; alternatively, apparent solar time minus mean solar time.

equator: the great circle on the surface of a body formed by the intersection of the surface with the plane passing through the center of the body perpendicular to the axis of rotation.

equinox: either of the two points at which the ecliptic intersects the celestial intermediate equator; also the time at which the Sun passes through either of these
intersection points; i.e. when the apparent longitude of the Sun is $0°$ or $180°$.
When required, the equinox can be designated by the ephemeris of the Earth
from which it is obtained (e.g. vernal equinox of DE 405).

equinox right ascensions: right ascensions that are measured from the equinox,
instead of from the CEO or CIO.

fixed ecliptic: ecliptic for any ephemeris at epoch J2000.0. The ecliptic has a
specified offset from the ICRS origin and the obliquity has a determined value.

Fizeau interferometer: interferometer whose two entry pupils are parts of a single
telescope, not necessarily a complete mirror. The interferences are uniform all
over the field of view of the instrument.

free core nutation: component of the nutation due to the free wobble of the Earth's
core. It is not modeled in the IAU 2000 precession–nutation formula and is provided by the IERS.

galactic coordinates: system of coordinates (galactic longitude and latitude) whose
principal plane is the mean plane of the Galaxy and the origin near the galactic
center. It is defined by its position in the ICRF.

Gaussian gravitational constant ($k = 0.017\,202\,098\,95$): the constant defining
the astronomical system of units of length (astronomical unit), mass (solar mass)
and time (day), by means of Kepler's third law. The dimensions of k^2 are those
of Newton's constant of gravitation: $L^3/(M^1T^2)$.

geocentric: with reference to, or pertaining to, the center of the Earth.

geocentric apparent right ascensions and declinations: positions measured in
the GTRF of a specific time. The topocentric positions have been corrected for
diurnal aberration and parallax, polar motion, UT1 irregularities, and TCG –
TT. Geocentric apparent right ascensions and declinations may be thought as
equivalent to previous designations of *apparent right ascensions and declinations*.

Geocentric Celestial Reference Frame (GCRF): the transformation of the
Barycentric Celestial Reference Frame (BCRF) to the geocenter. It is also a
realization of the GCRS, by extragalactic objects. More specifically, geocentric
ICRF.

Geocentric Celestial Reference System (GCRS): a global direction fixed coordinate system centered at the geocenter, defined by the metric tensor of the IAU
2000 resolutions. It is defined such that its spatial coordinates are kinematically
non-rotating with respect to the barycentric ones.

geocentric coordinates: the latitude and longitude or rectangular coordinates of
a point on the Earth's surface relative to the center of the Earth. Also, celestial
coordinates given with respect to the center of the Earth.

Geocentric Coordinate Time (TCG): coordinate time for GCRS a coordinate system at the geocenter, related to Terrestrial Time (TT) by a relativistic linear transformation.

Geocentric True Reference Frame (GTRF): another name for the Celestial Intermediate Reference Frame (CIRF). Observed positions are corrected for aberration.

geocentric true right ascensions and declinations: coordinates measured in a geocentric true reference system at a specific time in TCG or TT. They are geocentric apparent positions corrected for deflection of light and annual aberration.

geodesic: the path with minimum length between two points in a mathematically defined n-dimension space.

geodesic precession and nutation: the effect on a moving reference frame around a fixed reference frame when both frames have fixed coordinate directions. Specifically, the coordinate transformation between GCRS and BCRS. The geodesic precession is the secular rotational part of the transformation, and geodesic nutation is the periodic part.

geodetic coordinates: the latitude and longitude of a point on the Earth's surface determined from the geodetic vertical (normal to the specified spheroid).

geoid: the equipotential surface on which the geopotential is $U_G = 62\,636\,856\,\mathrm{m}^2/\mathrm{s}^2$. It roughly coincides with mean sea level in the open ocean.

geometric positions: the positions of an object in a three dimensional reference frame that can be either topocentric, geocentric, or barycentric with the $Oxyz$ axes parallel to the ICRF axes.

gnomonic projection: the geometric transform of the projection of an area of the celestial sphere onto a plane tangent to the celestial sphere. It transforms differential spherical coordinates into rectangular coordinates.

gravitational lens: a celestial body whose gravitational field bends light from a more distant source, so that one or several images of it are produced (*see* Einstein distance).

Greenwich sidereal date (GSD): the number of sidereal days and fraction of a day elapsed at the International meridian since the beginning of the Greenwich sidereal day that was in progress at Julian date 0.0.

Greenwich sidereal day number: the integral part of the Greenwich sidereal date.

Greenwich sidereal time (GST): the hour angle measure of the Terrestrial Ephemeris Origin (TEO), or TIO, meridian (Greenwich or International meridian) with respect to the ICRF.

height: elevation upwards from the geoid to a fixed point. More generally, distance upwards from a given level, particularly the geoid.

heliocentric: with reference to, or pertaining to, the center of the Sun.

horizon: a plane perpendicular to the line from an observer to the zenith. The great circle formed by the intersection of the celestial sphere with the horizon is called the astronomical horizon.

hour angle: angular distance measured westward along the celestial equator from the meridian to the hour circle that passes through a celestial object.

hour circle: a great circle on the celestial sphere that passes through the celestial poles and is therefore perpendicular to the celestial equator.

inclination: the angle between two planes or their poles; usually the angle between an orbital plane and a reference plane; one of the standard orbital elements that specifies the orientation of an orbit.

interferometry: astrometric technique producing interferences between two images of a celestial source for the derivation of geometric relations between the baseline of the instrument and the apparent direction of the source.

International Atomic Time (TAI): the continuous scale resulting from analyses by the Bureau International des Poids et Mesures (BIPM) of atomic time standards in many countries. The fundamental unit of TAI is the SI second (*see* second), and the epoch is 1958, January 1.0.

International Celestial Reference Frame (ICRF): radio sources realization of the barycentric, fixed, celestial reference system (ICRS), based on extragalactic sources. Other realizations have a specific name (e.g. HCRF: Hipparcos Celestial Reference Frame).

International Celestial Reference System (ICRS): a global direction fixed coordinate system, defined by the metric tensor of the IAU 2000 resolutions. It can be geocentric (GCRS) or barycentric (BCRS) in its restricted meaning.

International Terrestrial Reference Frame (ITRF): a realization of the ITRS by a set of coordinates and velocities of fiducial points on the Earth (e.g. ITRF 2000).

International Terrestrial Reference System (ITRS): the terrestrial reference system to which the positions on the Earth are referred. It is geocentric and has no global residual rotation with respect to horizontal motions at the Earth's surface.

invariable plane: the plane through the center of mass of the Solar System perpendicular to the angular momentum vector of the Solar System.

isoplanatic patch: region of the sky within which the refraction effect is relatively uniform.

Julian date (JD): the interval of time in days and fraction of a day since 4713 BC January 1, Greenwich noon, Julian proleptic calendar. In precise work the time scale, TCB, TCG, TT, or Universal Time, should be specified.

Julian date, modified (MJD): the Julian date minus 2 400 000.5.

Julian day number: the integral part of the Julian date.

Julian year: a period of 365.25 days. This period served as the basis for the Julian calendar, and is now used as a standard astronomical period.

latitude, celestial: angular distance on the celestial sphere measured North or South of the ecliptic along the great circle passing through the poles of the ecliptic and the celestial object.

latitude, terrestrial: angular distance on the Earth measured North or South of the equator along the meridian of a geographic location.

leap second: a second (*see* second) added between 60 s and 0 s at announced times to keep UTC within ±0.90 s of UT1. Generally, leap seconds are added to UTC at the end of June or December.

light-time: the interval of time required for light to travel from a celestial body to the Earth. During this interval the motion of the body in space causes an angular displacement of its apparent place from its geometric place.

light-year: the distance that light traverses in a vacuum during one year.

limb darkening: reduction of brightness when approaching the limb of stars (or Sun).

local sidereal time: the local hour angle of a catalog equinox.

longitude, celestial: angular distance on the celestial sphere measured eastward along the ecliptic from the dynamical equinox to the great circle passing through the poles of the ecliptic and the celestial object.

longitude, terrestrial: angular distance measured along the Earth's equator from the international (Greenwich) meridian or the TEO to the meridian of a geographic location.

Lorentz transformation: in special relativity, relationship between coordinates and time in one reference frame and those in another reference frame which is moving with respect to the first reference frame.

Love and Shida numbers: numbers that describe the proportionality of the local reaction of the deformable Earth to an external potential.

magnitude: a measure on a logarithmic scale of the brightness of a celestial object. There are many magnitude systems depending on the filters used (e.g. UBV or SLOAN).

magnitude, absolute: magnitude that a star would have if it were at a distance of ten parsecs from the Sun.

mean equator and equinox: the celestial reference system determined by ignoring short-period variations (nutation) in the motions of the celestial equator. Thus, the mean equator and equinox are affected only by precession. This requires using a separated precession expression and nutation theory.

mean place: the barycentric position, referred to the mean equator and equinox of a standard epoch, of an object on the celestial sphere. A mean place is determined by removing from the directly observed position the effects of refraction, geocentric and stellar parallax, diurnal and stellar aberration, deflection of light, and by referring the coordinates to the mean equator and equinox of a standard epoch. In compiling star catalogs it has been the practice not to remove the secular part of stellar aberration.

mean Sun: a fictitious body defined by Newcomb to define a mean solar time that averaged the Earth rotation. It was used to define the Besselian year until 1984.

meridian: a great circle passing through the celestial poles and through the zenith of a location on Earth. For planetary observations a meridian is half the great circle passing through the planet's poles and through any location on the planet.

metric: in the theory of relativity, the expression representing the square of the four-dimensional distance of two close generalized events; it is noted ds^2.

Michelson interferometer: two identical telescopes point at a given star, the images are recombined forming interference patterns based on the differences in the phases of the two images.

moving celestial reference frame: celestial reference frame defined by the true equator and equinox or the CEO. The coordinates are right ascension and declination with the origin to be specified.

nadir: the point on the celestial sphere diametrically opposite to the zenith.

node: either of the points on the celestial sphere at which the plane of an orbit intersects a reference plane. The position of a node is one of the standard orbital elements used to specify the orientation of an orbit.

nutation: the oscillations in the motion of the pole of rotation of a freely rotating body that is undergoing torque from external gravitational forces. The secular part of the motion is called precession. For the Earth, terms with periods less than two days are included in the polar motion. Nutation of the Earth's pole is discussed in terms of components in obliquity and longitude (*see* precession, polar motion).

obliquity: in general, the angle between the equatorial and orbital planes of a body or, equivalently, between the rotational and orbital poles. For the Earth the obliquity of the ecliptic is the angle between the planes of the equator and the ecliptic.

occultation: the obscuration of one celestial body by another of greater apparent diameter; especially the passage of the Moon in front of a star or planet, or the disappearance of a satellite behind the disk of its primary. If the primary source of illumination of a reflecting body is cut off by the occultation, the phenomenon is also called an eclipse. The occultation of the Sun by the Moon is a solar eclipse.

opposition: a configuration of the Sun, Earth and a planet in which the apparent geocentric longitude (*see* longitude, celestial) of the planet differs by 180° from the apparent geocentric longitude of the Sun.

orbit: the path in space followed by a celestial body.

parallactic correction: the difference in apparent direction of an object as seen from two different locations; conversely, the angle at the object that is subtended by the line joining two designated points.

parallax, annual: the difference between hypothetical geocentric and barycentric observations.

parallax, diurnal or geocentric: the difference in direction between a topocentric observation and a hypothetical geocentric observation.

parallax, horizontal: the difference between the topocentric and hypothetical geocentric observations of an object, when the object is on the astronomical horizon.

parallax, orbital: the difference in direction between an observation from a spacecraft and a hypothetical geocentric observation.

parallax, stellar: the angle subtended at the observed object by one astronomical unit.

parsec: the distance at which one astronomical unit subtends an angle of one second of arc; equivalently, the distance to an object having an annual parallax of one second of arc. A parsec is equal to 3.261 633 light-years.

period: the interval of time required to complete one revolution in an orbit or one cycle of a periodic phenomenon, such as the rotation of a body or a cycle of phases.

perturbations: deviations between the actual orbit of a celestial body and an assumed reference orbit; also, the forces that cause deviations between the actual and reference orbits. Perturbations, according to the first meaning, are usually calculated as quantities to be added to the coordinates of the reference orbit to obtain the precise coordinates.

phase: the ratio of the illuminated area of the apparent disk of a celestial body to the area of the entire apparent disk taken as a circle. For the Moon, phase designations are defined by specific configurations of the Sun, Earth and Moon. For eclipses, phase designations (total, partial, penumbral, etc.) provide general descriptions of the phenomena. More generally, for use with oddly shaped bodies, phase might be defined as $0.5(1 + \cos(\text{phase angle}))$.

phase angle: the angle measured at the center of an illuminated body between directions of the light source and the observer.

phase coefficient: expression for the observationally determined magnitude correction for variation in brilliancy of planets with phase angle.

photometry: a measurement of the intensity of light usually specified for a specific wavelength filter (*see* magnitude).

polar motion: the irregularly varying motion of the Celestial Intermediate Pole (CIP) with respect to the ITRF, including all periodic terms less than two days.

precession: the uniformly progressing motion of the pole of rotation of a freely rotating body undergoing torque from external gravitational forces. In the case of the Earth, the component of precession caused by the Sun and Moon acting on the Earth's equatorial bulge is called luni-solar precession; the component caused by the action of the planets is called planetary precession. The sum of luni-solar and planetary precession is called general precession.

precession–nutation: the ensemble of effects of external torques on the motion of the CIP.

precision: evaluation of the uncertainty due to the random errors of a measurement. With the accuracy, it is one of the two components of the uncertainty.

probability density function (pdf): a continuous function that takes on, for each possible value of a random variable, the associated probability.

proper motion: the projection onto the celestial sphere of the space motion of a star relative to the barycenter of the Solar System. Proper motion is usually tabulated in star catalogs as changes in right ascension and declination per year or century.

proper time: in the theory of relativity, the time linked to an object, which also applies to the local physics in a frame at rest with respect to the object.

pulsar: rapidly spinning bodies emitting very stable periodic pulses of radiowaves and often in other wavelengths.

radial velocity: the rate of change of the distance to an object.

reference frame: practical realization of a reference system, usually as a catalog of positions and proper motions of a certain number of fiducial points on the sky (stars or radio sources). For instance, ICRF is the realization of ICRS.

reference system: theoretical concept of a system of coordinates, including time and standards, necessary to specify the bases for representing the position and motion of bodies in space.

refraction, astronomical: the change in direction of travel (bending) of a light ray as it passes obliquely through the atmosphere. As a result of refraction the observed altitude of a celestial object is greater than its geometric altitude. The amount of refraction depends on the altitude of the object and on atmospheric conditions.

right ascension: angular distance measured eastward along the celestial equator from the CEO, or equinox, to the hour circle passing through the celestial object. Right ascension is given either in degrees, or in hours, minutes, and seconds.

right ascension α and declination δ: without qualification, right ascensions and declinations measured in ICRS. The epoch of the positions must be specified. By extension, catalog or mean position in a barycentric reference system from the catalog right ascension fiducial point and from the catalog equator, respectively. They are designated with the catalog reference system as necessary (e.g. Hipparcos right ascension and declination at epoch 1991.25 on TCB).

Roche limit: critical distance at which an approaching smaller body becomes structurally unstable and disintegrates.

second: the time unit in the Système International d'Unités (SI). It is the duration of 9 192 631 770 cycles of radiation corresponding to the transition between two hyperfine levels of the ground state of cesium-133.

seeing: size of a smeared star from the short-period perturbations of the atmosphere during an observation.

selenocentric: with reference to, or pertaining to, the center of the Moon.

semidiameter: the angle at the observer subtended by the equatorial radius of the Sun, Moon or a planet.

Shida number: *see* Love and Shida numbers.

sidereal day: time necessary for sidereal time to increase by 360°.

sidereal hour angle: angular distance on the celestial sphere measured westward along the celestial equator from the catalog equinox to the hour circle passing through the celestial object. It is equal to 360° minus right ascension in degrees.

sidereal time: the measure of the angle defined by the apparent diurnal motion of the catalog equinox; hence, a measure of the rotation of the Earth with respect to the stars rather than the Sun. It is often expressed in hours, minutes, and seconds, one hour being equal to 15°.

solar time: the measure of time based on the diurnal motion of the Sun. The rate of diurnal motion undergoes seasonal variation because of the obliquity of the ecliptic and because of the eccentricity of the Earth's orbit. Additional small variations result from irregularities in the rotation of the Earth on its axis.

solstice: either of the two points on the ecliptic at which the apparent longitude (*see* longitude, celestial) of the Sun is 90° or 270°; also, the time at which the Sun is at either point.

space motion: three-dimensional velocity vector of a star, expressed in kilometers per second. Its components are the radial velocity and the transverse velocity derived from the proper motion and the parallax.

space-time coordinate system: a four-dimensional set of three space and one time coordinates, based on a specific metric, that specify an event in relativity theory.

speckles: irregular patterns at the focus of the telescope produced by light from a point source crossing atmosphere cells.

spectral types or classes: categorization of stars according to their spectra, primarily due to differing temperatures of the stellar atmosphere. From hottest to coolest, the main spectral types are O, B, A, F, G, K and M.

standard deviation: an evaluation of the precision of a measurement assuming a random (Gaussian) distribution of errors, with a level of confidence of 68.27%. It is the square root of the variance, which is the sum of the squares of the measurement residuals divided by the number of the measurements minus one.

standard epoch: a date and time that specifies the reference system to which celestial coordinates are referred. Prior to 1984, coordinates of star catalogs were commonly referred to the mean equator and equinox of the beginning of a Besselian year. Beginning with 1984 the Julian year has been used, as denoted by the prefix J, e.g. J2000.0.

stellar angle, or Earth rotation angle: angle measured along the true equator of the Celestial Intermediate Pole (CIP) between the Terrestrial Ephemeris Origin (TEO), or TIO, and the Celestial Ephemeris Origin (CEO), or CIO, positively in the retrograde direction. It is proportional to UT1.

synodic period: for planets, the mean interval of time between successive conjunctions of a pair of planets, as observed from the Sun; for satellites, the mean interval between successive conjunctions of a satellite with the Sun, as observed from the satellite's primary.

synodic time: pertaining to successive conjunctions; successive returns of a planet to the same aspect as determined by Earth.

TAI: Temps Atomique International, *see* International Atomic Time

Terrestrial Ephemeris Origin (TEO): the origin of terrestrial longitudes of the International Terrestrial Reference System (ITRS) of date. Also called Terrestrial Intermediate Origin (TIO).

Terrestrial Intermediate Origin (TIO): *see* TEO.

Terrestrial Time (TT): the independent argument for apparent geocentric ephemerides, previously known as Terrestrial Dynamical Time (TDT). At 1977 January $1^{d}00^{h}00^{m}00^{s}$ TAI, the value of TT was exactly 1977 January 1.000 3725 days. The unit of TT is the SI second or a day of 86 400 SI seconds at the geoid. One realization of TT is TT(TAI) = TAI + 32.184 s. See Geocentric Coordinate Time (TCG).

time dilatation: the positive difference between a time interval measured in a moving frame and the same one measured in a frame fixed with respect to the events.

tip-tilt correction: two-dimensional corrections of the secondary mirror to compensate for the changes in the atmosphere during an observation.

topocentric: with reference to, or pertaining to, a point on the surface of the Earth.

topocentric position: the observed position corrected for refraction and instrumental parameters, on UTC or TT.

transit: the passage of the apparent center of the disk of a celestial object across a meridian; also, the passage of one celestial body in front of another of greater apparent diameter (e.g., the passage of Mercury or Venus across the Sun, or Jupiter's satellites across its disk); however, the passage of the Moon in front of the larger apparent Sun is called an annular eclipse. The passage of a body's shadow across another body is called a shadow transit; however, the passage of the Moon's shadow across the Earth is called a solar eclipse.

true equator: another name of the Celestial Intermediate Equator; it is the equatorial plane through the center of the Earth and perpendicular to the CIP pole at some specific time.

true equator and equinox: the classical moving celestial coordinate frame determined by the instantaneous positions of the celestial equator and ecliptic. The motion of this system is due to the progressive effect of precession and of nutation.

true positions: ICRF positions that have been transformed to the CEO and celestial intermediate, or true, equator of a specific epoch by the application of precession–nutation and geodesic precession and nutation. The same name applies for positions in the equator–equinox reference frame (*see* moving celestial reference frame).

true value: value that would be obtained by a perfect measurement. True values are by nature indeterminate and an expected value is often used as a true value.

uncertainty: parameter, associated with the result of a measurement, that characterizes the dispersion that can be attributed to include the expected value of the measured quantity with some given probability. Its components are accuracy and precision. Also, estimation of the difference between the measurement and the true value.

Universal Time (UT): a measure of time that conforms, within a close approximation, to the mean diurnal motion of the Sun and serves as the basis of all civil timekeeping. It is determined from observations of the diurnal motions of the stars. It has several realizations: UT0, UT1, and UTC.

UT0: the Universal Time as determined directly from observations; it is slightly dependent on the place of observation.

UT1: time scale deduced from the UT0 corrected for the shift in longitude of the observing station caused by polar motion. It was related by a mathematical formula to the sidereal time in the moving reference frame, or to the stellar angle in the CIRF.

UTC: *see* Coordinated Universal Time.

vernal equinox: the ascending node of the ecliptic on the celestial equator; also, the time at which the apparent longitude (*see* longitude, celestial) of the Sun is $0°$ (*see* equinox).

vertical: the apparent direction of gravity at the point of observation (normal to the plane of a free level surface).

year: a period of time based on the revolution of the Earth around the Sun. The calendar year is an approximation to the tropical year (*see* year, tropical). The anomalistic year is the mean interval between successive passages of the Earth through perihelion. The sidereal year is the mean period of revolution with respect to the background stars.

year, Besselian: the period of one complete revolution in right ascension of the fictitious mean sun, as defined by Newcomb. The beginning of a Besselian year, traditionally used as a standard epoch, is denoted by the suffix ".0". Since 1984 standard epochs have been defined by the Julian year rather than the Besselian year. For distinction, the beginning of the Besselian year is now identified by the prefix B (e.g., B1950.0).

year, tropical: the period of one complete revolution of the mean longitude of the Sun with respect to the dynamical equinox. The tropical year is longer than the Besselian year by $0.148T$ s, where T is centuries from B1900.0.

zenith: in general, the point directly overhead on the celestial sphere. The astronomical zenith is the extension to infinity of a plumb line. The geocentric zenith is defined by the line from the center of the Earth through the observer. The geodetic zenith is the normal to the geodetic ellipsoid at the observer's location (*see* deflection of the vertical).

zenith distance: angular distance on the celestial sphere measured along the great circle from the zenith to the celestial object. Zenith distance is $90°$ minus altitude.

References

Note: the chapters of this book where the references are cited are indicated in parentheses.

Adams, J. C., 1853, *Phil. Trans. R. Soc. London*, **CXLIII**, 397–406. *(Ch. 1)*

Altamimi, Z., Sillard, P. and Boucher, C., 2002, J. Geophys. Res., **107** NB10, 2214, ETG2. *(Ch. 9)*

Andersen, J., 1991, *Astron. & Astrophys. Review*, **3**, 91–126. *(Ch. 12)*

Andersen, J., 2001, *Transactions of the IAU*, vol. XXIV B. *(Ch. 5)*

Anderson, J. and King, I. R., 1999, *Publ. Astro. Soc. Pacific* **111**, 1095–1098. *(Ch. 14)*

Angel, R. and Woolf, N., 1997, *Astrophys. J.*, **475**, 373–379. *(Ch. 14)*

Aoki, S., Guinot, B., Kaplan, G. H., Kinoshita, H., McCarthy, D. D. and Seidelmann, P. K., 1982, *Astron. & Astrophys.*, **105**, 359–361. *(Ch. 8, 10, 14)*

Aoki, S. and Kinoshita, H., 1983, *Celest. Mech.*, **29**, 335–360. *(Ch. 10)*

Arias, E. F., Charlot, P., Feissel, M. and Lestrade, J.-F., 1995, *Astron. & Astrophys.*, **303**, 604–608. *(Ch. 7)*

Backer, D. C., 1993, in *Galactic High-Energy Astrophysics, High Accuracy Timing and Positional Astronomy*, J. van Paradijs and H. M. Maitzes (Eds.), *Lecture Notes in Physics*, **418**, Springer Verlag, Berlin, Heidelberg, 193–253. *(Ch. 12)*

Backer, D. C. and Hellings, R. W., 1986, *Ann. Rev. Astron. Astrophys.*, **24**, 537–575. *(Ch. 14)*

Bastian, U., Röser, S. *et al.*, 1993, *PPM Star Catalogue*, Spektrum Akademisher Verlag, Heidelberg, **III** and **IV**. *(Ch. 11)*

Bastian, U., Röser, S., Høg, E. *et al.*, 1996, *Astron. Nachr.*, **317**, 281–288. *(Ch. 2)*

Battin, R. H., 1987, *An Introduction to the Mathematics and Methods of Astrodynamics*, American Institute of Aeronautics and Astronautics Inc., New York. *(Ch. 3)*

Beichman, C. A., 1998, *Proceed. SPIE*, **3350**, 719. *(Ch. 14)*

Bertotti, B., Carr, B. J. and Rees, M. J., 1983, *Monthly Not. RAS*, **203**, 945–951. *(Ch. 14)*

Blaauw, A., Gum, C. S., Pawsey, J. L. and Westerhout, G., 1960, *Monthly Not. RAS*, **121**, 123–131. *(Ch. 7)*

Black, D. C., 1995, *Ann. Rev. Astron. Astrophys.*, **33**, 359–380. *(Ch. 14)*

Black, D. C. and Scargle, J. D., 1982, *Astrophys. J.*, **263**, 854–869. *(Ch. 14)*

Blandford, R. and Teukolsky, S. A., 1976, *Astrophys. J.*, **205**, 580–591. *(Ch. 14)*

Blandford, R., Romani, R. and Narayan, R., 1984, *J. Astrophys. Astron.*, **5**, 369–388. *(Ch. 14)*

Borkowski, K. M., 1989, *Bulletin Géodésique*, **69**, (1), 50–56. *(Ch. 9)*

Boss, B., Albrecht, S., Jenkins, H. *et al.*, 1937, *General Catalogue of 33342 stars for the Epoch 1950*, Carnegie Institution of Washington Publ., 5 volumes. *(Ch. 11)*

Boss, L., 1910, *Preliminary General Catalogue of 6188 Stars for Epoch 1900*, Carnegie Institution of Washington, Washington. *(Ch. 11)*

Boucher, C., 1990, in *Variations in Earth Rotation*, D. D. McCarthy and W. E. Carter (Eds.) , 197–201. *(Ch. 9)*

Boucher, C., Altamimi, Z. and Sillard, P., 1999, *The 1997 International Terrestrial Reference Frame (ITRF-97)*, IERS Technical note 27, May 1999, Observatoire de Paris. *(Ch. 9)*

Bouillé, F., Cazenave, A., Lemoine, J.-M. and Crétaux, J.-F., 2000, *Geophys. J. Int.*, **143**, 71–82. *(Ch. 9)*

Brosche, P., Wildermann, E. and Geffert, M., 1989, *Astron. & Astrophys.*, **211**, 239–244. *(Ch. 14)*

Brouwer, D. and Clemence, G.M., 1961, *Methods of Celestial Mechanics*, Academic Press, New York and London. *(Ch. 8, 9)*

Brumberg, V. A., 1991a, *Essential Relativistic Celestial Mechanics*, Adam Hilger, Bristol, Philadelphia, New York. *(Ch. 5, 7)*

Brumberg, V .A., 1991b, in *Reference Systems, Proceedings of IAU Coll. 127*, J. A. Hughes, C. A. Smith and G. H. Kaplan (Eds.), U.S. Naval Observatory, Washington D.C., 36–49. *(Ch. 7)*

Brumberg, V. A. and Groten, E., 2001, *Astron. & Astrophys.*, **367**, 1070–1077. *(Ch. 7)*

Brumberg, V. A. and Kopeikin, S. M., *Celestial Mech. and Dynam. Astron.*, **48**, 23–44. *(Ch. 5)*

Buffett, B. A., Mathews, P. M. and Herring, T. A., 2002, *J. Geophys. Res.*, **107** ETG 5-1-14. *(Ch. 8)*

Capitaine, N., 1990, *Celest. Mech. Dyn. Astron.*, **46**, 127–143. *(Ch. 8)*

Capitaine, N., Guinot, B. and Souchay, J., 1986, *Celest. Mech.*, **39**, 283–307. *(Ch. 8)*

Capitaine, N., Chapront, J. Lambert S. and Wallace, P., 2002, *Astron. & Astrophys.*, **400**, 1145–1154. *(Ch. 10)*

Capitaine, N. and Gontier, A.-M., 1993, *Astron. & Astrophys.*, **275**, 645–650. *(Ch. 10)*

Capitaine, N., Guinot, B. and McCarthy, D. D., 2000, *Astron. & Astrophys.*, **355**, 398–405. *(Ch. 10)*

Cazenave, A., Valette, J.-J. and Boucher, C., 1992, *J. Geophys. Res.*, **97**, 7109–7119. *(Ch. 9)*

Chapront, J., Chapront-Touzé, M. and Francou, G., 1999, *Astron. & Astrophys.*, **343**, 624–633. *(Ch. 8)*

Chapront, J., Chapront-Touzé, M. and Francou, G., 2002, *Astron. & Astrophys.*, **387**, 700–709. *(Ch. 10)*

Chapront-Touzé, M. and Chapront, J., 1988, *Astron. & Astrophys.*, **190**, 342–352. *(Ch. 13)*

Charbonneau, D., Brown, T., Latham, D. and Mayor, M., 2000, *Astrophys. J.*, **529**, L45–48. *(Ch. 12)*

Chen, J.L., Wilson, C.R., Eanes, R.J. and Tapley, B.D., 2000, *J. Geophys. Res.*, **105**, 16271–16277. *(Ch. 9)*

Cheng, M. K. and Tapley, B. D., 1999, *J. Geophys. Res.*, **104**, 2661–2681. *(Ch. 9)*

Cheng, M. K., Shum, C. K. and Tapley, B. D., 1997, *J. Geophys. Res.* **102**, 22377–22390. *(Ch. 9)*

Clausen, J. V., Gimenez, A. and Scarfe, C., 1986, *Astron. & Astrophys.*, **167** 287–296. *(Ch. 12)*

Clemence, G. M., 1948, *Astron. J.*, **53**, 169–179. *(Ch. 1)*

Colavita, M. M. and Shao, M., 1994, in *Planetary Systems: Formation, Evolution, and Detection*, B. F. Burke, J. H. Rahe, E. E.Roettger (Eds.), Kluwer Academic Publishers, Dordrecht, 385–390. *(Ch. 11)*

Corbin, T., 1992, *Transactions of the International Astronomical Union*, Vol. XXIB, 1991, 149–150. *(Ch. 11)*

Corbin, T. and Urban, S., 1988, in *IAU Symposium No. 133, Mapping the Sky*, S. Debarbat, J. A. Eddy, H. K. Eichhorn and A. P. Upgren (Eds.), Kluwer Academic Publishers, Dordrecht, 287–291. *(Ch. 14)*

Cummings, A., Marcy, G. W. and Butler, R. P., 1999, *Astrophys J.*, **526**, 890–915. *(Ch. 14)*

Danjon, A., 1929, *L'Astronomie*, **XLIII**, 13–22. *(Ch. 1)*

Defraigne, P., Dehant, V. and Paquet, P., 1995, *Celest. Mech. Dyn. Astron.*, **62**, 363–376. *(Ch. 8)*

Delaunay, C. E., 1865, *Comptes Rendus des Séances de l'Académie des Sciences*, **61**, 1023–1032. *(Ch. 1)*

DeMets, G., Gordon, R. G., Argus, D. F. and Stein, S., 1994, *Geophys. Res. Lett.*, **21**, 2191–2194. *(Ch. 9)*

Derr, J., 1969, *J. Geophys. Res.*, **74**, 5202–5220. *(Ch. 9)*

Deshpande, A. A., D'Allessandro, F. and McCulloch, P. M., 1996, *J. Astrophys. Astron.*, **17**, 7–16. *(Ch. 14)*

de Sitter, W., 1927, *Bull. of the Astron. Institutes of the Netherlands*, **IV**, 21–38. *(Ch. 1, 10)*

Dick, S. J., 1997, in *History of Astronomy, An Encyclopedia*, John Lankford (Ed.), Garland Publishing Inc., New York & London, 47–60. *(Ch. 1)*

Dieckvoss, W., 1971, in *Conference on Photographic Astrometric Techniques*, H. Eichhorn (Ed.), NASA publication, 161–167. *(Ch. 11)*

DMA, 1987, in *Department. of Defense World Geodetic System 1984* (DMA TR 83502, Washington D.C.). *(Ch. 9)*

Downs, G. S. and Reichley, P. E., 1983, *Astrophys. J. Suppl.*, **53**, 169–240. *(Ch. 14)*

Duncombe, R. L., Jefferys, W. H., Shelus, P. J. *et al.*, 1991, *Advan. Space Res.*, **11–2**, 87–96. *(Ch. 2)*

Eckhardt, D. H., 1981, *The Moon and the Planets*, **3**, 3–49. *(Ch. 13)*

Eddington, A. S., 1922, *The Mathematical Theory of Relativity*, Cambridge University Press, Cambridge. *(Ch. 5)*

Eichhorn, H., 1960, *Astron. Nachr.*, **285**, 233. *(Ch. 11)*

Eichhorn, H., 1974, *Astronomy of Star Positions*, F. Ungar Publishing Co., New York. *(Ch. 11)*

Eichhorn, H., 1988, *Astrophys. J.*, **334**, 465–469. *(Ch. 14)*

Eichhorn, H. and Jeffreys, W. H., 1971, *Publ. McCormick Obs.*, **16**. *(Ch. 14)*

Eichhorn, H. and Russell, J., 1976, *Mothly Not. RAS*, **174**, 679–694. *(Ch. 14)*

Eichhorn, H. and Williams, C. A., 1963, *Astron. J.*, **68**, 221–231. *(Ch. 14)*

Eichhorn, H., Googe, W. D. and Gatewood, G., 1967, *Astron. J.*, **72**, 626–630. *(Ch. 14)*

ESA, 1997, *The Hipparcos and Tycho Catalogues*, European Space Agency Publication, SP-1200, June 1997, 17 volumes. *(Ch. 2, 7, 11)*

ESA, 2000, *GAIA, Concept and Technology Report*, European Space Agency Publication, ESA–SCI (2000) 4. *(Ch. 2)*

Felli, M. and Spencer, R. E., 1989, *Very Long Baseline Interferometry, Techniques and Applications*, Kluwer Academic Publishers, Dordrecht. *(Ch. 2)*

Ferrel, W., 1864, *Proc. of the American Academy of Arts and Sciences*, **VI**, 379–383. *(Ch. 1)*

Fock, V., 1976, *The Theory of Space Time and Gravitation*, Pergamon Press, Oxford. *(Ch. 5)*

Folker, W. M., Charlot, P., Finger, M. H. *et al.*, 1994, *Astron. & Astrophys.* **287**, 279–289. *(Ch. 7)*

Fricke, W., 1982, *Astron. & Astrophys.*, **107**, L13–L16. *(Ch. 11)*
Fricke, W., Kopff, A. *et al.*, 1963, *Veröff. des Astron. Rechen-Institut Heidelberg*, **10**. *(Ch. 11)*
Fricke, W., Schwan, H., Lederle, T. *et al.*, 1988, *Veröff. des Astron. Rechen-Institut, Heidelberg*, **32**. *(Ch. 11)*
Fricke, W., Schwan, H. and Corbin, T., 1991, *Veröff. Astron. Rechen-Institut, Heidelberg*, **33**. *(Ch. 11)*
Fricke, W., Schwan, H., Corbin, T. *et al.*, 1994, *Veröff. Astron. Rechen-Institut, Heidelberg*, **34**. *(Ch. 11)*
Fukushima, T., 1991, *Astron. & Astroph.*, **244**, L11–L12. *(Ch. 7, 8)*
Fukushima, T., 1995, *Astron. & Astroph.*, **294**, 895–906. *(Ch. 5)*
Fukushima, T., 1999, *J. of Geodesy*, **73**, 603–610. *(Ch. 9)*
Fukushima, T., 2001, *Report on Astronomical Constants*, in *Highlights of Astronomy, Proceedings of the XXIV-th IAU General Assembly*, H. Rickman (Ed.), Astron. Soc. of the Pacific, 107–112. *(Ch. 7, 8)*
Gai, M., Carollo, D., Delbo, M. *et al.*, 2001, *Astron. & Astrophys.*, **367**, 362–370. *(Ch. 14)*
Garfinkel, B., 1944, *Astron. J.*, **50**, 169–179. *(Ch. 6)*
Gatewood, G. D., 1976, *Icarus*, **27**, 1–12. *(Ch. 14)*
Gatewood, G. D., Breakiron, L., Goebel, R. *et al.*, 1980, *Icarus*, **41**, 205–231. *(Ch. 14)*
Gatewood, G. D., 1987, *Astron. J.*, **94**, 213–224. *(Ch. 14)*
Gilbert, F. and Dziewonski, A. M., 1975, *Phil. Trans. Roy. Soc. London.*, **A278**, 187–269. *(Ch. 8)*
Gliese, W., 1963, *Veröff. Astron. Rechen-Institut, Heidelberg*, **12**. *(Ch. 11)*
Goad, C. C., 1980, *J. Geophys. Res.*, **85**, 2679–2683. *(Ch. 9)*
Googe, W. D., Eichhorn, H. and Lukac, C. F., 1970, *Monthly Not. RAS*, **150**, 35–44. *(Ch. 14)*
Grafarend, E. W. and Ardalan, A. A., 1999, *J. of Geodesy*, **73**, 611–623. *(Ch. 9)*
Guinot, B., 1979, in *Time and the Earth's Rotation*, IAU Symp. 82, D. D. McCarthy and J. D. H. Pilkington (Eds.), D. Reidel Publ. Co., Dordrecht, 7–18. *(Ch. 8, 10)*
Guinot, B., 1997, *Metrologia*, **34**, 261–290. *(Ch. 5)*
Guseva, I. S., 1987, *Proceedings of a Workshop on Refraction Determination*, G. Teleki (Ed.) Publication of the Belgrad Observatory, 69–74. *(Ch. 6)*
Hagihara, Y., 1970–76, *Celestial Mechanics*, MIT Press and the Japan Society for the Promotion of Science, 5 volumes. *(Ch. 8)*
Hale, A. and Doyle, L.R., 1994, *Astrophys. Space Science*, **212**, 335–342. *(Ch. 14)*
Hamilton, Sir W. R., 1866, *Elements of Quaternions*, Longmans, Green and Co., London. *(Ch. 3)*
Hartkopf, W. I., Mason, B. D. and Worley, C. E., 2001a, *Astron. J.*, **122**, 3472–3479. *(Ch. 12)*
Hartkopf, W. I., Mc Allister, H. A. and Mason, B. D., 2001b, *Astron. J.*, **122**, 3480–3481. *(Ch. 12)*
Hazard, C., Sutton, J., Argue, A. N. *et al.*, 1971, *Nature Phys. Sci.*, **233**, 89–91. *(Ch. 7)*
Heiskanen, W. A. and Moritz, H., 1967, *Physical Geodesy*, Freeman and Co., San Francisco, London. *(Ch. 9)*
Hellings, R. W., 1986, *Astron. J.*, **91**, 650–659. *(Ch. 14)*
Hemenway, P., 1980, *Celest. Mech.*, **22**, 89–109. *(Ch. 14)*
Herring, T. A., 1996, Private communication. *(Ch. 8)*
Herring, T. A., Mathews, P. M. and Buffett, B. A., 2002, *J. Geophys. Res.*, **107**, ETG 4-1-12. *(Ch. 8)*

Hilton, J. L., 2002, Private communication. *(Ch. 8)*

Hofmann-Wallenhof, B., Lichtenegger, H. and Colins, J., 1992, *Global Positioning System, Theory and Practice*, Springer Verlag, Wien, New-York. *(Ch. 2, 10)*

Høg, E., 1997, in *Hipparcos Venice'97*, B. Batrick (Ed.), ESA Publ. SP-402, July 1997, Noordwijk, 25–30. *(Ch. 11)*

Høg, E., Bastian, U., Egret, D. et al., 1992, *Astron. & Astrophys.*, **258**, 177–185. *(Ch. 11)*

Høg, E., Fabricius, C., Makarov, V. V. *et al.*, 2000a, *Astron. & Astrophys.*, **355**, L19–L22. *(Ch. 11)*

Høg, E., Fabricius, C., Makarov, V. V., Bastian, U. *et al.*, 2000b, *Astron. & Astrophys.*, **357**, 367–386. *(Ch. 7, 11)*

IAG, 1980, The Geodetic Reference System 1980, *Bulletin Géodésique*, **54**, 395–405. *(Ch. 9)*

IAU, 1992, *Proceedings of the XXI-st General Assembly*, J. Bergeron (Ed.) IAU Transactions, Kluwer Academic Publishers, Dordrecht, 41–52. *(Ch. 7)*

IAU, 1998, *Proceedings of the XXIII-rd General Assembly*, J. Andersen (Ed.) IAU Transactions, Kluwer Academic Publishers, Dordrecht 39–41. *(Ch. 7)*

IAU, 2001, *Proceedings of the XXIV-th General Assembly*, H. Rickman (Ed.) IAU Transactions, Astro. Soc. of the Pacific, 33–59. *(Ch. 7)*

IERS, 1996, *IERS Conventions (1996)*, IERS Technical Note 21, D. D. McCarthy (Ed.), Observatoire de Paris. *(Ch. 8)*

IERS, 2003, *IERS Conventions 2000*, International Earth Rotation Service. *(Ch. 9, 10)*

Ilyasov, Y. P., Kopeikin, S. M. and Rodin, A. E., 1998, *Astronomy Letters*, **24**, 228–236. *(Ch. 14)*

IUGG, 1992, *Bulletin Géodésique*, **66**, 128–129. *(Ch. 9)*

Iyer, B. R. and Will, C. M., 1995, *Phys. Rev.*, **D 52**, 6882–6893. *(Ch. 14)*

Jaech, J. L., 1985, *Statistical Analysis of Measurement Errors*, An Exxon Monograph, John Wiley & Sons, New York. *(Ch. 4)*

James, T. S. and Ivins, E. R., 1997, *J. Geophys. Res.*, **102**, 605–633. *(Ch. 9)*

Jaranowski, P., 1997, in *Mathematics of Gravitation, Part II, Gravitational Wave Detection*, A. Krolak (Ed.), Banach Center Publications of Inst. Math., Polish Acad. Sci., **41**, 55. *(Ch. 14)*

Jaranowski, P. and Schafer, G., 1997, *Phys. Rev.*, D **55**, 4712–4722. *(Ch. 14)*

Jefferys, W. H., 1963, *Astron. J.*, **68**, 111–113. *(Ch. 14)*

Jefferys, W. H., 1979, *Astron. J.*, **84**, 1775–1777. *(Ch. 14)*

Jefferys, W. H., 1987, *Astron. J.*, **93**, 755–759. *(Ch. 14)*

Johnston, K. J., Fey, A. L., Zacharias, N. *et al.*, 1995, *Astron. J.*, **110**, 880–915. *(Ch. 7)*

Johnston, K. J. and de Vegt, C., 1999, *Annual Rev. Astron. Astrophys.*, **37**, 97–125. *(Ch. 7)*

Kammeyer, P., 2000, *Celestial Mechanics and Dynamical Astronomy*, **77**, 241–272. *(Ch. 10)*

Kaplan, G. H., Josties, F. J., Angerhofer, P. E. *et al.*, 1982, *Astron. J.*, **87**, 570–576. *(Ch. 7)*

Kaspi, V. M., Taylor, J. H. and Ryba, M. F., 1994, *Astrophys. J.*, **428**, 713–728. *(Ch. 14)*

Kinoshita, H., 1977, *Celestial Mechanics*, **15**, 277–326. *(Ch. 8)*

Kinoshita, H. and Kozai, H., 1989, *Celestial Mechanics*, **45**, 231–244. *(Ch. 9)*

Kinoshita, H. and Souchay, J., 1990, *Celest. Mech. and Dyn. Astr.*, **48**, 187–265. *(Ch. 8)*

Klioner, S. A., 1993, *Astron. & Astrophys.*, **279**, 273–277. *(Ch. 7)*

Koch, D. G., Borucki, W., Webster, L. *et al.*, 1999, in *Space telescopes and Instruments V.*, P. Y. Bely and J. B. Breckinridge (Eds.), *Proc. SPIE*, **3356**, 599–607. *(Ch. 14)*

Kopeikin, S. M., 1997, *Monthly Not. RAS*, **288**, 129–137. *(Ch. 14)*

Kopeikin, S. M., 1999, *Monthly Not. RAS*, **305**, 563–590. *(Ch. 14)*

Kopeikin, S. M., 2001, in *Reference Frames and Gravitomagnetism*, Proceedings of the Spanish Relativity Meeting 2000, J.-F. Pascual-Sanchez (Ed.), World Science, Singapore. *(Ch. 5)*

Kopeikin, S. M. and Gwinn, C. R., 2000, *Towards Models and Constants for Sub-Microarcsecond Astrometry*, K. J. Johnston, D. D. McCarthy, B. J. Luzum and G. H. Kaplan (Eds.), IAU Colloquium 180, Publ. U.S. Naval Observatory, Washington, 303–307. *(Ch. 5)*

Kopeikin, S. M. and Ozenoy, L. M., 1999, *Astrophys. J.*, **523**, 771–785. *(Ch. 12)*

Kopeikin, S. M. and Schäffer G., 1999, *Phys. Review*, D **60**, 124002. *(Ch. 5)*

Kopff, A., 1937, *Veröff. Astron. Rechen-Institut, Heidelberg*, **54**. *(Ch. 11)*

Korff, D., 1973, *J. of Opt. Soc. of America*, **63**, 971–980. *(Ch. 2)*

Kovalevsky, J., 1967, *Introduction to Celestial Mechanics*, D. Reidel Publ. Co., Dordrecht. *(Ch. 8, 12)*

Kovalevsky, J., 2000, in *Towards Models and Constants for Sub-Microarcsecond Astrometry*, IAU Colloquium 180, K. J. Johnston, D. D. McCarthy, B. J. Luzum and G. H. Kaplan (Eds.), US Naval Observatory, Washington DC 3–9. *(Ch. 7)*

Kovalevsky, J., 2002, *Modern Astrometry*, second edition, Springer-Verlag, Berlin, Heidelberg. *(Ch. 2, 10, 11)*

Kovalevsky, J., 2003, *Astron. & Astrophys.*, **404**, 743–747. *(Ch. 6)*

Kovalevsky, J., Falin, J.-L., Pieplu, J.-L. *et al.*, 1992, *Astron. & Astrophys.*, **258**, 7–17. *(Ch. 11)*

Kovalevsky, J., Lindegren, L. and Froeschlé, M., 2000, in *Journées 1999, Systèmes de référence spatio-temporels*, M. Soffel and N. Capitaine (Eds.), Paris Observatory, 119–130. *(Ch. 2)*

Kovalevsky, J., Lindegren, L., Perryman, M. A. C. *et al.*, 1997, *Astron. & Astrophys.*, **323**, 620–633. *(Ch. 7, 11)*

Kovalevsky, J. and McCarthy, D., 1998, in *Highlights in Astronomy*, J. Andersen (Ed.), **11A**, 182–186. *(Ch. 10)*

Kuimov, K. V., Sorokin, F. D., Kuz'min, A. V. and Barusheva, N. T., 2000, *Astronomy Reports*, **44**, 474–480. *(Ch. 14)*

Labeyrie, A., 1970, *Astron. & Astrophys.*, **6**, 85–87. *(Ch. 2)*

Labeyrie, A., 1975, *Astrophys. J. Letters*, **196**, L71–75. *(Ch. 2)*

Labeyrie, A., 1995, *Astron. & Astrophys.*, **298**, 544–548. *(Ch. 14)*

Lambeck, K., 1980, *The Earth's Variable Rotation: Geophysical Causes and Consequences*, Cambridge University Press, Cambridge. *(Ch. 10)*

Laskar, J., 1986, *Astron. & Astrophys.*, **157**, 59–70. *(Ch. 13)*

Lattanzi, M. G. and Bucciarelli, B., 1991, *Astron. & Astrophys.*, **250**, 565–572. *(Ch. 14)*

Léger, A., Rouan, D., Schneider, J. *et al.*, 1999, in *Bioastronomy 99: A new Era in Bioastronomy*, Kohala Coast Meeting, Hawaii, August 2–6, 1999. *(Ch. 14)*

Léna, P., 1996, *Astrophysique, Méthodes physiques de l'observation*, CNRS Editions, Paris. *(Ch. 2)*

Le Provost, C., Genco, M. L., Lyard, F. *et al.*, 1994 *J. Geophys. Res.*, **99**, 24777–24797. *(Ch. 9)*

Lestrade, J.-F., Jones, D. L., Preston, R. A. *et al.*, 1995, *Astron. & Astrophys.*, **304**, 182–188. *(Ch. 7)*

Liebelt, P. B., 1967, *An Introduction to Optimal Estimation*, Addison-Wesley Publ. Co., Reading (Mass). *(Ch. 4)*

Lieske, J. H., 1979, *Astron. & Astrophys.*, **73** 282–284. *(Ch. 8)*

Lieske, J. H., Lederle, T., Fricke, W. and Morando, B., 1977, *Astron. & Astrophys.*, **58**, 1–16. *(Ch. 8, 11)*

Lindegren, L., Høg, E., van Leeuwen, F. *et al.*, 1992, *Astron. & Astrophys.*, **258**, 18–30. *(Ch. 11)*

Lindegren, L. and Kovalevsky, J., 1995, *Astron. & Astrophys.*, **304**, 189–201. *(Ch. 7)*

Luri, X. and Arenou, F., 1997, in *Hipparcos Venice'97*, B. Batrick (Ed.), ESA Publ., SP-402, 449–452. *(Ch. 11)*

Lutz, T. E. and Kelker, D. H., 1973, *Publ. Astron. Soc. Pacific*, **85**, 573–578. *(Ch. 11)*

Ma, C., Arias, E. F., Eubanks, T. M. *et al.*, 1998, *Astron. J.*, **116**, 516–546. *(Ch. 2, 7)*

Ma, C. and Feissel, M., 1997, *Definition and Realization of the International Celestial Reference System by VLBI Astrometry of Extragalactic Objects*, IERS Technical Note No 23, Observatoire de Paris. *(Ch. 7)*

Manchester, R. N. and Taylor, J. H., 1977, *Pulsars*, Freeman, San Francisco. *(Ch. 14)*

Marcy, G. W. and Butler, R. P., 1992, *Publ. Astron. Soc. Pacific*, **104**, 270–277. *(Ch. 14)*

Marcy, G. W. and Butler, R. P., 1993, *Astron. Soc. of the Pacific Conference Series*, **47**, 175–182. *(Ch. 14)*

Marcy, G. W. and Butler, R. P., 1998, *Ann. Rev. Astron. Astrophys.*, **36**, 57–97. *(Ch. 14)*

Marcy, G. W. and Butler, R. P., 1999, in *The Origin of Stars and Planetary Systems*, C. J. Lada and N. D. Kylafis (Eds.), Kluwer Academic Publishers, Dordrecht, 681–708. *(Ch. 14)*

Marcy, G. W. and Butler, R. P., 2000, *Publ. Astron. Soc. Pacific*, **112**, 137–140. *(Ch. 14)*

Marini, J. W. and Murray, C. W., 1973, *Correction of the Laser Range Tracking Data for Atmospheric Refraction at Elevations above* 10°, NASA-GSFC document X-591-73-35. *(Ch. 6)*

Markowitz, W., Hall, R. G., Essen, L. and Perry, J. V. L., 1958, *Phys. Rev. Letters*, **1**, 105. *(Ch. 1)*

Martinez, P. and Klotz, A., 1999, *A Practical Guide to CCD Astronomy*, Cambridge University Press, Cambridge. *(Ch. 2, 14)*

Mason, R. D., Wycoff, G. L., Hartkopf, W. I. *et al.* 2001, *Astron. J.*, **122**, 3466–3471. *(Ch. 12)*

Mathews, P. M., Herring, T. A. and Buffett, B. A., 2002, *J. Geophys. Res.*, **107**, ETG 3-1-26. *(Ch. 8)*

Matsakis, D. N. and Foster, R. S., 1996, in *Amazing Light, A Festschrift for Charles Townes*, R. Chiao (Ed.), Springer Verlag, Berlin, 445–462. *(Ch. 14)*

Matsakis, D. N., Taylor, J. H. and Eubanks, T. M., 1997, *Astron. & Astrophys.*, **326**, 924–928. *(Ch. 14)*

Mayor, M. and Queloz, D., 1995, *Nature*, **378** (6555), 355–359. *(Ch. 12, 14)*

Mc Allister, H. A., 1996, *Sky and Telescope*, November 1996, 29–35. *(Ch. 12)*

McCarthy, D. D. (Ed.), 1996, *IERS Conventions*, IERS Technical Note 21, Observatoire de Paris. *(Ch. 6, 9, 10)*

McCarthy, D. D. (Ed.), 2003, *IERS Conventions*, (in press). *(Ch. 6, 9, 10)*

McCarthy, D. D. and Babcock, A. K., 1986, *Physics of the Earth and Planetary Interiors*, **44**, 281–292. *(Ch. 10)*

McCarthy, D. D. and Luzum, B. J., 2003, *Celest. Mech. and Dyn. Astron.*, **85**, 37–49. *(Ch. 8)*

McLean, B., Lasker, B. and Lattanzi, M., 1998, *AAS Meeting*, **192**, 55.10. *(Ch. 11)*

Melchior, P., 1966, *The Earth Tides*, Pergamon Press, Oxford, London. *(Ch. 9)*

Mendenhall, W., Schaeffer, R. L. and Wackerly, D. D., 1986, *Mathematical Statistics and Applications*, 3rd edition, Duxbury Press, Boston. *(Ch. 4)*

Mignard, F., Froeschlé, M., Badiali, M. *et al.*, 1992, *Astron. & Astrophys.*, **258**, 165–172. *(Ch. 12)*

Mitrovica, J. X. and Peltier, W. R., 1993, *J. Geophys. Res.*, **98**, 4509–4526. *(Ch. 9)*

Moller, O. and Kristensen, L. K., 1976, *Trans. IAU.*, **XVI B**, 166. *(Ch. 14)*

Monet, D. G., 1988, *Ann. Review. Astron. Astrophys.*, **26**, 413–440. *(Ch. 2, 14)*

Monet, D. G. *et al.*, 1998, *USNO-A*, US Naval Observatory, Washington D.C. *(Ch. 14)*

Morgan, H. R., 1952, *Astron. Papers Am. Ephemeris and Naut. Alm.*, **13**, part 3. *(Ch. 11)*

Moritz, H., 1980, *Advanced Physical Geodesy*, Abacus Press, Tunbridge Wells, Kent, UK. *(Ch. 9)*

Morrison, J. E., Röser, S., Lasker, B. M. *et al.*, 1996, *Astron. J.*, **111**, 1405–1410. *(Ch. 14)*

Morrison, J. E., Röser, S., McLean, B. *et al.*, 2001, *Astron. J.*, **121**, 1752–1763. *(Ch. 11)*

Morrison, J. E., Smart, R. L. and Taff, L. G., 1998, *Monthly Not. RAS*, **296**, 66–76. *(Ch. 14)*

Mould, R. A., 1994, *Basic Relativity*, Springer-Verlag, New York, Berlin. *(Ch. 3, 5)*

Murray, C. A., 1983, *Vectorial Astrometry*, Adam Hilger, Bristol, England. *(Ch. 7)*

Murray, C. A., 1989, *Astron. & Astrophys.* **218**, 325–329. *(Ch. 7)*

Murray, C. D. and Dermott, S. F., 1999, *Solar System Dynamics*, Cambridge University Press, Cambridge, UK. *(Ch. 8)*

Nelson, B. and Davis, W. D., 1972, *Astrophys. J.*, **174**, 617–628. *(Ch. 12)*

Newcomb, S., 1878, *Washington Observations for 1875. (Ch. 1)*

Newcomb, S., 1912, *Astron. Papers Am. Ephemeris and Naut. Alm.*, **IX**, part I. *(Ch. 1)*

Nordtvedt, K., 1970, *Astrophys. J.*, **161**, 1059–1067. *(Ch. 5)*

Ohanian, H. C., 1980, *Gravitation and Spacetime*, second edition, W.W. Norton & Co., New York. *(Ch. 5, 7)*

Owen, W. M. and Yeomans, D. K., 1994, *Astron. J.* **107**, 2295–2298. *(Ch. 14)*

Paczyński, B., 1996, *Ann. Rev. Astron. Astrophys.*, **34**, 419–459. *(Ch. 6)*

Peale, S. J., 1997, *Icarus*, **127**, 269–289. *(Ch. 14)*

Peters, J., 1907, *Veröff. des König. Astron. Rechen-Instituts zu Berlin*, **33**. *(Ch. 11)*

Petit, G., 2000, in *Towards Models and Constants for Sub-microarcsecond Astrometry*, IAU Colloquium 180, K. J. Johnston, D. D. McCarthy, B. J. Luzum, and G. H. Kaplan (Eds.), US Naval Observatory, Washington DC, 275–282. *(Ch. 14)*

Popper, D. M., 1984, *Astron. J.*, **84**, 132–144. *(Ch. 12)*

Poveda, A., 1988, *Astrophys. Space Science*, **143**, N1/2, 67–78. *(Ch. 12)*

Pulkovo Observatory, 1985, *Refraction Tables of Pulkovo Observatory*, fifth edition, Pulkovo. *(Ch. 6)*

Rawley, L. A., Taylor, J. H. and Davis, M. M., 1988, *Astrophys. J.*, **326**, 947–953. *(Ch. 14)*

Rindler, W., 1977, *Essential relativity*, Springer-Verlag, New York, Heidelberg. *(Ch. 5)*

Roddier, C. and Roddier, F., 1975, *J. Opt. Soc. of America*, **65**, 664–667. *(Ch. 2)*

Röser, S., 1999, in *Review in Modern Astrometry*, R. Schielicke (Ed.), Astronomische Geselschaft, Hamburg, **12**, 97–106. *(Ch. 2)*

Röser, S. and Bastian, U., 1991, *PPM Star Catalogue*, Spektrum Akademisher Verlag, Heidelberg, **I** and **II**. *(Ch. 11)*

Röser, S. and Bastian, U., 1993, *Bulletin d'Information du CDS*, Observatoire de Strasbourg, **42**, 11–16. *(Ch. 11)*

Röser, S., Bastian, U. and Kuzmin, A., 1994, *Astron. & Astrophys. Suppl.*, **105**, 301–303. *(Ch. 11)*

Russell, J. L., Lasker, B. M., McLean, B. J., Sturch, C. R. and Jenkner, H., 1990, *Astron. J.*, 2059–2081. *(Ch. 11)*

Ryba, M. F. and Taylor, J. H., 1991a, *Astrophys J.*, **371**, 739–748. *(Ch. 14)*

Ryba, M. F. and Taylor, J. H., 1991b, *Astrophys J.*, **380**, 557–563. *(Ch. 14)*

Safizadeh, N., Dalal, N. and Griest, K., 1999, *Astrophys. J.*, **522**, 512–517. *(Ch. 14)*

SAO, 1966, *Star Catalog positions and proper motions of 258 997 stars for the epoch and equinox of 1950.0*, Publication of the Smithsonian Institution of Washington, n° 4562. *(Ch. 11)*

Schneider, P., Ehlers, J. and Falco, E. E., 1992, *Gravitational Lensing*, Springer Verlag, Berlin. *(Ch. 6)*

Seidelmann, P. K., 1982, *Celest. Mech.*, **27**, 79–106. *(Ch. 8, 10)*

Seidelmann, P. K., 1991, *Advan. Space Res.*, **11–2**, 103–111. *(Ch. 2)*

Seidelmann, P. K., (Ed.), 1992, *Explanatory Supplement to the Astronomical Almanac*, University Science Books, Mill Valley, California. *(Ch. 6, 8, 9, 10, 13)*

Seidelmann, P. K., Abalakin, V. K., Bursa, M. *et al.*, 2002, *Celest. Mech.*, **82**, 83–110. *(Ch. 13)*

Seidelmann, P. K. and Kovalevsky, J., 2002, *Astron. & Astrophys.*, **392**, 341–351. *(Ch. 8, 10)*

Serkowski, K., 1976, *Icarus*, **27**, 13–24. *(Ch. 14)*

Shaffer, D. B. and Marscher, A. P., 1987, in *Superluminal radio-sources*, J.-A. Zensus and T. J. Pearson (Eds.), Cambridge University Press, Cambridge, 67–71. *(Ch. 7)*

Shao, M., Colavita, M. M., Hines, B. E. *et al.*, 1988, *Astron. & Astrophys.*, **193**, 357–371. *(Ch. 2)*

Silvey, S. D., 1991, *Statistical Inference*, Chapman and Hall, London, New York. *(Ch. 4)*

Simon, J.-L., Bretagnon, P., Chapront, J. *et al.*, 1994, *Astron. & Astrophys.*, **282**, 663–683. *(Ch. 8)*

Simon, J.-L., Chapront-Touzé, M., Morando, B. and Thuillot, W., 1996, *Introduction aux éphémérides astronomiques*, Editions de Physique, Paris. *(Ch. 6)*

Skrutskie, M. F., 2001, *AAS Meeting*, **198**, 33.01. *(Ch. 11)*

Smith, E. K. Jr and Weintraub, S., 1953, *Proceedings IRE*, **41**, 1035–1037. *(Ch. 6)*

Smith, H., 1987, *Astron. & Astrophys.*, **171**, 336–341. *(Ch. 11)*

Smith, H. Jr and Eichhorn, H., 1996, *Monthly Not. RAS*, **281**, 211–218. *(Ch. 11)*

Söderhjelm, S., Evans, D. W., van Leeuwen, F. and Lindegren, L., 1992, *Astron. & Astrophys.*, **258**, 157–164. *(Ch. 12)*

Soffel, M. H., 1989, *Relativity in Astrometry, Celestial Mechanics and Geodesy*, A & A Library, Springer-Verlag, Berlin, New York. *(Ch. 5, 6)*

Soffel, M. H., 2000, *Towards Models and Constants for Sub-Microarcsecond Astrometry*, K. J. Johnston, D. D. McCarthy, B. J. Luzum and G. H. Kaplan (Eds.), IAU Colloquium 180, Publ. U.S. Naval Observatory, Washington DC, 283–292. *(Ch. 5, 7, 8)*

Souchay, J., Loysel, B., Kinoshita, H. and Folgeira, M., 1999, *Astron. & Astroph. Supp. Series*, **135**, 111–131. *(Ch. 8)*

Spencer Jones, H., 1939, *Monthly Not. R.A.S.*, **99**, 541. *(Ch. 1, 10)*

Standish, E .M., Newhall, X. X., Williams, J. G. and Folkner, W. F., 1995, *JPL Planetary and Lunar Ephemerides*, JPL IOM **314**, 10–127. *(Ch. 7)*

Standish, E. M., 1998, JPL Interoffice Memorandum, IOM **312**, F-98-048. *(Ch. 8)*

Stark, H. and Woods, J. W., 1986, *Probability, Random Processes, and Estimation Theory for Engineers*, Prentice-Hall, Englewood Cliffs, New Jersey. *(Ch. 4)*

Stephenson, F. R. and Lieske, J. H., 1988, *Astron. & Astrophys.*, **200**, 218–224. *(Ch. 10)*

Stephenson, F. R. and Morrison, L. V., 1984, *Phil. Trans. R. Soc. London*, **A313**, 47–70. *(Ch. 10)*

Stinebring, D. R., 1982, PhD thesis, Cornell University, Ithaca, N.Y. *(Ch. 14)*

Stone, R. C. and Dahn, C. C., 1995, in *Astronomical and Astrophysical Objectives of Sub-milliarcsecond Astrometry*, IAU Symp. 166, E. Høg and P. K.Seidelmann (Eds.), Kluwer, Dordrecht, 3–8. *(Ch. 7)*

Stoyko, N., 1937, *Comptes-rendus de l'Académie des Sciences*, **250**, 79–82. *(Ch. 10)*

Taff, L. G., 1988, *Astron. J.*, **96**, 409–411. *(Ch. 14)*

Taff, L. G., Lattanzi, M. G. and Bucciarelli, B., 1990a, *Astrophys. J.*, **358**, 359–369. *(Ch. 14)*

Taff, L. G., Bucciarelli, B. and Lattanzi, M. G., 1990b, *Astrophys. J.*, **361**, 667–672. *(Ch. 14)*

Taylor, G. E., 1955, *J. British Astron. Ass.*, **65**, 84. *(Ch. 13)*

Taylor, J. H., 1991, *Proc. IEEE*, **79**, 1054–1062. *(Ch. 14)*

Taylor, J. H., 1992, *Phil. Trans. R. Soc. London*, B **341**, 117–134. *(Ch. 14)*

Taylor, J. H., 1993, *Classical Quantum Gravity*, **10**, S135. *(Ch. 12)*

Taylor, J. H. and Weisberg, J. M., 1989, *Astrophys. J.*, **345**, 434–450. *(Ch. 5, 12, 14)*

Thomasson, P., 1986, *Quarterly Jour. RAS*, **27**, 413–431. *(Ch. 2)*

Thompson, A. R., Clark, B. G., Wade, C. M. and Napier, P. J., 1980, *Astrophys. J. Supp.*, **44**, 151–167. *(Ch. 2)*

Thompson, A. R., Moran, J. M. and Swenson, G. W. Jr, 1986, *Interferometry and synthesis in Radio Astronomy*, John Wiley and Sons, New York. *(Ch. 14)*

Triebes, K., Gilliam, L., Hilby, T. *et al.*, 2000, in *UV, Optical, and IR Space Telescopes and Instruments*, Proc. SPIE **4013**, 482–492. *(Ch. 14)*

Turon, C., Crézé, M., Egret, D. *et al.*, 1992a, *The Hipparcos Input Catalogue*, European Space Agency Publ., SP-1136, March 1992, 7 volumes *(Ch. 2, 11)*

Turon, C., Gómez, A., Crifo, F. *et al.*, 1992b, *Astron. & Astrophys.*, **258**, 74–81. *(Ch. 11)*

Urban, S. E., Corbin, T. E. and Wycoff, G. L., 1998, *Astron. J.*, **115**, 2161–2166. *(Ch. 11)*

Urban, S. E. and Corbin, T. E., 1998, Personnal communication and distributed CD-roms with the catalog. *(Ch. 11)*

Urban, S. E., Corbin, T. E., Wycoff, G. L. and Mason, B. D., 2000, in *Towards Models and Constants for Sub-Microarcsecond Astrometry*, IAU Colloquium 180, K. J. Johnston, D. D. McCarthy, B. J. Luzum and G. H. Kaplan (Eds.), US Naval Observatory, Washington DC, 97–103. *(Ch. 7)*

Urban, S. E., Seidelmann, P. K., Germain, M. *et al.*, 2000, in *Journées 1999, Systèmes de référence spatio-temporels*, M. Soffel and N. Capitaine (Eds.), Paris Observatory, 131–135. *(Ch. 2)*

Van Altena, W. F., 1983, *Ann. Rev. Astron. Astrophys.*, **21**, 131–164. *(Ch. 14)*

van Dam, T. M. and Wahr, J. M., 1987, *J. Geophys. Res.*, **92B**, 1281–1286. *(Ch. 9)*

van Dam, T. M. and Herring, T. A., 1994, *J. Geophys. Res.*, **99B**, 4505–4517. *(Ch. 9)*

Van de Kamp, P., 1967, *Principles of Astrometry*, W. H. Freeman & Co., San Francisco. *(Ch. 12)*

Vanivcek, P. and Krakiwsky, E., 1982, *Geodesy: the concepts*, North Holland Publ. Co., Amsterdam. *(Ch. 9)*

Van Leeuwen, F., 1997, *Space Sc. Rev.*, **81**, 201–409. *(Ch. 2)*

Vondrak, J. and Ron, C., 2000, in *Towards Models and Constants for Sub-microarcsecond Astrometry*, IAU Colloquium 180, K. J. Johnston, D. D. McCarthy, B. J. Luzum, and G. H. Kaplan (Eds.), U.S. Naval Observatory, Washington D.C., 248–253. *(Ch. 10)*

Wahr, J. M., 1981, *Geophys. J. R. Astron. Soc.*, **64**, 705–727. *(Ch. 8, 9)*

Wahr, J. M., 1985, *J. Geophys. Res.*, **90B**, 9363–9368. *(Ch. 9)*

Walter, H. G. and Sovers, O. J., 2000, *Astrometry of Fundamental Catalogues*, Springer Verlag, Berlin, Heidelberg *(Ch. 2, 10)*

Wertz, J., 1978, *Spacecraft Attitude Determination and Control*, Reidel Publ. Co., Dordrecht, Boston. *(Ch. 3)*

Will, C. M., 1974, in *Experimental Gravitation*, B. Bertotti, (Ed.), Academic Press, New York, 1–110. *(Ch. 8)*

Will, C. M., 1981, *Theory and experiment in gravitational physics*, Cambridge University Press, Cambridge. *(Ch. 5, 6)*

Will. C. M. and Nordvedt, K., 1972, *Astrophys. J.*, **177**, 757–774. *(Ch. 14)*

Wolszczan, A. and Frail, D. A., 1992, *Nature*, **355**, 145–147. *(Ch. 14)*

Woolard, E. W. and Clemence, G. M., 1966, *Spherical Astronomy*, Academic Press, New York. *(Ch. 6, 11)*

Worley, C. E. and Douglass, G. G., 1997, *Astron. & Astrophys. Suppl. Series*, **125**, 523. *(Ch. 12)*

Zacharias, N., 1996, *Proc. Astron. Soc. Pacific*, **108**, 1135–1138. *(Ch. 14)*

Zacharias, N., 2001, *Astrometry: telescopes and techniques, Encyclopedia of Astronomy and Astrophysics*, Nature Publishing Group, Institute of Physics Publishing, 111–120. *(Ch. 14)*

Zacharias, N., de Vegt, C. and Murray, C. A., 1997, in *Proc. ESA Symp. Hipparcos-Venice 1997*, B. Batrick (Ed.), ESA SP-402, 85–90. *(Ch. 7)*

Zacharias, N., Urban, S. E., Zacharias, M. I. *et al.*, 2000, *Astron. J.*, **120**, 2131–2147. *(Ch. 11)*

Zarrouati, O., 1987, *Trajectoires Spatiales*, Cepadues-Editions, Toulouse, France. *(Ch. 9)*

Zombeck, M. V., 1990, *Handbook of Space Astronomy and Astrophysics*, second edition, Cambridge University Press, Cambridge, 65. *(Ch. 11)*

Index